COMPUTER NUMERICAL CONTROL

SOCIETY OF MANUFACTURING ENGINEERS & DELMAR PUBLISHERS INC.

A PARTNERSHIP IN EDUCATIONAL EXCELLENCE

SME and DELMAR have proudly joined forces to form a partnership dedicated to educational excellence. We believe that quality manufacturing education is the key to keeping America competitive in the years ahead.

The Society of Manufacturing Engineers is an international technical society dedicated to advancing scientific knowledge in the field of manufacturing. SME has more than 80,000 members in 70 countries and serves as a forum for engineers and managers to share ideas, information, and accomplishments.

To be successful, today's engineers and technicians must keep pace with the torrent of information that appears each day. To meet this need, SME provides, in addition to the publication of books, many opportunities in continuing education for its members. These opportunities include: monthly meetings through five associations and more than 300 chapters; educational programs including seminars, clinics, and videotapes, as well as conferences and expositions.

Today's manufacturing technology students represent our future. Our goal is to provide these students with the finest manufacturing technology educational products. By pooling our many resources, SME and DELMAR are going to help teachers get the job done.

Together SME and DELMAR will provide outstanding educational materials to prepare students to enter the real world of manufacturing.

Thomas J. Drozda
Director of Publications
Society of Manufacturing Engineers

Gregory C. Spatz
President
Delmar Publishers Inc.

COMPUTER NUMERICAL CONTROL

CONCEPTS AND PROGRAMMING
2nd Edition

WARREN S. SEAMES

DELMAR PUBLISHERS INC.®

NOTICE TO THE READER

Dedication

This book is dedicated to my wife, Dolores, for her love, understanding, and patience throughout this project.

Cover Photos: Keyboard photo courtesy of Cincinnati Milacron Marketing Co. Industrial equipment photo courtesy of KGK International Corp.

Delmar Staff

Executive Editor: David C. Gordon
Associate Editor: Marjorie A. Bruce
Project Editor: Eleanor Isenhart
Production Coordinator: Larry Main
Design Coordinator: Susan C. Mathews

For information, address Delmar Publishers Inc.
2 Computer Drive West, Box 15-015
Albany, New York 12212

10 9 8 7 6 5 4 3 2

Printed in the United States of America
Published simultaneously in Canada
by Nelson Canada,
A Division of the Thomson Corporation

Library of Congress Cataloging in Publication Data

Seames, Warren S.
   Computer numerical control: concepts and programming/Warren S. Seames.—2nd ed.
      p.   cm.
   Includes index.
   ISBN 0-8273-3782-5. — ISBN 0-8273-3783-3
(instructor's guide)
   1. Machine-tools—Numerical control.   I. Title.
TJ1189.S36   1990   89-11700
621.9'023—dc20   CIP

CONTENTS

PREFACE

The second edition of *Computer Numerical Control: Concepts and Programming* offers an updated and improved introduction to the basic principles of computer numerical control, plus expanded programming examples and part problems for programming.

The programs presented in this text are not as complex as those found in industry. The examples were carefully selected to demonstrate the *basic* concepts of CNC programming. It was the author's goal to eliminate the learner's confusion that often accompanies the introduction of CNC programming. The learner will achieve an understanding of the programming principles through the programming applications and problems presented in the text. With a firm understanding of the basic principles, the learner can then move confidently ahead to apply the principles to actual industrial situations. Programming is an art, and like art in its generally visualized form, the programming practices used by programmers or NC instructors may vary. However, the basic principles of CNC programming are exhibited in each programmer's effort.

This new edition of *Computer Numerical Control* reflects the maturing of the industry with a complementary trend to standardization in codes and controllers. It has moved from a generic approach to one which reflects more programming for specific controllers to provide learners with real world practice.

FEATURES

The following features of this text aid the learner in understanding the principles of computer numerical control and mastering programming using the word address format.

- Learning objectives to identify the essential knowledge gained from a study of the chapter.
- Numerous illustrations to clarify and augment text discussions and highlight applications. Photos to identify equipment commonly found in industry and expand learner awareness of the real world of CNC.
- Solved programming examples with explanations of each step in the program to familiarize learners with the logical progression of commands and the use of the standard G and M codes.
- Emphasis on the word address format in programming examples and problems to mirror industrial usage.

- Emphasis on the specific algebra and trigonometry functions needed to solve cutter path geometry problems. Solved examples, clear explanations, and diagrams to clarify these math concepts.
- Introduction to numerical control principles to serve as a review or to make learners aware of the development of the industry to its current state.
- Clear, concise explanations of essential programming concepts including Cartesian coordinate system, part and machine coordinate systems, setting the machine origin and dimensioning (absolute and incremental).
- Explanation of the special requirements for tooling for NC, including tooling for hole operations and milling. Covers cutting tool materials (including carbide inserts) and calculations of speeds and feeds with solved problems.
- Coverage of programming functions, such as linear interpolation, circular interpolation, cutter diameter compensation, do loops, subroutines, nested loops, mirror imaging, polar rotation and helical interpolation.
- Discussion of the future of numerical control and job opportunities in this expanding industry.
- Sample part programs from industry for learner analysis.
- Illustrated glossary to reinforce understanding of essential terminology.

The major changes for the second edition include the following items.

- In Chapter 2, added content on setting machine origin and the difference between the part coordinate system and the machine coordinate system, with methods of transferring from one to the other.
- A new Chapter 3 on tooling now includes an introduction to process planning with an example showing the steps and documents utilized in planning the machining of a part prior to the actual programming of the machining process. The chapter also includes new and updated content on tools specifically designed for CNC, tooling materials and greater emphasis on the calculation of speeds and feeds.
- In Chapters 6, 7, 10, 11 and 15, added "Special Fanuc Section" consisting of a part drawing and instructions to program the cutting on a specific machine using a specific Fanuc controller. All special codes for the specific controller are defined.
- In Chapter 8, added a milling example and a lathe example requiring the computation of multiple cutter locations. Solutions are provided to show learners the steps to be taken.
- In Chapter 10, expanded the discussion of cutter diameter compensation to fine tune (compensate) for the difference between the programmed cutter diameter and the actual cutter diameter.
- Updated many programs and program explanations to reflect current usage of G codes in word address format.
- Chapters 13 and 14 on NC lathes, expanded with discussion of canned cycles and threading cycles.

- In Chapter 15, added a sample APT programming problem, including part drawing, part geometry figure, and postprocessor output.
- Added section on "Computer Graphics Programming" to stress the growing use of CAM software in manufacturing; also discusses CAD/CAM programming and computer integrated manufacturing (CIM).
- Added programming examples and problems in word address format.
- Added new Appendix 7, Lathe Canned Cycle Example, with part drawing, program, explanation of tooling used, sequence of operations, and program notes.

SOCIETY OF MANUFACTURING ENGINEERS (SME)

Delmar Publishers Inc. and the Society of Manufacturing Engineers have joined together to provide exceptional educational materials for the preparation of students to enter the manufacturing industries equipped with essential skills, and to further the education of those already employed in manufacturing. A list of complementary texts and other educational materials available from the Society of Manufacturing Engineers can be found following the Glossary.

ABOUT THE AUTHOR

Warren S. Seames is an experienced programmer with many years experience in industry and in teaching NC programming at the community college level. Mr. Seames is a former instrument maker and journeyman machinist. Presently he is an NC programmer, computer systems analyst, and resident applications programmer in industry. Mr. Seames is a member of the Society of Manufacturing Engineers.

ACKNOWLEDGMENTS

The author is indebted to the following individuals who reviewed the first edition and supplied many recommendations for the revised edition.

Alva B. Powell, Macomb Community College, Warren, MI
Jim Newman, Jefferson College, Hillsboro, MO
T. W. Sorrell, Northeast Missouri State University, Kirksville, MO

Lorin V. Waitkus, Cuyahoga Community College, Cleveland, OH
George T. Richardson, Mt. San Antonio College, Walnut, CA
Gary Dauer, Staples Technical Institute, Staples, MN
James H. Morris, Joliet Junior College, Joliet, IL
Melvin B. Gage, Linn Technical College, Linn, MO
David Dickie, Fresno City College, Fresno, CA
Ray Greb, Mesa State College, Grand Junction, CO
Robert G. Dixon, Chemeketa Community College, Salem, OR
Stanley Smolinsky, ITT Technical Institute, West Covina, CA

The following CNC instructors provided detailed critical reviews of the manuscript for the second edition of the text. Their contributions to the development of the new edition are gratefully acknowledged.

Mark E. Meyer, College of DuPage, Glen Ellyn, IL 60137
David Dickie, Fresno City College, Fresno, CA 93741
Ken S. Friend, Indiana Vocational Technical College, Richmond, IN 47274
Richard Colacino, National Tooling & Machining Association Training
    Center, Santa Clara, CA 95054
Melvin B. Gage, Linn Technical College, Linn, MO 65051
Ray Greb, Jr., Mesa State College, Grand Junction, CO 81501

CHAPTER 1

An Introduction to Numerical Control Machinery

OBJECTIVES Upon completion of this chapter, you will be able to:

* Describe the difference between direct and distributive numerical control.
* Describe the difference between a numerical control tape machine and a computer numerical control machine.
* Describe four ways that programs can be entered into a computer numerical controller.
* Explain two tape code formats in use with computer numerical control (CNC) machinery.
* Give the major objectives of numerical control.

Welcome to the world of numerical control. Numerical control (NC) has become popular in shops and factories because it helps solve the problem of making manufacturing systems more flexible. In simple terms, a *numerical control machine* is a machine positioned automatically along a preprogrammed path by means of coded instructions. The key words here are "preprogrammed" and "coded." Someone has to determine what operations the machine is to perform and put that information into a coded form that the NC control unit understands before the machine can do anything. In other words, someone has to program the machine.

Machines may be programmed manually or with the aid of a computer. Manual programming is called *manual part programming;* programming done by a computer is called *computer aided programming (CAP)*. Sometimes a manual program is entered into the machine's controller via its own keypad. This is known as *manual data input (MDI)*. This text will focus on manual part programming.

Advances in microelectronics and microcomputers have allowed the computer to be used as the control unit on modern numerical control machinery. This computer takes the place of the tape reader found on earlier NC machines. In other words, instead of reading and executing the program directly from punched tape, the program is loaded into and executed from the machine's computer. These machines, known as *computer numerical control* (CNC) machines, are the NC machines being manufactured today. The primary focus of this text is the MDI programming of *computer numerical control* (CNC) machinery.

THE HISTORY OF NC

In 1947, John Parsons of the Parsons Corporation, began experimenting with the idea of using three-axis curvature data to control machine tool motion for the production of aircraft components. In 1949, Parsons was awarded a U.S. Air Force contract to build what was to become the first numerical control machine. In 1951, the project was assumed by the Massachusetts Institute of Technology. In 1952, numerical control arrived when MIT demonstrated that simultaneous three-axis movements were possible using a laboratory-built controller and a Cincinnati Hydrotel vertical spindle. By 1955, after further refinements, numerical control became available to industry.

Early NC machines ran off punched cards and tape, with tape becoming the more common medium. Due to the time and effort required to change or edit tape, computers were later introduced as aids in programming. Computer involvement came in two forms: computer aided programming languages and direct numerical control (DNC). *Computer aided programming languages* allowed a part programmer to develop an NC program using a set of universal "pidgin English" commands, which the computer then translated into machine codes and punched into the tape. *Direct numerical control* involved using a computer as a partial or complete controller of one or more numerical control machines (see Figure 1−1). Although some companies have been reasonably successful at implementing DNC, the expense of computer capability and software and problems associated with coordinating a DNC system renders such systems economically unfeasable for all but the largest companies.

Recently a new type of DNC system called *distributive numerical control* has been developed (Figure 1−2). It employs a network of computers to coordinate the operation of a number of CNC machines. Ultimately, it may be possible to coordinate an entire factory in this manner. Distributive numerical control solves some of the problems that exist in coordinating a direct numerical control system. There is another type of distributive numerical control that is a spin-off of the system previously explained. In this system, the NC program is transferred in its entirety from a host computer directly to the machine's controller. Alternately, the program can be transferred from a mainframe host computer to a personal computer (PC) on the shop floor where it will be stored until it is needed. The program will then be transferred from the PC to the machine controller.

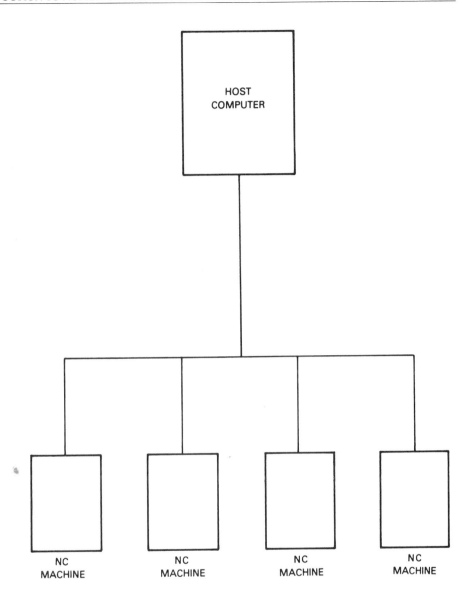

FIGURE 1–1
Direct numerical control

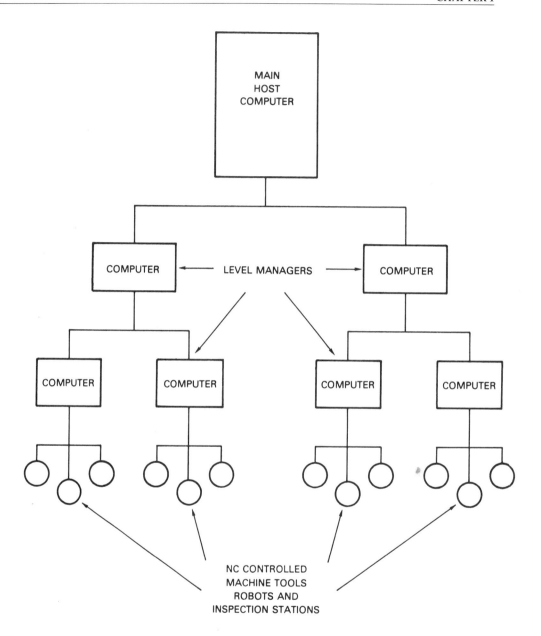

FIGURE 1-2
Distributive numerical control

CNC MACHINES

Figures 1–3 to 1–10 show modern CNC machines. CNC machines have more programmable features than older NC tape machinery and may be used as stand-alone units, or in a network of machines such as a flexible machining center (described in Chapter 16). They are easier to program, and most CNC machines may be programmed by more than one method.

All machines can be programmed via an on-board computer keyboard. In addition to the keyboard, there is a tape reader or electronic connector to allow the transfer of a program written elsewhere to the CNC machine.

A CNC machine is a *soft-wired controller*; that is, once the NC program is loaded into the computer's memory, no hardware is necessary to transfer the numerical control codes to the controller. The controller uses a permanent resident program, called an *executive program,* to process the codes into the electrical pulses that control the machine. The executive program is often referred to as the executive "software," but technically speaking, software is a misnomer. The executive program is more appropriately called *firmware.* In any CNC machine, the executive program resides in ROM memory, and the NC code resides in RAM memory.

ROM stands for *read only memory.* The information in ROM memory is written into the electronic chip and cannot be erased without special equipment. ROM can be accessed by the computer, but cannot be altered. This is why the executive program cannot be erased and is always active when the machine is on.

RAM stands for *random access memory.* RAM can be accessed and altered (written to) by the computer. The NC code is written into RAM by either the keyboard or an outside source. The contents of RAM are lost when the controller is turned off. Many CNC controllers utilize a battery backup system that powers the computer long enough for the program to be transferred (saved) to some storage media in the event of a power loss; other CNC controllers use a special type of RAM called *CMOS memory,* which retains its contents even when the power to the computer is turned off.

INPUT MEDIA

Input media are used to electronically or mechanically store the NC programs until they are needed. A program is simply read from the input medium when it is loaded into the machine. As mentioned previously, there are different methods of inputting the NC code into a controller. Whereas old NC machinery could only read programs from punched tape or a direct numerical control system, CNC machines may possess multiple means of program input.

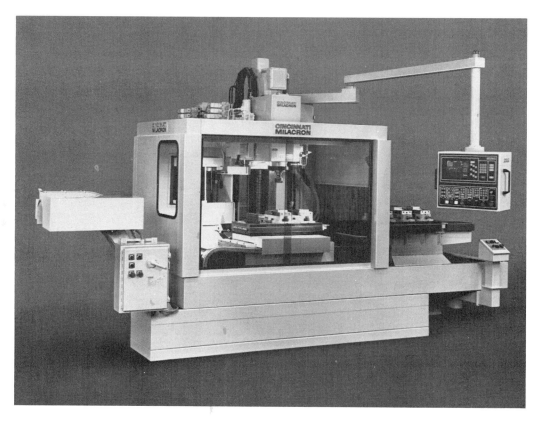

FIGURE 1–3
A vertical spindle machining center, featuring twin pallets *(Photo courtesy of Cincinnati Milacron)*

The most popular medium for program storage is still *punched tape* made of paper or mylar plastic (mylar is most commonly used as it is stronger than paper and less likely to tear). The NC program code is entered into the tape by use of a *tape puncher* which punches a series of holes that represent the NC codes. A tape reader employing electrical, optical, or mechanical means senses the holes in the tape and transfers the coded information into the machine computer.

A tape puncher may be attached to a teletype machine or a Flexowriter (a typewriterlike machine). With either of these two pieces of equipment, a code character is punched into the tape as it is being typed onto a sheet of paper. It is more common, however, to see a tape puncher attached to a microcomputer. The NC code is entered into either a CAM (computer-aided manufacturing) or wordprocessor type of program and punched into the tape after all editing of the program is completed.

FIGURE 1–4
A horizontal machining center utilizing twin matrix tool storage magazines. Note the workpiece and tool delivery systems. Safety guards are removed to show clarity. *(Photo courtesy of Cincinnati Milacron)*

Magnetic tape is another popular storage medium. Early experiments with magnetic tape for program storage were not very successful, because the shop environment was not conducive to the delicate tapes of yesteryear. Today's high-quality tapes can survive the rigors of the shop environment with reasonable care in their handling. The most commonly used style of magnetic tape is ¼-inch computer grade cassette tape. The cassette case affords good protection and the small size is convenient for storage. Standards for tape format and tape coding have been developed by the Electronics Industries Association (EIA).

BINARY NUMBERS

An understanding of how the controller processes information is helpful in learning to program computer numerical control machinery. Computers and computer-controlled machinery do not deal in Arabic symbols or numbers. All of

FIGURE 1–5
A horizontal machining center featuring a Fanuc controller *(Photo courtesy of Bendix Corp.)*

the internal processing is done by calculating or comparing binary numbers. Binary numbers contain only two digits, zero and one, as illustrated in Figure 1–11.

Within the CNC controller, each binary digit "one" may represent a positive charge and a "zero" a negative charge, or a "one" may be the presence of charge and a "zero" the absence of charge. The method used depends on the particular controller. In either case, the CNC program code in binary form must be loaded into the computer. Programming formats and languages allow the NC code to be written using alphabetic characters and base-ten decimal numbers. When the NC program is punched or recorded on tape, the information is translated into binary form.

TAPE FORMATS

Various tape coding formats have been developed and tried since the beginning of numerical control. Today, tapes are primarily made using the EIA standard RS-274 format, also called *word address format.* Program information is contained in program lines, called *blocks,* which are punched into the tape in one of two tape code standards. RS-274 is a *variable* block coding for-

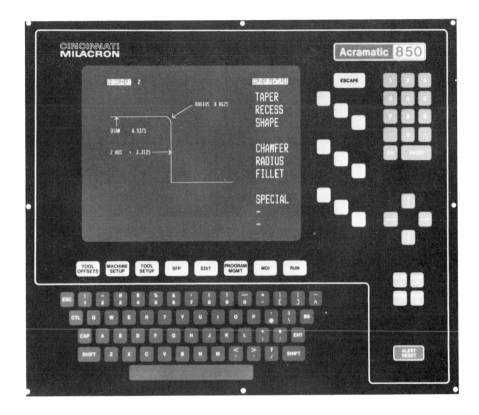

FIGURE 1-6
A CNC controller featuring interactive graphics *(Photo courtesy of Cincinnati Milacron)*

mat, meaning that the information contained in a block may be arranged in any order. Discussion of MDI programming using word address format begins in Chapter 6.

RS-244 Binary Coded Decimal

The EIA RS-244 standard, illustrated in Figure 1 – 12, is one of two tape codes used for NC tapes. It became a standard early in the development of numerical control. Notice that the code utilizes lowercase letters and limited punctuation. Each hole punched into the tape represents the binary digit "one," while a blank space represents the digit "zero." The tape code allows alphabetic characters and base-ten numbers to be translated into the binary code that the controller requires. For this reason, RS-244 is also known as *binary coded decimal* (BCD).

FIGURE 1–7
A modern vertical spindle machining center *(Photo courtesy of Bridgeport Machines Division of Textron Inc.)*

RS-358

At the time that NC was being implemented in industry, other industries were also using punched tape. Government, telephone, and computer industries all required a tape code that contained both upper and lowercase letters and more punctuation than the limited NC tape code used. What was adequate for numerical control use was not sufficient for other applications. The standard that was adopted for tape coding in these industries was the American Standard Code for Information Interchange (ASCII). To expand the role of computers in NC programming and strive for one standard tape code, EIA RS-358 was adopted for use. This code, known as both ISO (International Standards Organization) and ASCII, is a subset of the ASCII code used in other applications. It is illustrated in Figure 1–13. Both codes are used to prepare tape for use with CNC machines today. Many CNC controllers can detect which of the two formats is being sent and accept either one.

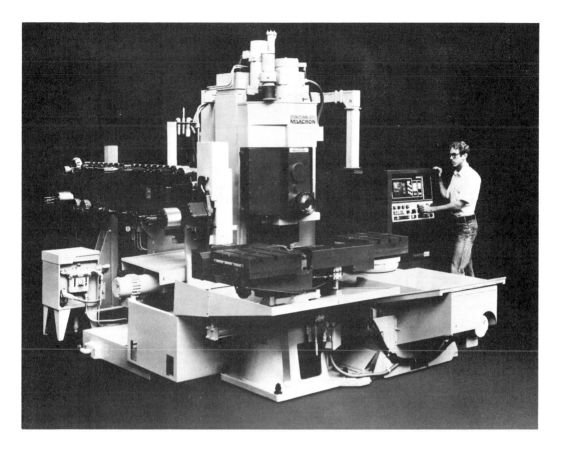

FIGURE 1–8
A modern horizontal spindle machining center *(Photo courtesy of Cincinnati Milacron)*

OBJECTIVES OF NUMERICAL CONTROL

Numerical control (NC) was developed with these goals in mind:

1. To increase production
2. To reduce labor costs
3. To make production more economical
4. To do jobs that would be impossible or impractical without NC
5. To increase the accuracy of duplicate parts

Before deciding (in light of NC objectives) to utilize an NC or CNC machine for a particular job, the requirements and economics of the job must be weighed against the following advantages and disadvantages of the machinery. Such an

FIGURE 1–9
A horizontal machining center equipped with an eight-pallet automatic workchanger. Safety guards have been removed for clarity. *(Photo courtesy of Cincinnati Milacron)*

evaluation is necessary to determine if such a machine is practical for the particular job. (Note: NC is a general term used for numerical control. It is also used to describe numerical control machinery that runs directly off of tape. CNC refers specifically to computer numerical control. CNC machines are all NC machines, but not all NC machines are CNC machines.)

Advantages

1. Increased productivity
2. Reduced tool/fixture storage and cost
3. Faster set-up time
4. Reduced parts inventory
5. Flexibility that speeds changes in design
6. Better accuracy of parts
7. Reduction in parts handling
8. Better uniformity of parts
9. Better quality control
10. Improvement in manufacturing control

FIGURE 1–10
A CNC centerless grinding machine. This machine features an epoxy granite bed. Safety guards have been removed for clarity. *(Photo courtesy of Cincinnati Milacron)*

Disadvantages

1. Increase in electrical maintenance
2. High initial investment
3. Higher per-hour operating cost than traditional machine tools
4. Retraining of existing personnel

This is not a complete listing of the various advantages and disadvantages of numerical control machines; however, it should give a general idea of the types of jobs for which NC machines are suited.

ARABIC	BINARY	ARABIC	BINARY
0	0	18	1 0 0 1 0
1	1	19	1 0 0 1 1
2	1 0	20	1 0 1 0 0
3	1 1	21	1 0 1 0 1
4	1 0 0	22	1 0 1 1 0
5	1 0 1	23	1 0 1 1 1
6	1 1 0	24	1 1 0 0 0
7	1 1 1	25	1 1 0 0 1
8	1 0 0 0	26	1 1 0 1 0
9	1 0 0 1	27	1 1 0 1 1
10	1 0 1 0	28	1 1 1 0 0
11	1 0 1 1	29	1 1 1 0 1
12	1 1 0 0	30	1 1 1 1 0
13	1 1 0 1	31	1 1 1 1 1
14	1 1 1 0	32	1 0 0 0 0 0
15	1 1 1 1	64	1 0 0 0 0 0 0
16	1 0 0 0 0	128	1 0 0 0 0 0 0 0
17	1 0 0 0 1		

FIGURE 1–11
Binary numbers compared to Arabic numbers

APPLICATIONS IN INDUSTRY

Developed originally for use in aerospace industries, NC is enjoying widespread acceptance in manufacturing. The use of CNC machines continues to increase, becoming visible in most metalworking and manufacturing industries. Aerospace, defense contract, automotive, electronic, appliance, and tooling industries all employ numerical control machinery. Advances in micro-electronics have lowered the cost of acquiring CNC equipment. It is not unusual to find CNC machinery in contract tool, die, and moldmaking shops. With the advent of low cost OEM (original equipment manufacturer) and retrofit CNC vertical milling machines, even shops specializing in one-of-a-kind prototype work are using CNCs.

Although numerical control machines traditionally have been machine tools, bending, forming, stamping, and inspection machines have also been produced as numerical control systems. Since this text is written with the student machinist in mind, only CNC machines will be considered.

TRACK NUMBER

8	7	6	5	4		3	2	1	
		●			•				0
					•			●	1
					•		●		2
			●		•		●	●	3
					•	●			4
			●	●	•			●	5
			●	●	•		●		6
					•	●	●	●	7
				●	•				8
			●	●	•			●	9
	●	●			•			●	a
	●	●			•		●		b
	●	●	●		•		●	●	c
	●	●			•	●			d
	●	●	●		•	●		●	e
	●	●	●		•	●	●		f
	●	●			•	●	●	●	g
	●	●		●	•				h
	●	●	●	●	•			●	i
	●		●		•			●	j
	●		●		•		●		k
	●				•		●	●	l
	●		●		•	●			m
	●				•	●		●	n
	●				•	●	●		o
	●		●		•	●	●	●	p
	●		●	●	•				q
	●			●	•			●	r
		●	●		•		●		s
		●			•		●	●	t
		●	●		•	●			u
		●			•	●		●	v
		●			•	●	●		w
		●	●		•	●	●	●	x
		●	●	●	•				y
		●		●	•			●	z
	●	●		●	•		●	●	. (Period)
		●	●	●	•		●	●	, (Comma)
		●	●		•			●	/
	●	●	●		•				+
	●				•				-
			●		•				Space
	●	●	●	●	•	●	●	●	Delete
●									CARR. RET. (EOB)
		●		●	•		●		Back Space
		●	●	●	•	●	●		Tab
				●	•		●	●	End of Record
					•				Tape Feed Hole
									Blank Tape

● = Hole in Tape

• = Tape Feed Hole

FIGURE 1–12
EIA RS-244 tape code

TRACK NUMBER

8	7	6	5	4		3	2	1	
		●	●		•				0
●		●	●		•			●	1
●		●	●		•		●		2
		●	●		•		●	●	3
●		●	●		•	●			4
		●	●		•	●		●	5
		●	●		•	●	●		6
●		●	●		•	●	●	●	7
●		●	●	●	•				8
		●	●	●	•		●		9
	●				•			●	A
	●				•		●		B
●	●				•		●	●	C
	●				•	●			D
●	●				•	●		●	E
●	●				•	●	●		F
	●				•	●	●	●	G
	●			●	•				H
●	●		●		•			●	I
●	●		●		•		●		J
	●		●		•		●	●	K
●	●		●		•	●			L
	●		●		•	●		●	M
	●		●		•	●	●		N
●	●		●		•	●	●	●	O
	●	●			•				P
●	●	●			•			●	Q
●	●	●			•		●		R
	●	●			•		●	●	S
●	●	●			•	●			T
	●	●			•	●		●	U
	●	●			•	●	●		V
●	●	●			•	●	●	●	W
●	●		●	●	•				X
	●		●	●	•			●	Y
	●		●	●	•		●		Z
●	●	●	●	●	•	●	●	●	Delete
●			●		•				Back Space
			●		•		●		Horiz. Tab
			●		•	●			Line Feed
●			●		•	●		●	Carr. Ret. (EOB)
●		●			•				Space
●		●			•	●		●	%
		●		●	•				((Open Paren.)
●		●		●	•			●	) (Close Paren.)
		●		●	•		●	●	+
		●		●	•	●		●	—
●		●		●	•	●	●	●	/
		●	●	●	•		●		: (Colon)
					•				Tape Feed Hole
									Virgin Tape

FIGURE 1–13
EIA RS-358 tape code

● = Hole in Tape
• = Tape Feed Hole

SUMMARY

The important concepts presented in this chapter are:

- A numerical control machine is a machine that is positioned automatically along a preprogrammed path by way of coded instructions.
- Direct numerical control involves a computer that acts as a partial or full controller to one or more NC machines. Distributive numerical control is a network of computers and numerical control machinery coordinated to perform some task.
- CNC machines use an on-board computer as a controller.
- Offline programming is the programming of a part away from the computer keyboard (hence the term *offline*). This is usually done with a microcomputer.
- There are four ways to input programs into CNC machinery: MDI (manual data input), punched tape, magnetic tape, and DNC (direct numerical control/distributive numerical control).
- Computers work with binary numbers. The CNC program must be loaded into the controller in binary form.
- RS-244 and RS-358 are tape codes used to place information on punched tape. The information is punched into the tape in binary form.
- Before deciding on a numerical control machine for a specific job, the advantages and disadvantages of NC must be weighed in view of the primary objectives of numerical control.

REVIEW QUESTIONS

1. What is a numerical control machine?
2. What is the difference between an NC tape machine and a CNC machine?
3. What is direct numerical control?
4. What is distributive numerical control?
5. Name four ways to enter a program into a computer numerical control machine.
6. What does CNC mean? What does MDI mean?
7. What are RS-244 and RS-358?
8. What are the five major objectives of numerical control?
9. What are the advantages of NC? What are the disadvantages?

CHAPTER 2

Numerical Control Systems

OBJECTIVES Upon completion of this chapter, you will be able to:

- Describe the two types of control systems in use on NC equipment.
- Name the four types of drive motors used on NC machinery.
- Describe the two types of loop systems used.
- Describe the Cartesian coordinate system.
- Define a machine axis.
- Describe the motion directions on a three-axis milling machine.
- Describe the difference between absolute and incremental positioning.
- Describe the difference between datum and delta dimensioning.

COMPONENTS

A CNC machine consists of two major components: the *machine tool* and the *controller,* or *machine control unit* (MCU), which is an on-board computer. These components may or may not be manufactured by the same company. General Numeric, Fanuc, General Electric, Bendix, Cincinnati Milacron, and G & L Electronics are among those manufacturers of CNC controllers that supply units to makers of machine tools. Figure 2–1 shows a typical controller. Each controller is manufactured with a standard set of built-in codes. Other codes are added by the machine tool builders. For this reason, program codes vary somewhat from machine to machine. Every CNC machine, regardless of manufacturer, is a collection of systems coordinated by the controller.

TYPES OF CONTROL SYSTEMS

There are two types of control systems used on NC machines: *point-to-point systems* and *continuous-path systems. Point-to-point* machines move only in straight lines. They are limited in a practical sense to hole operations

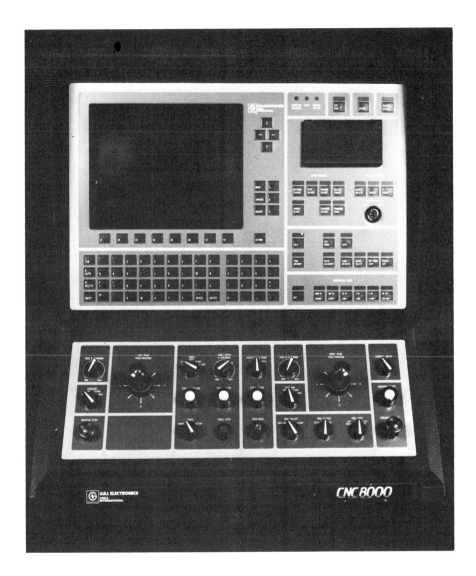

FIGURE 2–1
A modern CNC controller *(Photo courtesy of Giddings & Lewis/Davis Corp.)*

(drilling, reaming, boring, etc.) and straight milling cuts parallel to a machine axis. When making an axis move, all affected drive motors run at the same speed. When one axis motor has moved the instructed amount, it stops while the other motor continues until its axis has reached its programmed location. This makes the cutting of 45-degree angles possible, but not arcs or angles other than 45 degrees. Angles and arc segments must be programmed as a series of straight line cuts (see Figure 2–2).

A *continuous-path* machine (or *contouring system*) has the ability to move its drive motors at varying rates of speed while positioning the machine; the

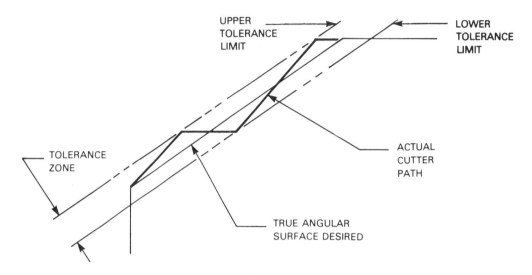

POINT-TO-POINT ANGLE MADE BY MACHINE
CAPABLE OF 45 DEGREE ANGLES.

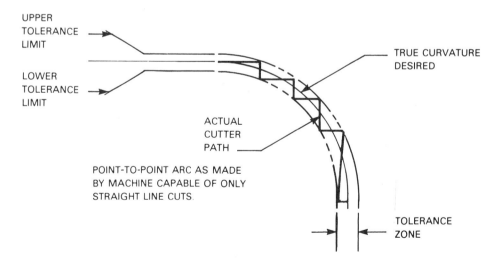

POINT-TO-POINT ARC AS MADE
BY MACHINE CAPABLE OF ONLY
STRAIGHT LINE CUTS.

FIGURE 2–2
Point-to-point angles and arcs

cutting of arc segments and any angle may be easily accomplished (see Figure 2–3). At one time, point-to-point machines were common; their electronics were less expensive to produce and they were, therefore, less expensive to acquire. Technological advancements, however, have narrowed the cost difference between point-to-point and continuous-path machines to where most CNC machines now manufactured are of the continuous-path type.

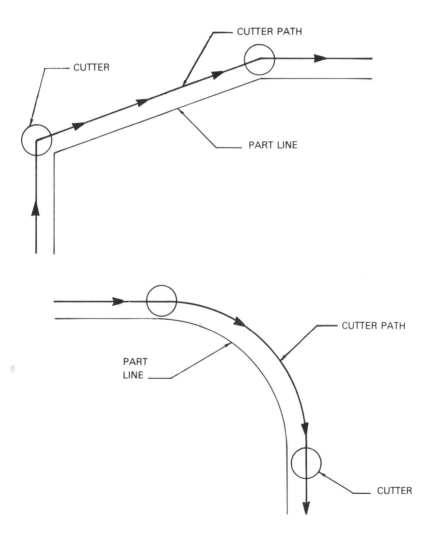

FIGURE 2–3
Continuous-path angles and arcs

SERVOMECHANISMS

It is helpful to understand the drive systems used on NC machinery. The *drive motors* on a particular machine will be one of four types: stepper motors, DC servos, AC servos, or hydraulic servos. Stepper motors move a set amount of rotation (a step) every time the motor receives an electronic pulse. DC and AC servos are widely used variable speed motors found on small and medium continuous-path machines. Unlike a stepper motor, a servo does not move a set distance; when current is applied, the motor starts to turn; when the current is removed, the motor stops turning. The AC servo is a fairly recent development. It can develop more power than a DC servo and is commonly found on CNC machining centers. Hydraulic servos, like AC or DC servos, are variable speed motors. Because they are hydraulic motors, they are capable of producing much more power than an electrical motor. They are used on large NC machinery, usually with an electronic or pneumatic control system attached.

LOOP SYSTEMS

Loop systems are electronic feedback systems that send and receive electronic information from the drive motors. Two types of loop systems are currently in use: *open* and *closed loop.* The type of system used affects the overall accuracy of the machine. This is valuable to know before selecting a machine to be used for a close tolerance part. Open loop systems use stepper motors; closed loop systems usually use hydraulic, AC, or DC servos.

Figure 2–4 is a block diagram of an open loop system. The machine gets its information from the reader and stores it in the storage device. When the information is needed, it is sent to the drive motor(s). After the motor has completed its move, a signal is sent back to the storage device that the move has been completed, indicating that the next instruction may be received. Notice that there is no process to correct for error induced by the drive system. (There is no such thing as a perfect positioning drive system or motor.)

A closed loop system block diagram is shown in Figure 2–5. As in an open loop system, the machine gets its information from the reading device and stores it in the storage device. When the information is sent to the drive motor(s), the position of the motor is monitored by the system and compared to what was sent. If an error is detected, the necessary correction is sent to the drive system.

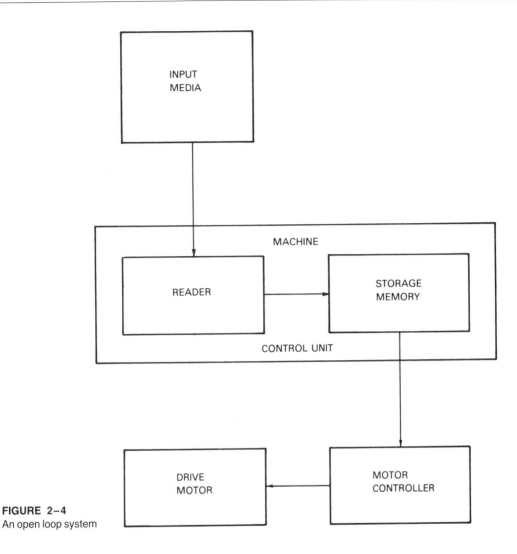

FIGURE 2–4
An open loop system

If the error is large, the machine may simply stop executing the program until the inaccuracy is corrected. This type of system eliminates most errors in position produced by the drive motors.

Recent advances in stepper motor technology have made the manufacture of extremely accurate open loop systems possible. These systems also eliminate the extra hardware and electronics required for closed loop systems.

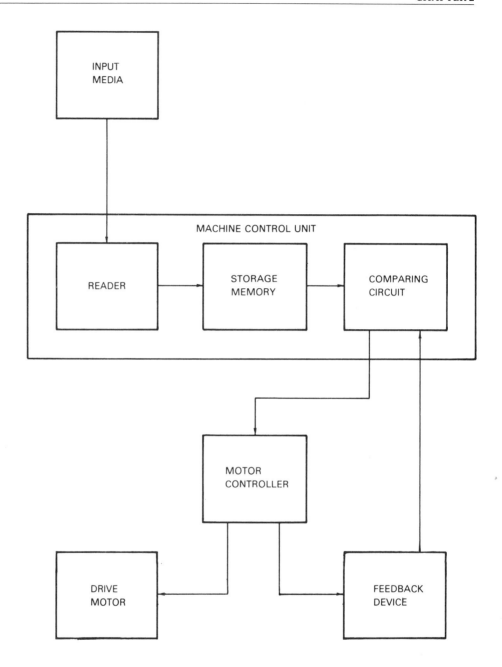

FIGURE 2–5
A closed loop system

THE CARTESIAN COORDINATE SYSTEM

The basis for all machine movement is the Cartesian coordinate system. Figure 2−6 illustrates two- and three-axis coordinate systems. On a machine tool, an *axis* is a direction of movement. The X and Y axes on the coordinate system shown in Figure 2−6(a) can be likened to a two-axis milling machine, where X is the direction of the table travel, and Y is the direction of the cross (or saddle) travel. Figure 2−6(b) illustrates a three-axis coordinate system. Using a vertical mill for example, X would be the table travel, Y the cross (saddle) travel, and Z the spindle travel (up and down). Figure 2−7 illustrates the three-axis system on a vertical mill. Machines are also available in four- and five-axis arrangements. A six-axis layout is shown in Figure 2−8. The milling machines programmed in this text will all use this EIA standard axis arrangement.

Cartesian coordinate systems are divided into quarters (quadrants). In Figure 2−9, the quadrants have been labeled I, II, III, and IV, respectively, in a counterclockwise direction. This is the universal way of labeling axis quadrants. Note that the signs of X and Y change when moving from quadrant to quadrant.

Figure 2−10 shows a number of points on a two-axis Cartesian system. Each of the points can be defined by a set of coordinates. The X-axis value is given first; the Y-axis value second. In mathematics this set of points is called an *ordered pair.* In numerical control programming, the points are referred to as *coordinates.* In later chapters, Cartesian coordinates will be used in writing numerical control programs.

POSITIVE AND NEGATIVE MOVEMENT

Machine axis direction is defined in terms of *spindle movement.* On some axes, the machine slides actually move; on other axes, the spindle travels. For purposes of standardization, the positive and negative direction of each axis is always defined as if the spindle did the traveling. The arrows in Figure 2−7 show the positive and negative direction of spindle movement along each axis. On a vertical mill, the table would move in the direction opposite to the sign indicated. For example, to make a move in the + X direction (spindle right), the table would move to the left. To make a move in the + Y direction (spindle toward the column), the saddle would move away from the column. The Z axis movement is always positive when the spindle moves toward the machine head and negative when it moves toward the workpiece.

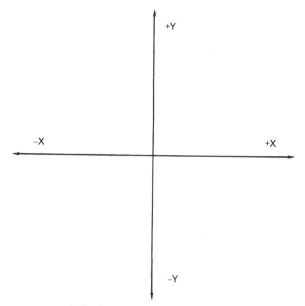

A. TWO-AXIS COORDINATE SYSTEM

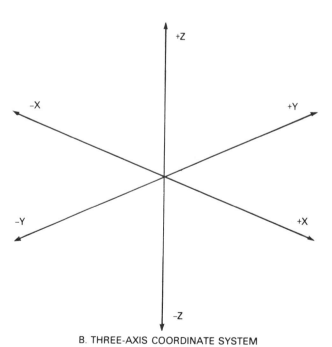

B. THREE-AXIS COORDINATE SYSTEM

FIGURE 2–6
Cartesian coordinate system

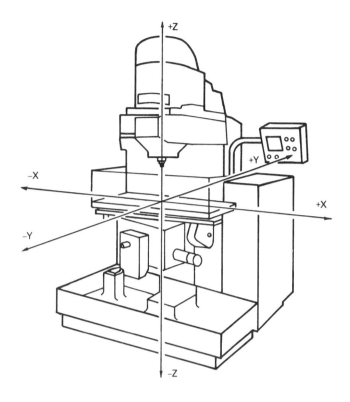

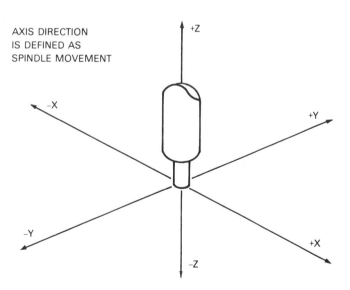

AXIS DIRECTION
IS DEFINED AS
SPINDLE MOVEMENT

FIGURE 2–7
Three-axis vertical mill

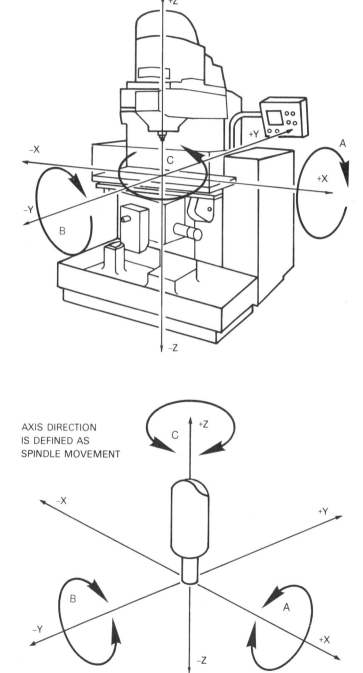

FIGURE 2–8
Six-axis machine layout

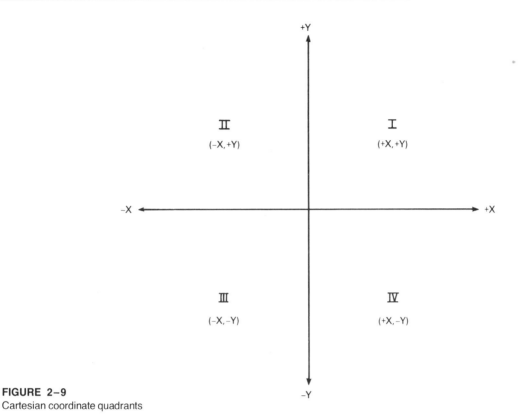

FIGURE 2-9
Cartesian coordinate quadrants

POSITIONING SYSTEMS

There are two ways that machines position themselves with respect to their coordinate systems. These two systems are called *absolute positioning* and *incremental positioning.*

Absolute Positioning

In absolute positioning (Figure 2–11), all machine locations are taken from one fixed zero point. Note that all positions on the part are taken from the X0/Y0 point at the lower left corner of the part. The first hole would have coordinates of X1.000, Y1.000; the second hole coordinates are X2.000, Y1.000; the third hole coordinates are X3.000, Y1.000. Every time the machine moves, the controller references the original zero point at the lower left corner of the part.

Incremental Positioning

In incremental positioning (see Figure 2–12) the X0/Y0 point moves with the machine spindle. Note that each position is specified in relation to the previous one. The first hole coordinates are X1.000, Y1.000; the second hole coordinates are X1.000, Y0.000. The third hole coordinates are again X1.000, Y0.000. After each machine move, the current location is reset to X0/Y0 for the next move. Figures 2–13 and 2–14 illustrate absolute and incremental positioning and their relationship to the Cartesian coordinate system. Notice that with incremental positioning, the coordinate system "moves" with the location. The machine controller does not reference any common zero point.

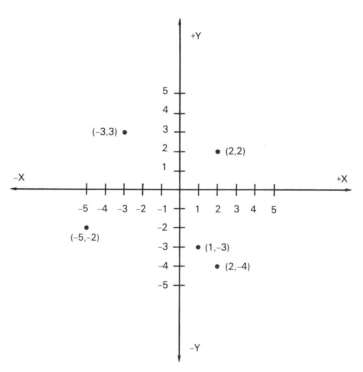

FIGURE 2–10
Cartesian coordinates

SETTING THE MACHINE ORIGIN

Most CNC machinery has a default coordinate system the machine assumes upon power-up, known as the *machine coordinate system*. The origin of this system is called the *machine origin* or *home zero location*. Home zero is usually, but not always, located at the tool change position of a machining center. A part is programmed independent of the machine coordinate system. The programmer will pick a location on the part or fixture. This location becomes the origin of the coordinate system for that part. The programmer's coordinate system is called the *local* or *part coordinate system*. The machine coordinate system and the part coordinate system will almost never coincide. Prior to running the part program, the coordinate system must be transferred from the machine system to the part system. This is known as *setting a zero point*.

There are three ways a zero point can be set on CNC machines: manually by the operator, by a programmed absolute zero shift, or by using work coordinates.

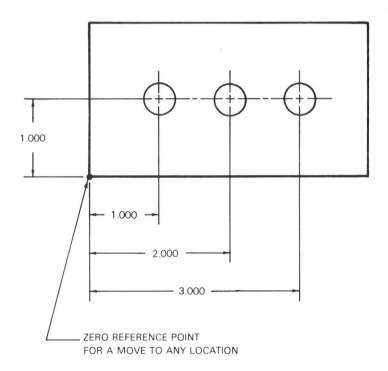

FIGURE 2–11
Absolute positioning

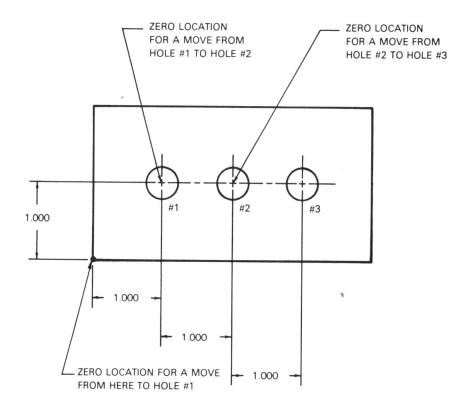

ZERO LOCATION
FOR A MOVE FROM
HOLE #1 TO HOLE #2

ZERO LOCATION
FOR A MOVE FROM
HOLE #2 TO HOLE #3

#1   #2   #3

1.000

1.000

1.000

ZERO LOCATION FOR A MOVE
FROM HERE TO HOLE #1

1.000

FIGURE 2–12
Incremental positioning

Manual Setting

When *manual zero setting* is used, the setup person positions the spindle over the desired part zero and zeros out the coordinate system on the MCU console. The actual keystroke sequence for accomplishing this varies from controller to controller.

Absolute Zero Shift

An *absolute zero shift* is a transferring of the coordinate system which is done inside the NC program. The programmer first commands the spindle to the home zero location. Next, a command is given that tells the MCU how far from the home zero location the coordinate system origin is to be located. An absolute zero shift is given as follows:

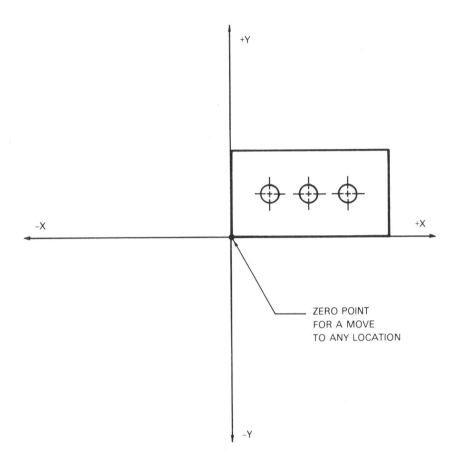

+Y

−X   +X

−Y

ZERO POINT
FOR A MOVE
TO ANY LOCATION

FIGURE 2–13
Relationship of the Cartesian coordinate system to the part when using absolute positioning

    (Send the spindle to home zero)
    N010 G28 X0 Y0 Z0
    (Set the current spindle position)
    (To X5.000 Y6.000 Z7.000)
    N020 G92 X5.000 Y6.000 Z7.000

In line N010 the spindle moves to home zero. Following line N020, even though the spindle did not physically move from home zero, the location of the spindle became X5.0 Y6.0 Z7.0 as far as the MCU is concerned. The machine will now reference the part coordinate system. G92 is a fairly standard command for an absolute zero shift. The term G92 line is often used to describe an absolute zero shift.

If more than one fixture is to be used on a machine, the programmer will want to use more than one part coordinate system. By sending the spindle back to home zero using a G28 X0 Y0 Z0 command, another G92 line can be used in the program to set the second part coordinate system.

Work Coordinates

A *work coordinate* is a modification of the absolute zero shift. Work coordinates are registers in which the distance from home zero to the part zero can be stored. The part coordinate system does not take effect until the work coordinate is commanded in the NC program.

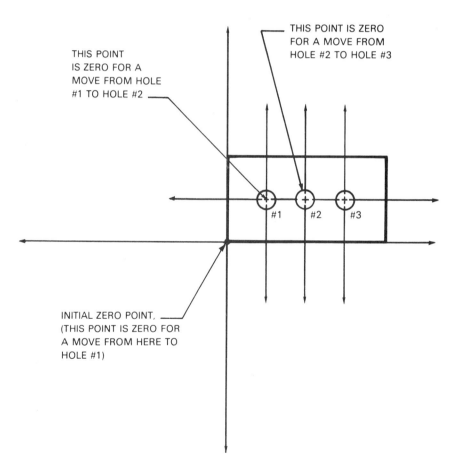

FIGURE 2–14

Relationship of the Cartesian coordinate system to the part when using incremental positioning

When using G92 zero shifts, the coordinate system was changed to the part coordinate system when the G92 line was issued. When using work coordinates, a register can be set at one place in the program and called at another. If more than one fixture is used on a machine, a second part zero can be entered in a second work coordinate, and called up when needed. The work coordinate registers can be set either manually by the operator, or in the program by the NC programmer, without having to send the spindle to the home zero location. This saves program cycle time by eliminating the moves to home zero in the program.

Work coordinates are set and called up in a program by commands called *G-codes*. G54, G55, and G56 would be examples of G-codes to call up different work coordinate registers. The following is an example of using work coordinates:

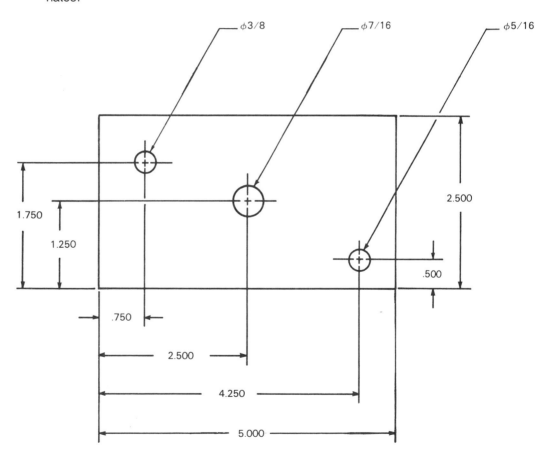

FIGURE 2-15
A datum dimensioned drawing

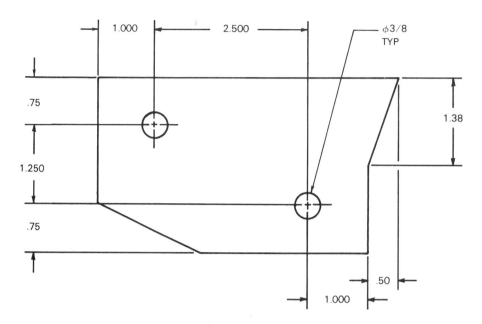

FIGURE 2–16
A delta dimensioned drawing

(Set work coordinate P1 — which is G54)
(and work coordinate P2 — which is G55)
N010 G10 L2 P1 X5.000 Y6.000 Z7.000
N020 G10 L2 P2 X10.000 Y3.000 Z15.000

(Call work coordinate G54 and move)
(To X1.000 Y1.000 Z.500)
N100 G54 X1.000 Y1.000 Z.500

(Call work coordinate G55 and move)
(To X2.000 Y2.000 Z3.000)
N110 G55 X2.000 Y2.000 Z3.000

In line N010, the G54 work coordinate is set to X5.0, Y6.0, Z7.0 from the home zero location. In line N020, the G55 work coordinate is set to X10.0 Y3.0 Z15.0 from home zero. In line N100, the G54 work coordinate is called, activating the part coordinate system. The spindle is moved to X1.0 Y1.0 Z.5 as referenced from the activated part coordinate system. In line N110, the G55 work coordinate is called, activating the second part coordinate system. The spindle is moved to X2.0 Y2.0 Z3.0 as referenced from the second part coordinate system.

Work coordinates remain active once called until cancelled by another work coordinate. They may be called on a line by themselves, or in a line with motion commands as in the example.

DIMENSIONING

In conjunction with NC (or N/C) machinery, there are two types of dimensioning practices used on part blueprints: *datum* and *delta.* These two dimensioning methods are related to absolute and incremental positioning. (Note: although this text uses the NC abbreviation, N/C is equally accepted and is beginning to become the more prominent form.)

Datum Dimensioning

In datum dimensioning, all dimensions on a drawing are placed in reference to one *fixed* zero point. Datum dimensioning is ideally suited to absolute positioning equipment. Figure 2–15 shows a datum dimensioned drawing; notice how all dimensions are taken from the corner of the part.

Delta Dimensioning

Dimensions placed on a delta dimensioned drawing are "chain-linked." Each location is dimensioned from the *previous* one, as shown in Figure 2–16. Delta drawings are suited for programming incremental positioning machines.

In many cases, the drafting practice does not suit the available machines. It is often necessary to calculate program coordinates from print dimensions because a delta dimensioned drawing is being used to program an absolute positioning machine, and vice versa. It is not uncommon to find the two methods mixed on one drawing.

SUMMARY

The important concepts presented in this chapter are:

- There are two types of NC control systems: point-to-point and continuous-path.

- There are four types of drive motors used on NC equipment: stepper motors, AC servos, DC servos, and hydraulic servos.

- Loop systems are electronic feedback systems used to help control machine positioning. There are two types of loop systems: open and closed. Closed loop systems can correct errors induced by the drive system; open loop systems cannot.
- The basis of machine movement is the Cartesian coordinate system. Any point on the Cartesian coordinate system may be defined by X/Y or X/Y/Z coordinates.
- An absolute positioning system locates machine coordinates relative to a fixed datum reference point.
- In an incremental positioning system, each coordinate location is referenced to the previous one.
- The machine coordinate system can be transferred to the part coordinate system manually, by an absolute zero shift, or by use of work coordinates.
- The positive or negative direction of an axis movement is always thought of as spindle movement.
- Machine movements occur along axes which correspond to the direction of travel of the various machine slides. On a vertical mill, the Z axis of a machine is always the spindle axis. The X and Y axes of a machine are perpendicular to the Z axis, with X being the axis of longer travel.
- There are two dimensioning systems used on part drawings intended for numerical control: datum and delta. Datum dimensioning references each dimension to a fixed set of reference points; delta dimensioning references each dimension to the previous one.

REVIEW
QUESTIONS

1. What is the difference between point-to-point and continuous-path systems?
2. What are the four types of drive motors used on numerical control machines?
3. What is a loop system?
4. What is the difference between an open and closed loop system? Why is the choice of loop system important?
5. What is a machine axis?
6. What machine feature determines the positive or negative direction of an axis?
7. What are the two types of positioning systems? What are the differences between them?
8. What three methods are used to origin a part on a CNC machine?
9. What two types of dimensioning systems are used on NC part prints?
10. Give the coordinates of the points shown in Figure 2–17.

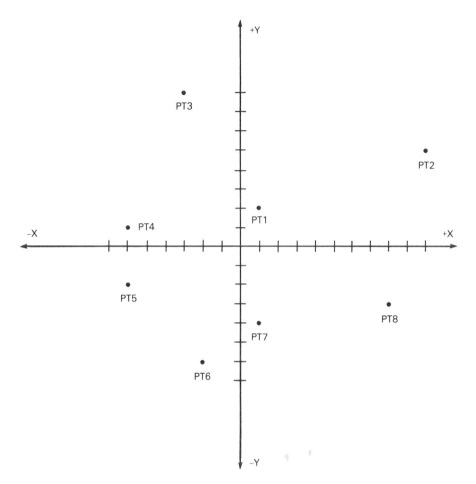

FIGURE 2–17
Coordinate system for review question #10

CHAPTER 3

Process Planning and Tool Selection

OBJECTIVES   Upon completion of this chapter, you will be able to:

- List the steps involved in process planning.
- List the factors that influence the selection of an NC machine, workholding devices, and tooling.
- Describe the types of tools available for hole operations.
- Describe the types of tools available for milling operations.
- Determine the proper grade of carbide insert for a given material.
- Describe some common NC turning tool types.
- Determine the proper spindle RPM to obtain a given cutting speed.
- Explain the importance of proper feedrates.

PROCESS PLANNING

Process planning is the term used to describe the development of an NC part program. A number of decisions must be made by the NC programmer to successfully program a part.

- Which NC machine should be used?
- How will the part be held in the machine?
- What machining operations and strategy will be used?
- What cutting tools will be used?

This process is known as *methodizing* — developing the entire method of producing the part.

Machine Selection

A programmer must first decide which machine will be used. This decision is based on a number of factors.

- What is the programmer's experience?
- What machines are available?

- How many parts are in the order? Are there enough to justify the setup time and higher per hour run cost on a more complex machine?
- Is the particular part best suited for a lathe or a milling machine application?
- Is a vertical or horizontal spindle preferred? Vertical spindles are advantageous for hole drilling and boring operations. Horizontal spindles are best for heavy milling operations. The horizontal orientation of the spindle causes the chips to fall away from the tool, whereas vertical spindles tend to keep the chips packed around the tool.

Fixturing

The next decision to be made is how will the workpiece be held? Again this decision is based on a number of factors, many of them economic.

- Will standard holding devices (clamps, mill vises, chucks, etc.) suffice, or will special fixturing need to be developed?
- What quantity of parts will be run? A large number of parts means special fixturing to shorten the machining cycle may be feasible, even if conventional workholding methods would otherwise be used.
- How elaborate does the fixturing need to be? If many part runs are foreseen, a more durable fixture must be designed. If only one or two part runs are projected, a simpler fixture can be used.
- What will make the best quality part?

Machining Strategy

The machining strategy must be developed before the NC program can be written. Machining sequences used in a part program are determined by the following decisions.

- What is the programmer's experience?
- What is the shape of the part and the blueprint tolerance?
- What tooling is available?
- How many parts are in the order?

Tool Selection

Tool selection is the final important step in process planning. The selection is based on the following decisions.

- What tools are available?
- What machining strategy is to be used?
- How many parts are in the order? If a large number of parts are in the order, special timesaving tools can be made or purchased.

- What are the blueprint tolerances?
- What machine is being used?

The NC Setup Sheet

Once the process planning is finished and the program written, the programmer must communicate to the setup personnel in the shop what tools and fixtures are to be used in the NC program. This information is often placed on setup sheets such as those shown in Figures 3 – 1 and 3 – 2. The setup sheet should contain all necessary information to prepare for the job. Any special instructions to the setup personnel or machine operators should be communicated. Any special notes regarding tooling should also be included.

TOOLING FOR NUMERICAL CONTROL

Tooling is a vital consideration in efficient NC programming. This section is intended to show the prospective NC programmer various standard tooling options available. It is assumed the student has had exposure to cutting tools commonly used in a machine shop. A brief look through any beginning machining text will supply any necessary review of cutting tools.

Cutting Tool Materials

Cutting tools are available in three basic material types: high speed steel, tungsten carbide, and ceramic. The type of tool should be carefully chosen.

High speed steel (HSS) is one type of tool material. It has the following advantages over carbide:

- HSS costs less than carbide or ceramic tooling.
- HSS is less brittle and not as likely to break during interrupted cuts.
- The tools can be resharpened easily.

High speed steel tools have the following disadvantages:

- HSS does not hold up as well as carbide or ceramic at the high temperatures generated during machining.
- HSS does not cut hard materials well.

Tungsten carbide (known simply as carbide) is another material often used for cutting tools. Carbide tools come in one of three basic types. Solid carbide tools are made from a solid piece of carbide. Brazed carbide tools use a carbide cutting tip brazed on a steel shank. Inserted carbide tooling utilizes indexable inserts made of carbide which are held in steel tool holders. Tungsten carbide has the following advantages over high speed steel:

STA. NO.	CRO REG.	TOOL DESCRIPTION
1	—	3.0 DIA. INSERTED FACE MILL W/ .015 R GRADE 883 INSERTS
2	D12	.500 DIA. 4-FLUTE SOLID CARB. END MILL
3	—	NO. 4 × 90° C'DRILL
4	—	1/4 DRILL (.250 DIA.)
5	—	.262 DIA. BORING BAR

NOTES: DRILL POINT ANGLES TO BE 118° INCL.

TAPE NUMBER: 1053

FIXTURE: 6 IN. MILL VISE

TABLE LAYOUT:

X

VISE STOP

PART

Y

Z

.100

VISE

PART

TABLE

NC SETUP SHEET FOR:

MACHINE: VERTICAL MACHINING CENTER

DRWN: WSS

PROG: WSS

DATE: 1-10-89

B/P REV: C

FIGURE 3–1
NC setup sheet for a CNC machining center

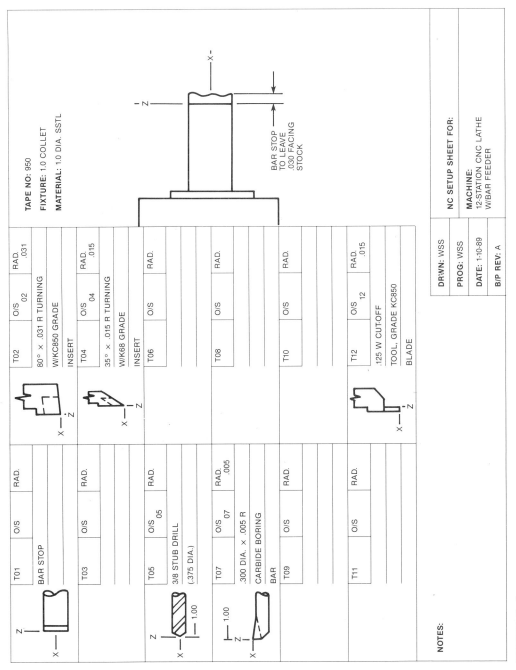

FIGURE 3–2
NC setup sheet for a CNC lathe

- Carbide holds up well at elevated temperatures.
- Carbide can cut hard materials well.
- Solid carbide tools absorb workpiece vibration and reduce the amount of "chatter" generated during machining.
- When inserted cutters are used, the inserts can be easily changed or indexed, rather than replacing the whole tool.

Carbide also has the following disadvantages:

- Carbide costs more than high speed steel.
- Carbide is more brittle than HSS, and has a tendency to chip during interrupted cuts.
- Carbide is harder to resharpen and requires diamond grinding wheels.

Ceramic tooling has made great advances in the past several years. While once very expensive, some ceramic inserts can now be purchased for less than the cost of carbide. Ceramic has the following advantages:

- Ceramic is sometimes less expensive than carbide when used in insert tooling.
- Ceramic will cut harder materials at a faster rate and has superior heat hardness.

Ceramic has the following disadvantages:

- Ceramic is more brittle than HSS or carbide.
- Ceramic must run within its given surface speed parameters. If run too slowly, the insert will break down quickly. Many machines do not have the spindle RPM range needed to use ceramics.

High speed steel is generally used on aluminum and other nonferrous alloys, while carbide is used on high silicon aluminums, steels, stainless steels, and exotic metals. Ceramic inserts are used on hard steels and exotic metals. Inserted carbide tooling is becoming the preferred tooling for many NC applications.

Some carbide inserts are coated with special substances, such as titanium nitride to improve the insert life. These coatings can increase tool life by up to 20 times when used in accordance with the manufacturer's recommended cutting speeds and feedrates.

TOOLING FOR HOLE OPERATIONS

There are four basic hole operations which are performed on NC machinery: drilling, reaming, boring, and tapping.

Drilling

Drills are available in different styles for different materials. Figure 3 – 3 shows a standard twist drill. Even with all the new tooling technology, twist drills remain one of the most common tools for making holes. Drills have a tendency to walk as they drill, resulting in a hole that is not truly straight. Centerdrills such as shown in Figure 3 – 4 are often used to predrill a pilot hole to help twist drills start straight. Drills also produce triangular shaped holes.

If a hole tolerance is closer than .003 inch, a secondary hole operation should be used to size the hole, such as boring or reaming. Large holes are sometimes produced by spade drills (Figure 3 – 5). The flat blades allow good chip flow and economical replacement of the drill tip.

Drill point angle must be considered when selecting a drill. The harder the material to be cut, the greater the drill point angle needs to be to maintain satisfactory tool life. Mild steel is usually cut with a 118-degree included angle drill point. Stainless steels often use a 135-degree drill point.

Drills are available in different types. HSS drills are the most common, but brazed carbide and solid carbide are also used. Carbide drills have a tendency to chip when drilling holes. When drilling hard materials cobalt drills (HSS with cobalt added to the alloy) are used. Cobalt drills have greater heat hardness than HSS drills.

Special drills utilizing carbide inserts have been developed for NC applications (Figure 3–6). The economics of using these tools should be considered by the programmer when hard materials or high run quantities are involved.

FIGURE 3–3
Tapered shank twist drill *(Photo courtesy of Morse Cutting Tools Division)*

FIGURE 3–4
Center drill *(Photo courtesy of DoALL Manufacturing)*

Reaming

Reaming is used to remove a small amount of metal from an existing hole as a finishing operation. Reaming is a precision operation which will hold a tolerance of + or − .0002 easily.

Reamers are made with two basic flute designs: straight fluted (Figure 3−7) and spiral fluted (Figure 3−8). Spiral fluted reamers produce better surface finishes than straight flutes, but are more difficult to resharpen. Reamers are available in three basic tool materials: high speed steel, brazed carbide, and solid carbide.

Boring

Boring removes metal from an existing hole with a single point boring bar. Boring heads are available in two designs: offset boring heads, in which the boring bar is a separate tool inserted into the head, and cartridge type. Cartridge boring heads use an adjustable insert in place of a boring bar.

Boring bars are available in the four material types: high speed steel, solid carbide, brazed carbide, and inserted carbide. Inserted carbide bars are used for large holes, whereas brazed and solid carbide bars are usually supplied in smaller sizes (up to ½−inch diameter).

Tapping

Tapping is used to produce internally threaded holes. They are available in several flute designs. Standard machine screw taps (Figure 3−9) are widely used, especially when tapping blind holes. Spiral pointed taps (known as gun taps) are preferred for thru hole operations. These taps shoot the chips forward and out the bottom of the hole. High spiral taps (Figure 3−10) are used for soft stringy material such as aluminum.

FIGURE 3−5
Spade drill *(Photo courtesy of DoALL Manufacturing)*

**INSERT
POCKET A**
EDGES 1 & 2 USED
IN POCKET "A"

**INSERT
POCKET B**
EDGES 3 & 4 USED
IN POCKET "B"

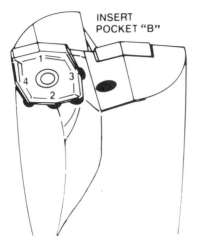

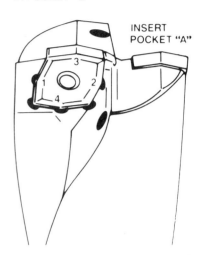

**4 CUTTING EDGES
FROM EACH INSERT**

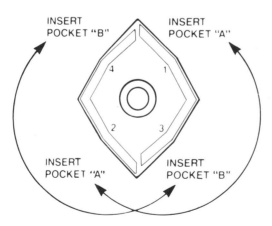

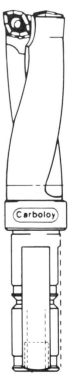

FIGURE 3–6
(Courtesy Carboloy Inc., A Seco Tools Company)

FIGURE 3–7
Straight flute chucking reamer *(Photo courtesy of Cleveland Twist Drill Company)*

FIGURE 3–8
Spiral flute chucking reamer *(Photo courtesy of DoALL Manufacturing)*

A special milling cutter called a thread hob (Figure 3 – 11) is sometimes used to mill a thread in a workpiece. Thread hobs make use of an NC machine's helical interpolation capabilities. Helical interpolation is presented in Chapter 12.

MILLING CUTTERS

The greatest advances in tooling for NC have taken place in the area of inserted milling cutters. *Milling* allows the contouring capabilities of the NC machine to be used to efficiently perform operations that would require special tooling if done manually. Milling cutters can be placed in two basic categories: solid milling cutters and inserted milling cutters. They can be further classified as end mills and face mills.

End Mills

End mills are available in HSS and solid carbide from .032 inch to 2 inches in diameter in two or four flute. Inserted end mills are available from .500 inch to 3 inch diameters. Figure 3–12 shows a four flute HSS end mill. Figure 3–13 shows a two flute solid carbide end mill. Two flute cutters with their deeper gullets are well suited for roughing operations. Four flute end mills, however, are more rigid because of their thicker core. The programmer's experience will determine when to use a two or four flute cutter.

Figures 3–14 and 3–15 illustrate two different types of inserted end mills. Inserted cutters are preferred for NC applications. Inserts are less expensive to replace than an entire tool. By indexing the inserts, four or six cutting edges can be used on one insert. When the insert is used up, it is thrown away rather than resharpened. Inserted cutters may also be used on many different types of workpiece materials by simply changing the inserts from one designed for aluminum, for example, to one designed for stainless steel.

Figures 3–16 and 3–17 show two different styles of inserted ball end mills. Ball end mills are also available in HSS and solid carbide. Ball mills are used for three-, four-, or five-axis contouring work, where the Z-axis will be used. They are also used to produce a given radius on a part.

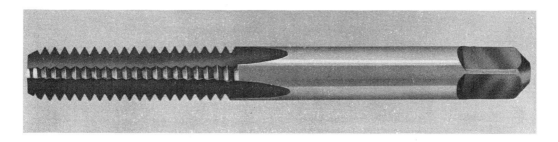

FIGURE 3–9
Machine screw tap *(Photo courtesy of Morse Cutting Tools Division)*

FIGURE 3–10
High spiral coated tap *(Photo courtesy of DoALL Company)*

Figure 3–18 shows a special type of inserted end mill called a cyclo mill, designed by Valenite GTE. It uses a series of round inserts staggered in a helical pattern. This mill can remove large amounts of material at fairly high speeds. It is just one example of inserted tooling that is being developed for NC use.

Face Mills

Face mills differ from end mills in their major application. Face mills are designed to remove large amounts of material from the face of a workpiece. They are manufactured in HSS, brazed carbide, and inserted carbide types.

Face mills are available in sizes from 2 inches to over 8 inches in diameter. Inserted carbide is the most common type of facing tool. The costs of large brazed carbide and HSS mills limit their application to special situations.

Figure 3–19 shows a common type of inserted face mill. Figure 3–20 shows a large diameter face mill. Note the number of inserts used. In Figure 3–

FIGURE 3–11
Thread hob *(Photo courtesy of GTE Valenite)*

FIGURE 3–12
Single end, multiple flute end mill, standard length flutes *(Photo courtesy of Sharpaloy Division, Precision Industries, Inc.)*

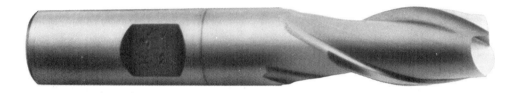

FIGURE 3–13
Solid carbide, two-flute, end mill *(Photo courtesy of DoALL Company)*

21, a special type of mill cutter is shown. This cutter is called a plunge and profile cutter. It is designed to plunge into the material first and then begin the cutting path. This design is a cross between an end mill and a face mill.

SPECIAL INSERTED CUTTERS

A number of special tools have been developed for uses with NC. The NC programmer is always confronted with new ideas to improve productivity. Prospective and experienced programmers should spend time looking at tooling catalogs to become acquainted with current tooling developments.

Figures 3–22 through 3–24 illustrate some of the current tooling ideas developed specifically for NC applications.

Carbide Inserts and Their Selection

Carbide inserts are manufactured in a variety of types and grades. The type of insert describes the shape of the insert. Some common shapes include

Style F
3/4''—1''—1-1/4''

Style G
1-1/2''—2''

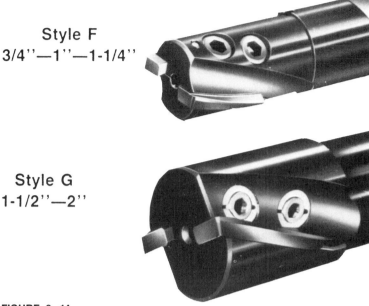

FIGURE 3–14
Inserted carbide end mills *(Photo courtesy of GTE Valenite)*

triangular, 80-degree diamond, 55-degree diamond, and round. The grade of insert refers to the hardness of the insert and application for which it was developed. The NC programmer must be aware of the type and grade of insert available when making tooling selections. Each type of insert is identified by a designation code. The identification system used on an insert will vary depending on the manufacturer. Figure 3–25 shows one such system.

Figure 3–26 illustrates some of the carbide grades available and their applications. Each grade of carbide is designated by an ANSI "C" number designation from C–1 to C–8. In addition, each grade of carbide has also been classified by the ISO. The ISO designation uses a "K" or "P" number, depending on insert hardness. In the United States, the ANSI system is generally used. In other countries, the ISO system is followed. Manufacturers develop their own grade system based on the ANSI or ISO rating. A C–2 ANSI grade insert would be called CQ2 by RWT corporation, K1 by Kennametal Inc., and VC-1 by Valenite GTE Inc. It is necessary therefore, for the programmer to consult the individual manufacturer's catalog to arrive at the proper grade number.

Lathe Tooling

Carbide inserted tools dominate tool selection for CNC turning applications. One of the more popular insert styles is the diamond insert. Figure 3–27 illustrates a series of inserted turning tools. Figure 3 – 28 shows a number of boring bars and toolholders.

A PROCESSING EXAMPLE

In a large company, the formal processing (determining the machine routing through the shop) is done by a processing engineer. The process is then sent to the programming department where the tooling concepts and machining strategy is done. In a small company, the NC programmer does both the processing and programming. It is important that the department processing a job work closely with the programming department to efficiently and economically produce a manufacturing process.

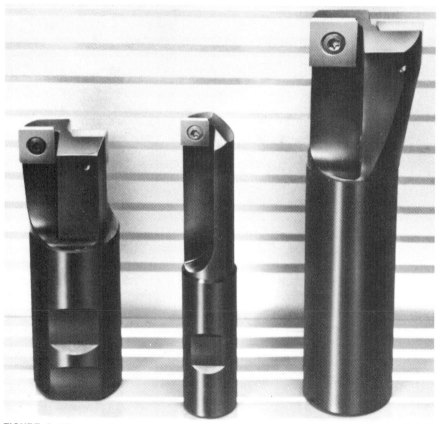

FIGURE 3–15
"Centerdex" two-flute inserted end mills *(Photo courtesy of GTE Valenite)*

Figure 3–29 is a part that is to be machined from an aluminum casting. The casting has .250 dia. of stock to be removed from the 4.000 and 3.000 diameters. The center of the casting was cored to 1.000 inch, and the 1.00 height was cast at 1.250. After consultation with the NC programming department, the process illustrated in Figure 3–30 was developed. The 4.000 inch diameter and .38 dimension are to be done on a conventional lathe. The part will then be routed to a vertical spindle CNC machining center where the balance of the part is to be completed.

The fixture concept to hold the part was developed by the NC programmer. The concept drawing is shown in Figure 3–31. The part will be nested in the 4.0015 diameter fixture bore. It will be clamped with four swiveling clamps, available as a purchased item from a tooling component supplier. This fixture design was based on the following factors.

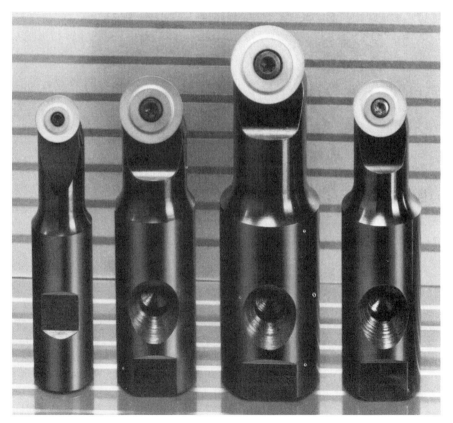

FIGURE 3–16
Ball nose end mills featuring round inserts *(Photo courtesy of GTE Valenite)*

FIGURE 3–17
Ball nose end mills featuring triangular inserts *(Photo courtesy of GTE Valenite)*

- The 4.000 diameter and .38 dimensions were completed in the previous operation, making this feature the logical choice for locating the part.
- The run quantity is only 200 parts. The fixture design is simple, making it economical to build.
- The design is easy to load.

The sequence of the machining operation at the machining center was planned as follows.

FIGURE 3–18
"Cyclo Mill" special multi-inserted milling cutter *(Photo courtesy of GTE Valenite)*

- Face the 1.000 and .25 dimensions using a 3¼ carbide inserted face mill.
- Center drill the .188 and .250 diameter holes. A 90-degree center drill was chosen. The 90 degree-chamfer will provide an edge break at the drilled hole, reducing the amount of deburr time.

- Drill the .188 diameter holes using a ³⁄₁₆ drill. Since drills almost always drill .001 or more oversize, the hole will be comfortably within tolerance.
- Drill the .250 diameter hole using a ¼ drill.
- Mill the 3.000 diameter using a 1 ¼ diameter inserted helical end mill. The end mill has inserts up the sides of the insert, allowing side cutting up to 2.00 deep.
- Using the same end mill, mill the 1.500 diameter bore.

The setup sheet for the NC operation is shown in Figure 3–32.

FIGURE 3–19
Carbide inserted face mill *(Photo courtesy of GTE Valenite)*

SPEEDS AND FEEDS

The efficiency and life of a cutting tool depend upon the cutting speed and the feedrate at which it is run.

Cutting Speed

The *cutting speed* is the edge or circumferential speed of a tool. In a machining center or milling machine application, the cutting speed refers to the edge speed of the rotating cutter. In a turning center or lathe application, the cutting speed refers to the edge speed of the rotating workpiece. Cutting speed

FIGURE 3–20
Large inserted face mill—note number of inserts on cutter *(Photo courtesy of GTE Valenite)*

FIGURE 3–21
Plunge and profile inserted milling cutter *(Photo courtesy of GTE Valenite)*

(CS) is expressed in surface feet per minute (SFM). It is the number of feet a given point on a rotating part or cutter moves in one minute.

Proper cutting speed varies from material to material. Generally, the softer the material, the higher the cutting speed. Recommended cutting speeds for various materials can be found in tables contained in machinists' handbooks, and tooling manufacturers' catalogs. Appendix 6 of this text contains one such chart.

It should be understood that cutting speed and spindle RPM are two different things. A .250-inch diameter drill turning at 1200 RPM has a cutting speed of approximately 75 surface feet per minute. A .500-inch diameter drill turning 1200 RPM has a cutting speed of approximately 150 SFM. The spindle RPM necessary to achieve a given cutting speed can be calculated by the formula:

$$RPM = \frac{CS \times 12}{D \times \pi}$$

Where: CS = cutting speed in surface feet per minute
$\quad\quad\quad D$ = diameter in inches of the tool (workpiece diameter for lathes)
$\quad\quad\quad \pi$ = 3.1416

FIGURE 3–22
Special small diameter inserted end mill *(Photo courtesy of GTE Valenite)*

The cutting speed of a particular tool can be determined from the RPM, using the formula:

$$CS = \frac{D \times \pi \times RPM}{12}$$

On the shop floor, the formulas are often simplified. The following formulas will yield results similar to the formulas just given.

$$RPM = \frac{CS \times 4}{D}$$

$$CS = \frac{RPM \times D}{4}$$

For turning applications, the diameter of the workpiece, rather than the tool diameter, is used to determine the cutting speed and spindle speed. For milling applications, the diameter of the tool is used.

Feedrates

Feedrate is the velocity at which a tool is fed into a workpiece. Feedrates are expressed two ways: inches per minute of spindle travel and inches per revolution of the spindle. For milling applications, feedrates are generally given in inches per minute (IPM). For turning they are expressed most often in inches per revolution (IPR).

Feedrates are critical to the effectiveness of a job. Too heavy a feedrate will result in premature dulling and burning of tools. Feedrates which are too light will result in tools chipping. This chipping will rapidly lead to tool burning and breakage.

Turning Feedrates

The vast majority of turning tools used with NC are inserted tools. The feedrates used vary with material type and insert type. Tables found in manu-

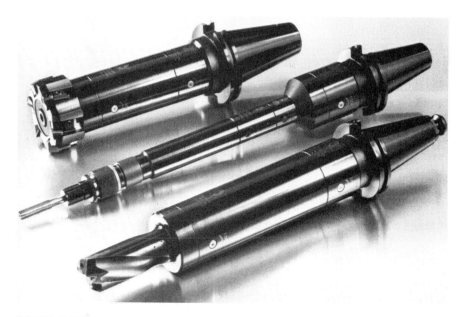

FIGURE 3–23
Special inserted tooling for use with NC. From top to bottom: an inserted milling cutter with interchangeable tooling extensions, a machine tap in a tap holder with interchangeable tooling extensions, and an inserted drill mounted in a holder with interchangeable extensions *(Photo courtesy of GTE Valenite)*

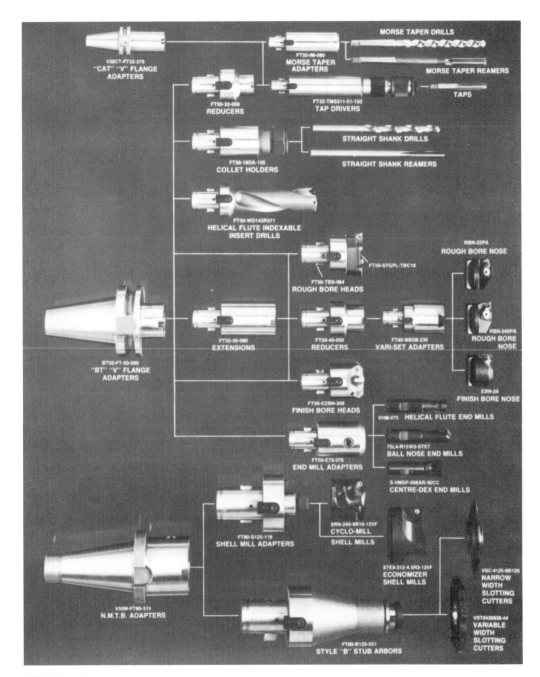

FIGURE 3–24

An NC tooling system featuring tool adapters, interchangeable extensions, tool bodies, boring heads, and arbors
(Photo courtesy of GTE Valenite)

Identification System

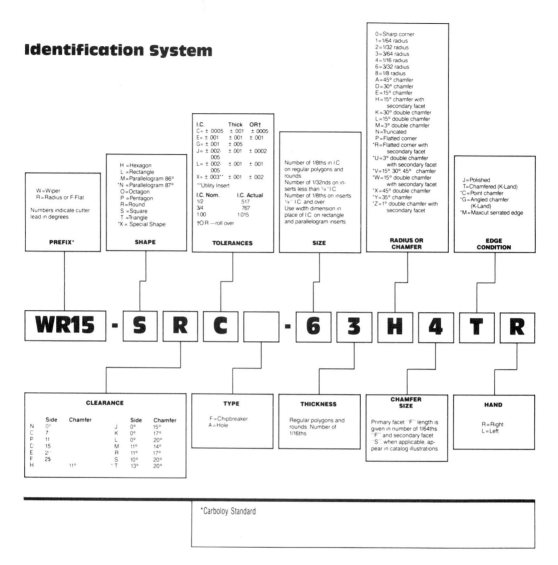

		0 = Sharp corner
		1 = 1/64 radius
		2 = 1/32 radius
		3 = 3/64 radius
		4 = 1/16 radius
		6 = 3/32 radius
		8 = 1/8 radius
		A = 45° chamfer
		D = 30° chamfer
		E = 15° chamfer
		H = 15° chamfer with secondary facet
		K = 30° double chamfer
		L = 15° double chamfer
		M = 3° double chamfer
		N = Truncated
		P = Flatted corner
		*R = Flatted corner with secondary facet
		*U = 3° double chamfer with secondary facet
		*V = 15°, 30°, 45° chamfer
		*W = 15° double chamfer with secondary facet
		*X = 45° double chamfer
		*Y = 35° chamfer
		*Z = 1° double chamfer with secondary facet

PREFIX*

W = Wiper
R = Radius or F-Flat

Numbers indicate cutter lead in degrees

SHAPE

H = Hexagon
L = Rectangle
M = Parallelogram 86°
*N = Parallelogram 87°
O = Octagon
P = Pentagon
R = Round
S = Square
T = Triangle
*X = Special Shape

TOLERANCES

I.C.	Thick	OR†
C = ± 0005	± 001	± 0005
E = ± 001	± 001	± 001
G = ± 001	± 005	
J = ± 002-005	± 001	± 0002
L = ± 002-005	± 001	± 001
X = ± 003**	± 001	± 002

**Utility Insert

I.C. Nom.	I.C. Actual
1/2	517
3/4	767
1.00	1015

†O.R —roll over

SIZE

Number of 1/8ths in I.C. on regular polygons and rounds

Number of 1/32nds on inserts less than ¼" I.C.

Number of 1/8ths on inserts ¼" I.C. and over

Use width dimension in place of I.C. on rectangle and parallelogram inserts

RADIUS OR CHAMFER

EDGE CONDITION

J = Polished
T = Chamfered (K-Land)
*C = Point chamfer
*G = Angled chamfer (K-Land)
*M = Maxcut serrated edge

WR15 - S R C □ - 6 3 H 4 T R

CLEARANCE

	Side	Chamfer		Side	Chamfer
N	0°		J	0°	15°
C	7		K	0°	17°
P	11		L	0°	20°
D	15		M	11°	14°
E	20		R	11°	17°
F	25		S	10°	20°
H		11°	*T	13°	20°

TYPE

F = Chipbreaker
A = Hole

THICKNESS

Regular polygons and rounds. Number of 1/16ths

CHAMFER SIZE

Primary facet "F" length is given in number of 1/64ths. "F" and secondary facet "S" when applicable, appear in catalog illustrations

HAND

R = Right
L = Left

*Carboloy Standard

FIGURE 3–25
Carbide insert identification system *(Courtesy of Carboloy Inc., A Seco Tools Company)*

facturers' catalogs and machining data handbooks are the best sources for turning feedrates. It must be noted that conditions such as part geometry, machine rigidity, and rigidity of the setup will affect both speeds and feedrates. The values given in the tables are starting points. The actual speed and feedrate used during the run will ultimately be determined during the job setup, when the first piece is run.

INSERT GRADE APPLICATION CHART

Cast iron and nonferrous materials	Alloy and tool steels Stainless steels
C-1: Roughing	C-5: Roughing
C-2: General Purpose	C-6: General Purpose
C-3: Finishing	C-7: Finishing
C-4: Precision Finishing	C-8: Precision Finishing

MANUFACTURER'S GRADE DESIGNATION

ANSI Class	ISO Class	Carboloy	Iscar	Kennametal	Sandvik	Valenite
C-8	P-01 P-05	210	IC-80t	K7H	F02	VC-8
C-7	P-10 P-25	350	IC-70	K45	S1P	VC-7
C-6	P-25 P-35	370	IC-50	KC850	S4	VC-55
C-5	P-40 P-50	518	IC-54	—	S35	VC-5
C-4	K-01 K-05	999	IC-4	K11	—	VC-4
C-3	K-10 K-15	905	IC-20	K68	H10	VC-3
C-2	K-20 K-25	883	IC-2	K6	H20	VC-2
C-1	K-30 K-20	820	IC-28	K1	H	VC-1

Note: Most manufacturers produce more than one grade per insert class. Consult the manufacturer's catalog for a complete listing.

FIGURE 3–26
Carbide insert grades

Drilling Feedrates

Drilling feedrates are dependent on the drill diameter. Tables in machinists' handbooks will list recommended feedrates in IPR for given diameters of a given tool material. For example, HSS drills from ⅛ to ¼ inch use feedrates of .002 to .004 IPR. Drills from ¼ to ½ inch use feedrates of .004 to .007 IPR. Drills from ½ to 1 inch use feedrates from .007 to .015 IPR. The final feedrate used will depend upon these factors.

For machining center use, the feedrates given in the tables will have to be converted to IPM values. To accomplish this, the following formula is used:

$$\text{IPM} = \text{RPM} \times \text{IPR}$$

Where: IPM = the required feedrate expressed in inches per minute
RPM = the programmed spindle speed in revolutions per minute
IPR = the drill feedrate to be used expressed in inches per revolution

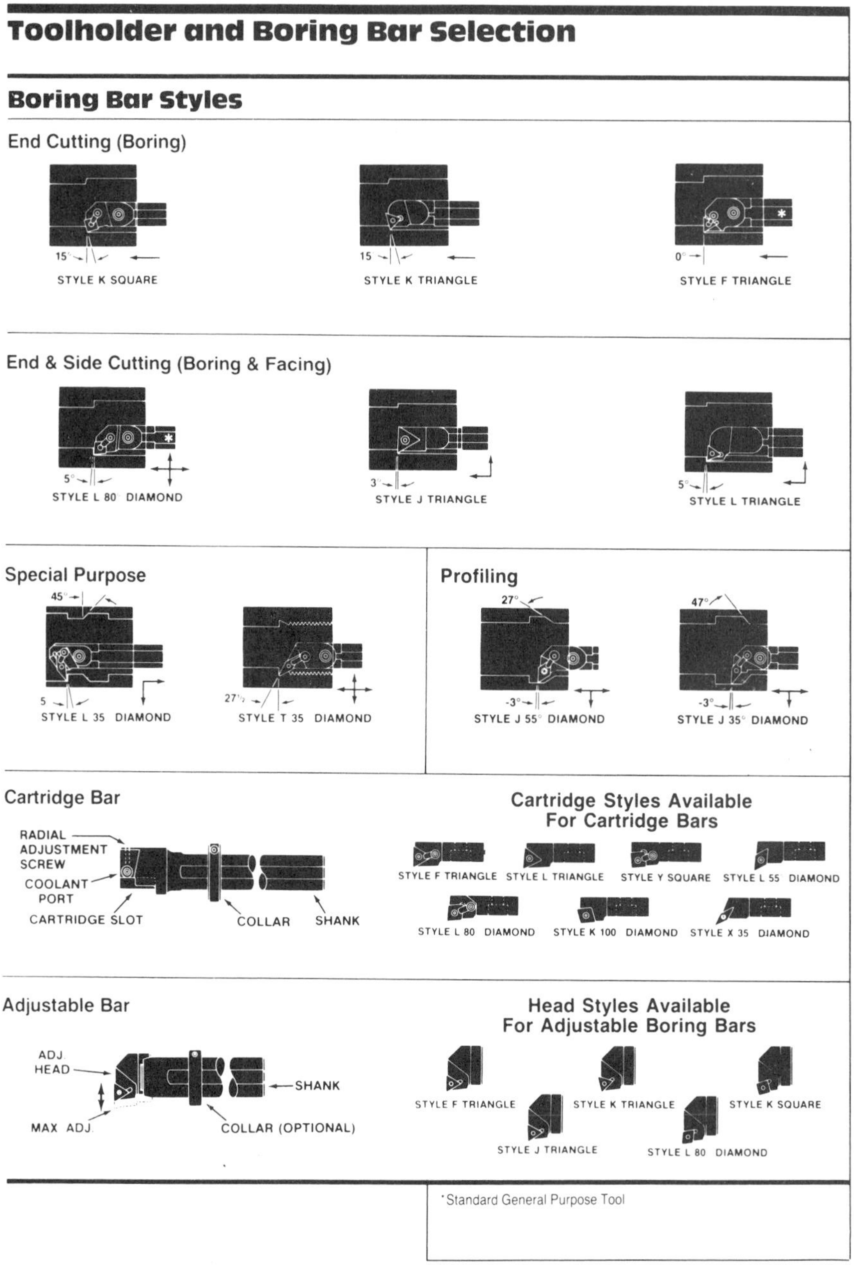

Toolholder and Boring Bar Selection

Boring Bar Styles

End Cutting (Boring)

15° — STYLE K SQUARE

15° — STYLE K TRIANGLE

0° — STYLE F TRIANGLE

End & Side Cutting (Boring & Facing)

5° — STYLE L 80° DIAMOND

3° — STYLE J TRIANGLE

5° — STYLE L TRIANGLE

Special Purpose

45° —
5 — STYLE L 35° DIAMOND

27½° — STYLE T 35° DIAMOND

Profiling

27° —
-3° — STYLE J 55° DIAMOND

47° —
-3° — STYLE J 35° DIAMOND

Cartridge Bar

RADIAL ADJUSTMENT SCREW
COOLANT PORT
CARTRIDGE SLOT
COLLAR
SHANK

Cartridge Styles Available For Cartridge Bars

STYLE F TRIANGLE   STYLE L TRIANGLE   STYLE Y SQUARE   STYLE L 55° DIAMOND

STYLE L 80° DIAMOND   STYLE K 100° DIAMOND   STYLE X 35° DIAMOND

Adjustable Bar

ADJ HEAD
MAX ADJ.
SHANK
COLLAR (OPTIONAL)

Head Styles Available For Adjustable Boring Bars

STYLE F TRIANGLE   STYLE K TRIANGLE   STYLE K SQUARE

STYLE J TRIANGLE   STYLE L 80° DIAMOND

*Standard General Purpose Tool

FIGURE 3–27

Toolholder and boring bar sections—boring bar styles *(Courtesy of Carboloy Inc., A Seco Tools Company)*

Toolholder and Boring Bar Selection

Toolholder Styles

Side Cutting Tools (Turning)

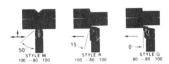

STYLE D SQUARE STYLE R SQUARE STYLE R TRIANGLE STYLE W TRIANGLE STYLE G TRIANGLE STYLE A TRIANGLE STYLE E TRIANGLE

STYLE M
100 — 80 100

STYLE R
100 — 80 100

STYLE G
80 — 80 100

End Cutting Tools (Facing)

STYLE K SQUARE STYLE F TRIANGLE

STYLE K
100 — 80 100

STYLE F
80 — 80 100

Side & End Cutting Tools (Turning & Facing)

STYLE G ROUND STYLE S SQUARE

STYLE L
80 — 80 100

Profiling Tools

STYLE L TRIANGLE STYLE J TRIANGLE STYLE J 55 DIAMOND STYLE P 55 DIAMOND STYLE J 35 DIAMOND STYLE L 35 DIAMOND

Special Purpose Tools

A   G   O
TEE LOCK

STYLE C
TRIANGLE

STYLE T
35° DIAMOND

O   G
ROUND

* Standard General Purpose Tool

FIGURE 3–28

Toolholder and boring bar selection—toolholder styles *(Courtesy of Carboloy Inc., A Seco Tools Company)*

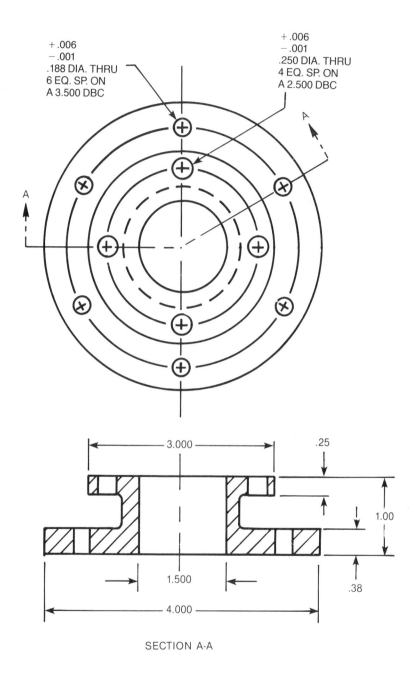

SECTION A-A

FIGURE 3–29
Part drawing

Milling Feedrates

Feeds used in milling depend not only on the spindle RPM, but also on the number of teeth on the cutter. The milling feedrate is calculated to produce a desired chip load on each tooth of the cutter. In end milling, for example, chip load should be .002 to .006 inch per tooth. The recommended chip loads for various mill cutters are given in machinists' handbooks. For inserted cutters, the insert manufacturer's catalog will list recommended chip loads for a given insert. To calculate the feedrate for a mill cut, the following formula is used:

$$F = R \times T \times RPM$$

Where: F = the milling feedrate expressed in inches per minute
          R = the chip load per tooth
          T = the number of teeth on the cutter
       RPM = the spindle speed in revolutions per minute.

Milling feedrates are also affected by machine and setup rigidity, and by part geometry.

In the case of inserted milling cutters, there is another factor which affects feedrates: chip thickness. This is not the chip load on the tooth, but the actual thickness of the chip produced at a given feedrate. Chip thickness will vary

MANUFACTURING PROCESS

Part Number: Adapter   **Job Number:** 000-000-001
Run Quantity: 200   **Material:** Alum. Casting.

OPERATION NUMBER	OPERATION CODE	DESCRIPTION OF OPERATION
010	issue	Issue 356 alum. castings
020	manual lathes	Chuck on 3.250 as cast dia. • Turn 4.000 ± .010 b/p dim to 4.000 ± .001 dia. (tooling dimension). • Face .38 b/p dim.
030	vert. mach. center	Locate parts in fixture NCF-000-100 • Drill .188 + .006 − .001 dia. thru 6 plcs. • Drill .250 + .006 − .001 dia. thru 4 plcs. • Bore 1.500 ± .010 dia. thru 1 plc. • Mill the 3.000 ± .010 dia., hole the 1.000 and .25 dims.
040	burr	• Deburr parts as required.
050	insp	• Inspect parts for b/p conformance.

FIGURE 3–30
Manufacturing process for part shown in Figure 3–29

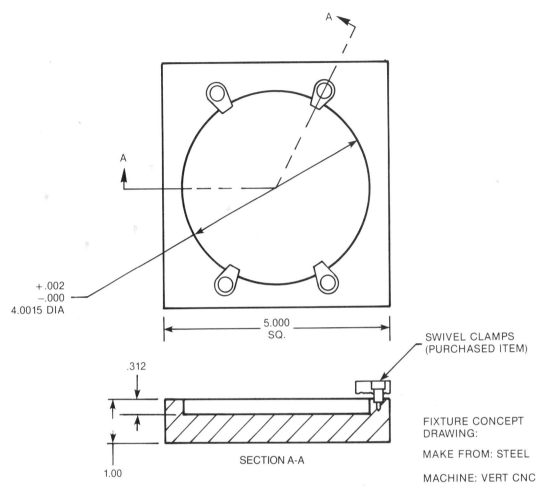

FIGURE 3–31
Fixture concept

some with the geometry of the cutter (positive rake, negative rake, neutral rake), but should be maintained in the range of .004 to .008 inch. Chip thickness less than or greater than these values will place either too little or too great a pressure on the insert for efficient machining. Once a feedrate has been calculated, the chip thickness it produces should be derived. If chip thickness is out of the recommended range, the feedrate should be adjusted to bring it to acceptable limits.

Chip thickness can be calculated by the following formula:

$$CT = \sqrt{\frac{W}{D}} \times R$$

STA. NO.	CRO REG.	TOOL DESCRIPTION
1	D11	3 1/4 INSERTED CARBIDE FACE MILL
2		NO. 4 × 90° C'DRILL
3		3/16 DRILL (.1875 DIA.)
4		1/4 DRILL (.250 DIA.)
5	D15	1 1/4 INSERTED CARBIDE HELICAL END MILL

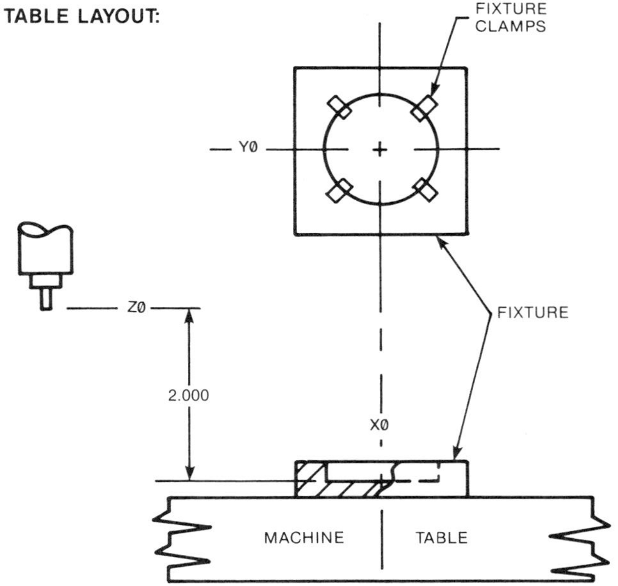

NOTES: TOOL NO. 2 REQUIRES 1.125 MIN EFF. LENGTH

TAPE NUMBER: 1000

FIXTURE: NCF-000-100

TABLE LAYOUT:

FIXTURE CLAMPS

FIXTURE

Y0

X0

Z0

2.000

MACHINE

TABLE

SETUP SHEET FOR CNC MACHINING CENTER	
DRWN: WSS	**MACHINE:** UNIVERSAL VERT. MACH. CENTER
PROG: WSS	
DATE: 3-4-89	**OPER. NO:** 030
B/P REV: A	

FIGURE 3–32
NC setup sheet for CNC machining center

Where: CT = the chip thickness
       W = the width of the cut
       D = the diameter of the cutter
       R = the feed per tooth

If the chip thickness is found to be too small, this modification of the preceding formula can be used to determine an acceptable feedrate:

$$f = \sqrt{\frac{D}{W}} \times CT$$

Where: f = the feed per tooth being calculated
         D = the diameter of the cutter
        CT = the desired chip thickness

This new calculated value of the feed per tooth can then be substituted back into the feedrate formula, and a new feedrate calculated.

Speed and Feed Example

An aluminum workpiece is to be milled using a carbide inserted mill cutter. The cutter is 1.750 diameter × 4 flute. What would be the appropriate spindle RPM and milling feedrate for the workpiece?

An appropriate cutting speed (SFM) for aluminum is 1000 surface feet per minute. Using this value in the spindle speed formula with a cutter diameter of 1.75:

$$RPM = \frac{1000 \times 12}{1.75 \times 3.1416}$$

$$RPM = \frac{12000}{5.4978}$$

$$RPM = 2{,}183$$

The feedrate can now be determined using the feedrate formula. The machinist handbook's tables give a recommended chip load of .002 to .006 inch. A value of .004 per tooth is selected. Using these values in the feedrate formula, the feedrate is calculated.

$$F = 2{,}183 \times 4 \times .004$$

$$F = 34.91 \text{ inches per minute}$$

The chip thickness is then calculated to insure the inserts will not break down prematurely. For this example, it will be assumed the width of the cut to be taken is 1.000 inch wide. Using this value, the chip thickness is determined as follows:

$$CT = \sqrt{\frac{1.000}{1.750}} \times .004$$

$$CT = .755 \times .004$$

$$CT = .00302$$

The chip thickness is less than the recommended minimum of .004. The feed per tooth is therefore calculated as using the feed per tooth formula. A chip thickness of .008 is used.

$$f = \sqrt{\frac{1.75}{1.000}} \times .008$$

$$f = 1.3229 \times .008$$

$$f = .010$$

The new value for the chip load per tooth is substituted in the feedrate formula, and the feedrate recalculated.

$$F = 2183 \times 4 \times .010$$

$$F = 87.32 \text{ inches per minute}$$

The 2183 RPM spindle speed and 87.32 inches per minute feedrate are "book value" rates. They will have to be adjusted up or down depending on the machine, fixture, tool, and workpiece rigidity.

SUMMARY

The important concepts presented in this chapter are:

- Process planning is the term used to describe the steps the programmer uses to develop and implement a part programming.
- The steps in process planning are: determine the machine, determine the workholding, determine the machining strategy, select the tools to be used.
- Tool selection is important to the efficiency of the NC program.
- Cutting tools for NC are made in high speed steel, tungsten carbide, and ceramic.
- Inserted cutters are the preferred tools for NC use.
- Inserts are manufactured in different grades with different applications intended.
- Cutting speed is the edge speed of the tool; it is a function of the spindle RPM and the tool diameter.

- Feedrates that are too heavy will result in excess tool wear and premature tool failure.
- Feedrates that are too light will result in chipping of tools and premature tool failure.
- When calculating milling feedrates, chip thickness must be considered.

REVIEW QUESTIONS

1. What is process planning?
2. List three factors that influence NC machine selection.
3. List three factors used by the programmer in deciding how to hold a workpiece.
4. List two factors that influence the development of machining strategy.
5. List three factors that determine tool selection.
6. What is an NC setup sheet?
7. What are the three basic cutting tool materials?
8. Give one advantage of HSS tools? Give one disadvantage.
9. Give one advantage of carbide tools. Give one disadvantage.
10. On what types of materials are ceramic inserts used?
11. What types of tools are preferred for NC use? Why?
12. What are the four common hole operations performed on NC machines?
13. Why are drilled holes not perfectly straight?
14. Is drill point angle important? Why?
15. Which reamer flute design produces better finishes: straight or spiral?
16. What are the two types of boring head?
17. What is the difference between a tap and a hob?
18. What are the two categories of milling cutters?
19. Into what further classifications can the categories in #18 be put?
20. Why should a programmer be familiar with insert grades?
21. What is cutting speed? Is it the same as the spindle RPM?
22. What happens when a feedrate is too light? too heavy?
23. What two ways are feedrates expressed?
24. How is a milling feedrate calculated?
25. Why should chip thickness be checked after a feedrate is determined?

CHAPTER 4

Tool Changing and Tool Registers

OBJECTIVES Upon completion of this chapter, you will be able to:

- Explain why the speed, repeatability, and accuracy of tool changing are important factors in numerical control.
- Name the two types of tool changes.
- Explain why quick-change tooling is used on NC mills.
- Explain how tooling is used in automatic tool change functions.
- Name the five types of automatic tool changers and briefly describe the operation of each.
- Describe the two basic methods of tool storage.
- Explain what tool registers are and what they are used for.
- Describe what tool offset length is and how it is determined.
- Explain how tool offsets may be entered by the operator during setup and how the programmer allows for this.

This chapter deals with CNC tool changing and tool registers. A good general understanding of these subjects is required for three-axis CNC programming.

TOOL CHANGES

There are two types of tool changes: *manual* and *automatic*. When referring to CNC mills, tool changing is understood to be manual unless otherwise stated. A machining center, on the other hand, incorporates automatic tool change (ATC). It is the tool-changing capability that separates the CNC machining center from CNC milling machines. Machining centers, like milling machines, have the capability to do numerous machining operations (drilling, tapping, spotfacing, and milling, among others). This is opposed to a machine capable of a single function only, such as an NC drilling machine. Figure 4–1 illustrates a CNC milling machine; Figures 4–2 and 4–3 illustrate CNC ma-

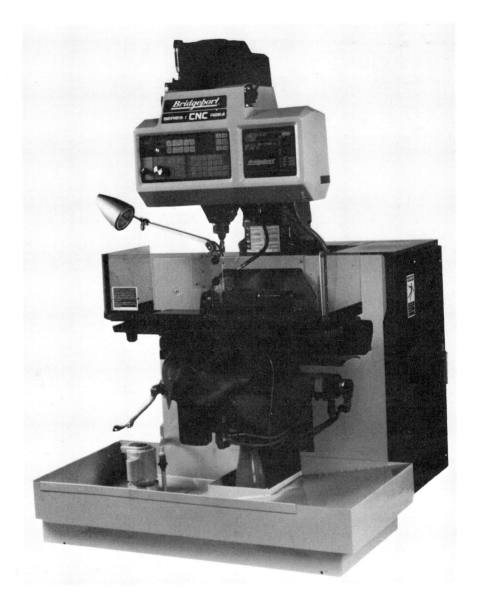

FIGURE 4-1
A vertical spindle CNC milling machine. Note the quick-change tooling system installed in the spindle. *(Photo courtesy of Bridgeport Machines Division of Textron Inc.)*

chining centers. Notice the presence of the tool changer on the machining centers. Also note that a machining center may have either a horizontal or vertical spindle just like a milling machine.

Tooling for Manual Tool Change

What is to be gained by the speed with which a CNC machine can position itself for hole drilling if the tool changes are so lengthy as to cancel the time and accuracy gained by using numerical control? Tool changing greatly influences the efficiency of numerical control, so tool changes should take place as quickly as is safely possible. The tool must not only be (1) accurately located in the spindle to assure proper machining of the workpiece, but (2) the tool must be located as accurately as possible in the same location and (3) in the same relationship to the workpiece each time it is inserted in the spindle. This is known as *repeatability* of a tool—the ability to locate, or repeat, its position in the spindle each time it is used.

Numerical control mills (manual tool change) usually are supplied with or have had added to them some type of quick-change tooling system to accomplish this task.

Most small vertical turret mills are manufactured with what is known as an R-8 spindle taper, which will accept R-8 collets. Figure 4–4 depicts an R-8 spindle and collet. The CNC milling machine in Figure 4–1 has an R-8 spindle employing a quick-change tool-changing system. The R-8 collet is a standard

FIGURE 4–2
A horizontal CNC machining center employing automatic tool change. Note the pivot insertion tool changer on the side. Tools are stored in a matrix magazine. Safety guards have been removed for clarity. *(Photo courtesy of Cincinnati Milacron)*

collet on Bridgeport vertical mills. Since most vertical turret mills are spin-offs of this design, the R-8 spindle has become psuedo-standard on these machines.

R-8 collets and R-8 tool holders require the use of a drawbar. For CNC use, either an automatically tightening drawbar is suppied with the machine, or a quick-change tool system is added.

The quick-change tool system consists of a quick release chuck (which is held in the machine spindle) and a set of tool holders that hold the individual tools needed for a particular part program. The chuck is a separate tool-holding mechanism that stays in the spindle. During a tool change, the tool holder is removed from the chuck (sometimes called the tool changer), and a tool holder containing the next required tool is installed in its place. The tools placed in the tool holders are securely held by means of set screws. Many varieties of these quick-change tool systems are available on the market. Figure 4–5 illustrates a quick-change tooling system in action.

Larger vertical mills and most horizontal mills use another type of spindle taper, called the American Standard Milling Machine Taper (Figure 4–6). Like the R-8 taper, this taper requires the use of a drawbar. If no automatic drawbar is supplied with a machine, a quick-change tooling system is added to the machine to improve tool changing.

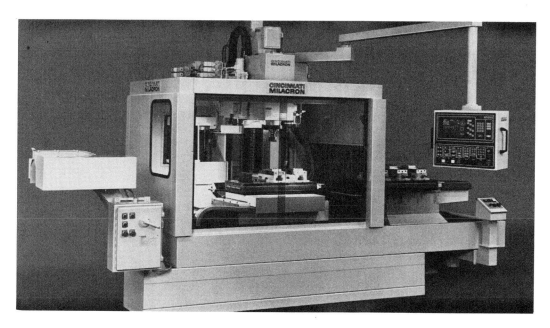

FIGURE 4–3
A vertical spindle CNC machining center *(Photo courtesy of Cincinnati Milacron)*

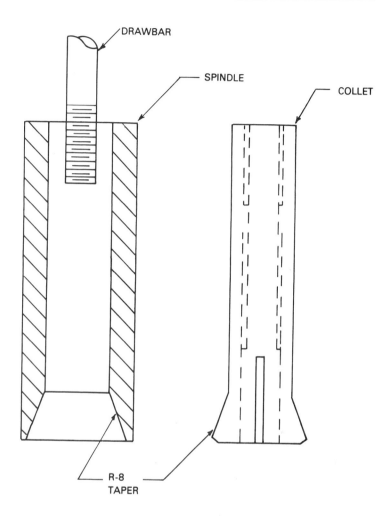

FIGURE 4–4
R-8 spindle and collet

Tooling for Automatic Tool Change

When automatic tool change is used, the requirements for speed and repeatability are even more critical. The machine's tool changer cannot think and correct for misalignment of tooling or tool setup errors like a human being. The tool changer will faithfully carry out its tool-changing cycle and nothing else (that's all it was programmed to do). Tooling used with a tool changer, therefore, must be (1) easy to center in the spindle; (2) easy for the tool changer to grab; and (3) have some means providing for the safe disengagement of the tool changer from the tool once secured in the spindle. Figure 4–7 depicts a common type of tool holder used with ATC (automatic tool change).

The tool changer grips the tool at point A in Figure 4–7 and places the tool in position, aligned with the spindle. The tool changer will then insert the tool

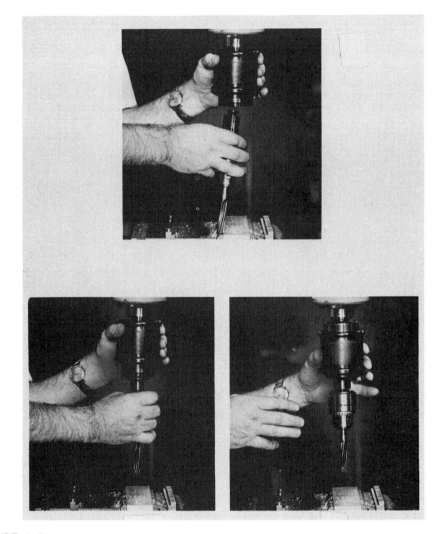

FIGURE 4–5
A quick-change tooling system used for manual tool change *(Photo courtesy of Immotion Quick Change Tool Systems)*

into the spindle. In some cases, insertion of the tool is accomplished by the spindle descending over the tool. As the tool engages the spindle, a split bushing in the spindle will close on the *tool retention knob* (point B in Figure 4–7). This split bushing holds the tool so that the tool changer can release its grip on the tool. The tool is then drawn completely up into the spindle and tightened. Using this procedure insures proper alignment of the tool with the spindle and prevents damage from occurring to the spindle or tool holder taper. Figure 4–8 shows tool insertion using a split bushing.

Another insertion method can be used with a different type of tool holder (Figure 4–9). Here, the tool changer grips the tool in slot A. After the tool is inserted into the spindle, the tool changer moves toward the spindle as the tool is drawn up into the spindle. When the tool is secured in the spindle, the tool changer slides off the tool holder from the side.

AUTOMATIC TOOL CHANGERS

Automatic tool changers, while varied, are made in five basic types: turret head, 180-degree rotation, pivot insertion, multi-axis, and spindle direct. Tools used in automatic tool change are secured in tool holders designed for that pur-

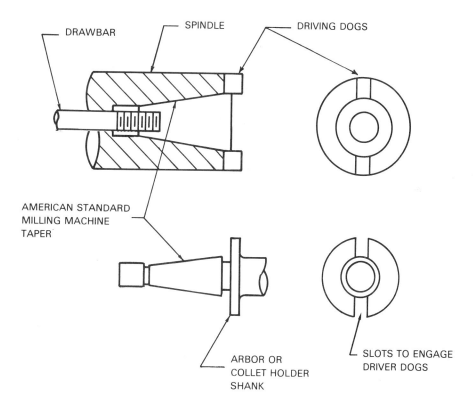

FIGURE 4–6
American Standard Milling Machine Taper used on spindle and arbor (or collet holder shank)

pose. These tool holders are installed directly in the spindle at each tool change by the tool changer. An assortment of tools and tool holders used with CNC machining centers is shown in Figure 4-10.

Turret Head

Tool changing accomplished through the use of a turret head is perhaps the oldest form of automatic tool change. A *turret head* is a number of spindles linked to the same milling machine head, as depicted in Figure 4-11. The tools are placed in the spindles prior to running the program. When another tool is needed, the head *indexes* (moves) to the desired position.

The main disadvantage of this system is the limited number of tool spindles available. In order to use a greater number of tools than available spindles, the operator must remove tools that have already been used and insert those called for later in the program. While other tool-changing methods require less machine operator attention once the program is running, no tool removal is actually performed during the tool change. This results in a very quick tool change. Turret heads are still being used today on certain types of NC machinery such as drilling machines.

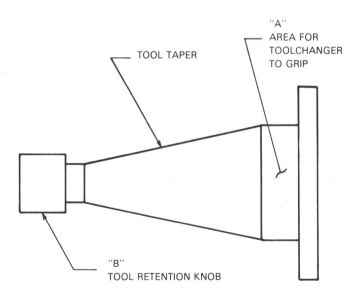

FIGURE 4-7
Typical toolholder used with ATC

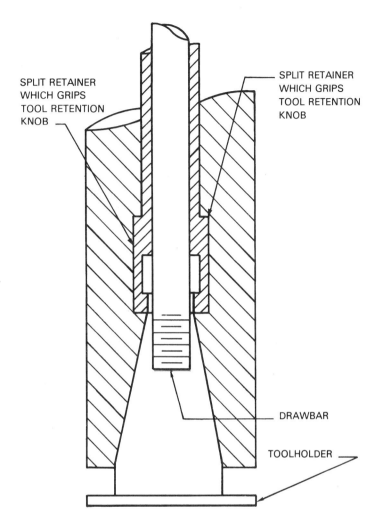

SPLIT RETAINER
WHICH GRIPS
TOOL RETENTION
KNOB

SPLIT RETAINER
WHICH GRIPS
TOOL RETENTION
KNOB

DRAWBAR

TOOLHOLDER

FIGURE 4–8
Split bushing closes over the retention knob to secure the tool as it is drawn into the spindle

180-Degree Rotation

The simplest of the true tool-changing mechanisms is the *180-degree rotation* tool changer (see Figure 4–12). Upon receiving a tool change command, the machine control unit sends the spindle to its fixed tool change coordinates. At the same time, the tool magazine is indexed to the proper position. The tool changer then rotates and engages both the tool in the spin-

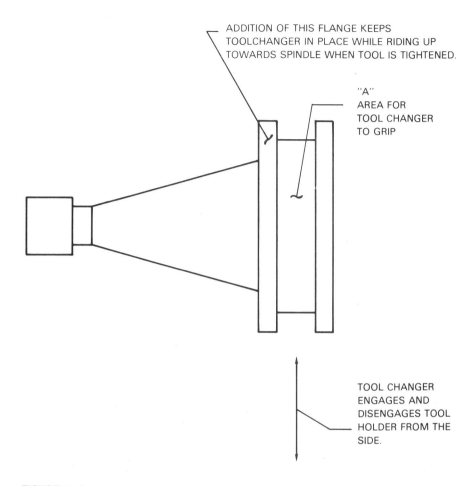

ADDITION OF THIS FLANGE KEEPS
TOOLCHANGER IN PLACE WHILE RIDING UP
TOWARDS SPINDLE WHEN TOOL IS TIGHTENED.

"A"
AREA FOR
TOOL CHANGER
TO GRIP

TOOL CHANGER
ENGAGES AND
DISENGAGES TOOL
HOLDER FROM THE
SIDE.

FIGURE 4−9
Tool changer moves in from the side to grip the toolholder in area A while the tool is secured in
the spindle.

dle and the tool in the magazine at the same time. The drawbar is removed
from the tool in the spindle, and the tool changer removes both tools from their
respective places. The tool changer then rotates 180 degrees and swaps the
tool that was in the spindle with the one that was in the magazine. While the
tool changer is rotating, the magazine repositions itself to accept the old tool
that was removed from the spindle. The tool changer then installs the new tool
in the spindle and the old tool in the magazine. Finally, the tool changer rotates
back to its "parked" position where it remains until needed. The tool change is
thus complete and the program continues.

The principal advantage of this type of changer is its simplicity. The amount of motion involved is minimal and tool changes are fast. The principal disadvantage is that the tools must be stored in a plane parallel to the spindle. The chances of chips and coolant getting on the tool holders are greatly increased compared to those in *side-* or *back-mounted magazines.* Extra protection for the tools must, therefore, be provided. Chips on the tool holder taper will also cause an inaccurate tool change, possibly damaging both the tool holder and the spindle. Some machining centers employ a transfer arm that allows the tool magazine to be stored on the side of the machine. When the tool change command is issued, the transfer arm removes the tool from the magazine and pivots to the front of the machine, positioning the tool to be engaged by the tool changer. The 180-degree rotation tool changer may be used on either horizontal or vertical spindle machines.

Pivot Insertion

An adaptation of the 180-degree rotation tool changer is the *pivot insertion* tool changer (one of the most popular types in use). A pivot insertion system combines the functions of the tool changer and transfer arm. The operation of a pivot insertion tool changer is depicted in Figure 4 – 13. Figure 4 – 14 shows a

FIGURE 4–10
An assortment of tools and toolholders used with CNC machining center *(Photo courtesy of Command Corporation International)*

pivot insertion tool changer on a horizontal machining center. This tool changer has the same physical design as that of the 180-degree rotation tool changer.

When a tool change command is given, the spindle is sent to the tool change location, and the tool magazine is rotated to the proper location for the tool changer to remove the new tool from its slot. The tool changer rotates and removes the new tool from the magazine, which is located on the side of the machine. The tool changer then pivots around to the front of the machine where it engages and removes the tool from the spindle, rotates 180 degrees, and inserts the new tool in the spindle. During this time, the tool magazine has indexed to the proper position to receive the old tool. The tool changer then

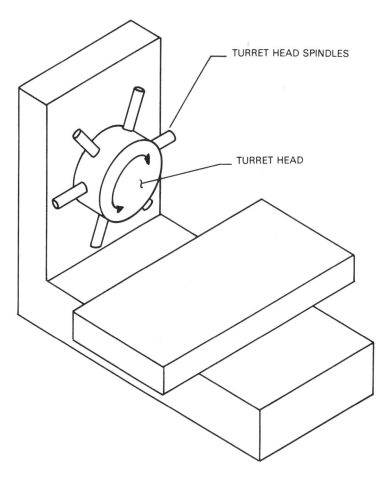

FIGURE 4–11
Turret head tool changer

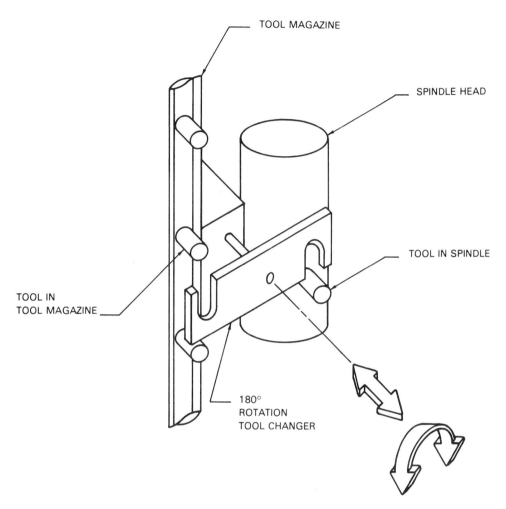

FIGURE 4–12
180-degree rotation tool changer

pivots around to the side of the machine and places the old tool in its slot in the tool magazine. Finally, the tool changer "parks," and the NC program continues.

The main advantage of this system is that the tools may be stored on the side of the machine away from potentially damaging chips. Its disadvantage as compared to the 180-degree rotation tool changer is that pivot insertion requires more motion and therefore results in a more time-consuming tool change.

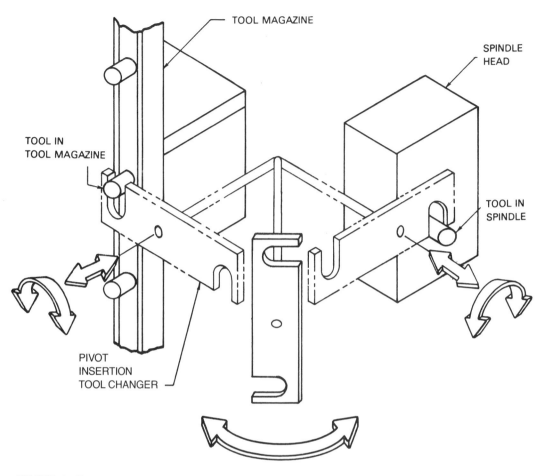

FIGURE 4–13
Pivot insertion tool changer

Multi-Axis

Multi-axis tool change operation is depicted in Figure 4 – 15. This type of tool changer can be used with either side-mounted or back-mounted tool magazines. Its design lends itself very well to use with vertical spindle machining centers. When given a tool change command, the tool changer moves from its "parked" position, grabs the tool that is in the spindle, and removes it. The tool changer then swings (or sweeps) back to the tool magazine and places the old tool into the magazine. The changer then removes the desired tool from the magazine, swings around to the spindle again, and installs the tool in the spindle. Finally, the tool changer returns to "park," and the tool change is completed.

The main advantage of this system is the placement of the tool magazine on the back or side of the machine, where maximum protection can be afforded to the tools. Its disadvantage is the amount of tool handling and motion that must be employed. Today, multi-axis tool changers are giving way to other tool-changing mechanisms such as the 180-degree rotation, and, on vertical spindle machining centers, the spindle direct tool changer.

Spindle Direct

Spindle direct tool changing differs from other types of tool changing in that the tool magazine (carousel) moves directly to the machine spindle or vice versa. Figure 4–16 depicts the operation of a spindle direct tool change. Figures 4–17 and 4–18 illustrate vertical machining centers employing spindle direct tool changers. When a tool change is initiated, the spindle is directed to

FIGURE 4–14
A pivot insertion tool changer on a horizontal machining center using twin matrix tool storage magazines. Guards have been removed for clarity. *(Photo courtesy of Cincinnati Milacron)*

the tool change location. The tool carousel indexes to the required tool slot, moves out of its "parked" position to the tooling position, and engages the toolholder that is in the spindle. The drawbar is then removed from the toolholder,

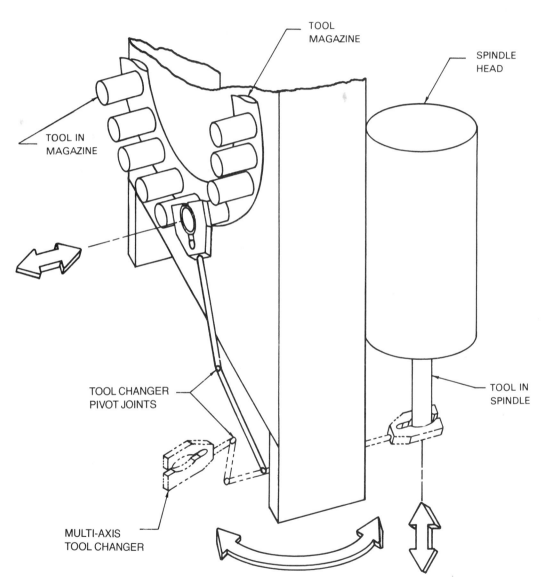

FIGURE 4–15
Multi-axis tool changer

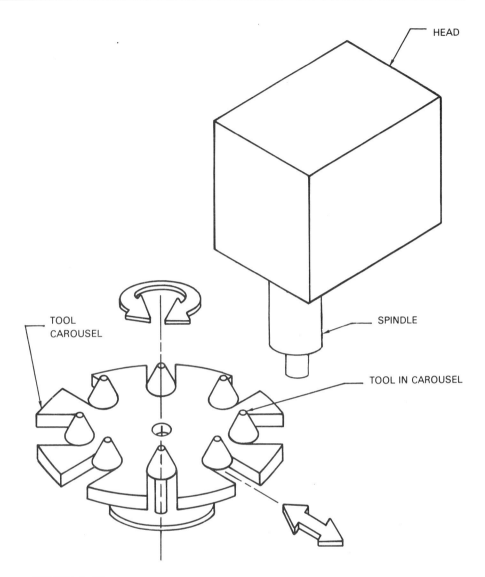

FIGURE 4–16
Spindle direct tool change

and the tool carousel moves downward, removing the tool. The carousel then indexes to align the required tool with the spindle, and moves upward, inserting the tool into the spindle where the tool is secured. Finally, the carousel moves sideways away from the spindle, thus disengaging itself from the tool holder, and returns to its "parked" position. The tool change is now complete.

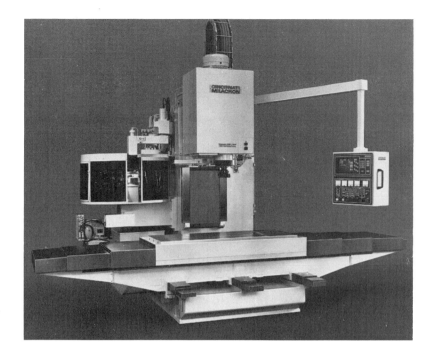

FIGURE 4-17
A vertical spindle machining center. Note the tool changer and carousel tool storage magazine. *(Photo courtesy of Cincinnati Milacron)*

On some large vertical spindle machinery the procedure varies from the one just described. Tool carousels on very large machinery are too large to manipulate easily. Rather than move the carousel, the spindle is moved to the carousel and lowered over it to remove and insert the tools.

TOOL STORAGE

As with tool changers, there are as many tool storage systems as there are manufacturers. However, tool storage systems may be loosely grouped into two types: *carousel* and *matrix*.

Carousel Magazine

A carousel magazine stores the tools in a circular fashion. The machining centers pictured in Figures 4-3, 4-17, and 4-18 employ a tool carousel for tool storage. When a particular tool is called up, the carousel indexes to position the correct tool in the proper location for the tool changer to grab it. In addition to their use for spindle direct tool change on vertical spindle machining centers, carousels may be mounted on carts and moved to the proper spot as needed, such as when spindle direct tool change is employed on large equipment. They may also be mounted on the sides or backs of machines, depending on the type of tool changer used.

FIGURE 4–18
A vertical spindle machining center using carousel tool storage *(Photo courtesy of Bridgeport Machines Division of Textron Inc.)*

Matrix Magazine

Figures 4–2 and 4–14 picture machining centers employing matrix tool magazines. Figure 4–2 shows a single matrix magazine; Figure 4–14 shows a double. In either case, tool holder sockets are incorporated into long chains. When a tool is needed, the chain of sockets moves to position the correct tool socket in line with the tool changer. The advantage of the matrix magazine is that it is not limited to a circular configuration. In an oval configuration, for example, a matrix magazine can store a large number of tools in a limited amount of space.

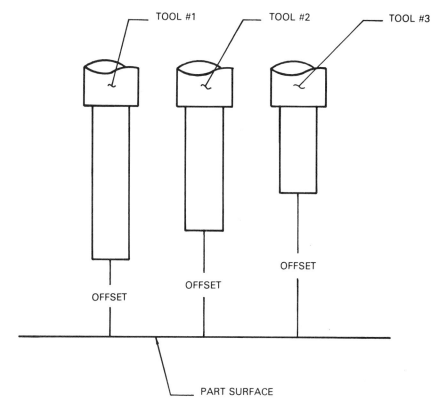

FIGURE 4–19
Tool length offset

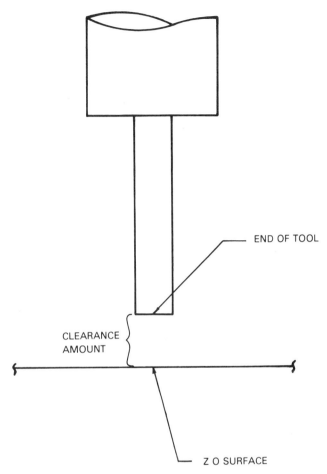

FIGURE 4-20
Tool clearance

TOOL LENGTH AND OFFSET

Tools used for machining vary in length. When using three-axis NC machinery, some means to compensate for the differing tool lengths must be employed. One method of dealing with this problem is to measure the tools prior to writing the program, so the programmed coordinates on any given tool movement will not interfere with the part, a clamp, or the machine table. Typically, the tool lengths are specified in an instruction sheet developed by the programmer that is sent to the shop floor for use in setting up the machine for a particular run.

Sometimes a tool setup drawing is used. Special tool setting equipment is needed to measure the tools accurately. The cost of this equipment, and the labor necessary to set the tools, must be included in the cost of any numerical control system utilizing premeasured tooling. This method of tool length compensation also makes the replacement of broken or dull tools complicated, as such tools must be set to a specific length to function properly. With tape machinery, however, measuring tools is usually the only way to accommodate the various tool lengths.

The advent of CNC machinery has revolutionized tool setting by introducing the *programmable tool register.* A tool register is a memory spot in the computer where the length of a tool may be stored. When a particular tool is called up, the computer checks the tool register to see how much *offset* has been programmed for that tool (see the discussion on *tool offset* that follows). These offset figures are usually entered by the operator at the time the machine is set up for the program run.

Tool Offset

Tool length offset is not the length of a tool but the distance from the part to the bottom of the tool (see Figure 4–19). After a Z0 point has been set, the longest tool to be used is installed in the machine. The table or machine head is then positioned with a specific distance between the tool and the workpiece. This distance is determined by either the programmer or setup man and must be sufficient to clear any clamps or other projections when the spindle is retracted (see Figure 4–20). The programmer may have to leave empty lines in the program for the setup operator to enter tool length offsets. To determine a particular tool offset, the tool is installed in the spindle and the spindle lowered until the tool is at the desired Z0 point on the part. The amount of offset for that tool will be displayed in the axis readout on the MCU. This offset amount, the distance from the tool to the part, is then entered in the MCU. The spindle can then be raised back to Z0, the tool removed, and the procedure repeated for the next tool.

Each time a tool is called up by the program, the offset value for that tool is used to shift the original Z0 point to the position on the part that the programmer desires as the Z0 point for that tool. To fully retract the spindle, the tool offset is cancelled, shifting the Z0 point back to its original position.

There are two basic types of tool offset methods being used on CNC machinery. Some controllers separate the offset from the tool; that is, when a particular tool is called up, the offset to be used with that tool must be called up separately within the CNC program. On other controllers, the offset is associated with a particular tool when it is entered in the machine control unit (MCU). When that particular tool is called up, the offset is automatically included.

SUMMARY

The important concepts presented in this chapter are:

- The speed, repeatability, and accuracy of a tool change greatly influence the efficiency of numerical control.
- There are two types of tool changes: manual and automatic.
- Machinery utilizing manual tool change generally incorporates some type of quick-change tooling system to facilitate the speed and accuracy of tool changes.
- Automatic tool changers are grouped into five categories: turret head, 180-degree rotation, pivot insertion, multi-axis, and spindle direct.
- Tool storage magazines are grouped into two types: carousel or matrix.
- Tool registers are places in the computer's memory to program tool offsets.
- A tool offset is the distance from the bottom of the tool to the desired Z0 point on the part.
- Tool offsets may be entered during setup. In this case the programmer leaves empty blocks in the program in which the tool offsets are placed by the setup man (or operator).

REVIEW QUESTIONS

1. Why is tool changing so important in numerical control?
2. What are the two types of tool changes?
3. On what type of machinery are R-8 spindle tapers found?
4. On what type of machinery is an American Standard Machine Taper used?
5. What type of device is used on manual tool change machines to increase the speed of the tool change?
6. What are the five basic types of tool changers?
7. How does a 180-degree rotation tool changer work? How does a pivot insertion tool changer work?
8. What type of machinery is a spindle direct tool-changing system best suited for?
9. What are the two types of tool storage magazines?
10. What is a tool register?

11. What is a tool length offset?
12. Why are tool registers an improvement over other types of tool length solutions?
13. How does a programmer allow for tool length offsets in a part program?
14. What procedure is used by the operator to determine the tool length offsets?

Programming Coordinates

OBJECTIVES Upon completion of this chapter, you will be able to:

- Explain what a hole operation is.
- Program hole operation coordinates using absolute and incremental positioning.
- Program milling coordinates using absolute and incremental positioning.

HOLE OPERATIONS

To understand how to program coordinates for hole operations, such as drilling, reaming, boring, and tapping, assume that the holes shown on the part drawing in Figure 5–1 are to be drilled using an absolute positioning machine. For hole #1, the coordinates are X0.7500, Y1.7500; for hole #2, the coordinates are X2.0000, Y0.2500; for hole #3, the coordinates are X3.0000, Y1.0000. Note that no plus or minus signs are given with any of these coordinates. If a coordinate is positive, no sign need be given; the machine will assume a positive coordinate unless otherwise indicated. Looking at Figure

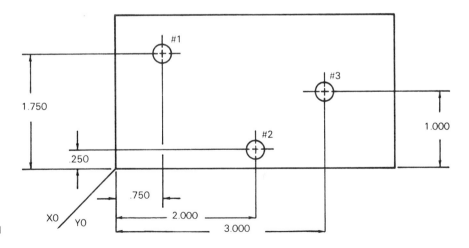

FIGURE 5–1

5–2, try to arrive at the coordinates to drill this part on an absolute positioning machine. The proper coordinates are as follows:

#1	X1.0000, Y0.5000	#5	X − 1.0000, Y − 0.5000
#2	X0.5000, Y1.0000	#6	X − 0.5000, Y − 1.0000
#3	X − 0.5000, Y1.0000	#7	X0.5000, Y − 1.0000
#4	X − 1.0000, Y0.5000	#8	X1.0000, Y − 0.5000

The same principles apply to the parts in Figures 5–1 and 5–2. The difference is that X0/Y0 is located at the center of the part in Figure 5–2. Notice that the signs of X and Y change as the coordinate locations move from quadrant to quadrant.

Figure 5–3 shows the same part as that in Figure 5–1 but delta dimensioned rather than datum dimensioned. Try to derive the proper coordinates to drill the holes in Figure 5–3, using an incremental positioning machine. The coordinates for the holes are as follows:

#1	X0.7500, Y1.7500
#2	X1.2500, Y − 1.5000
#3	X1.0000, Y0.7500

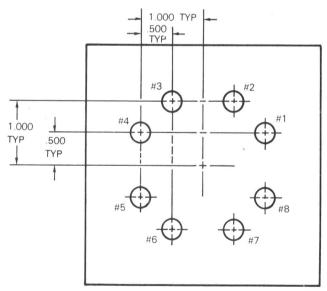

NOTES: 1) PART X0/Y0 IS CENTER OF PART
2) FOR INCREMENTAL MOVES, THE SPINDLE IS ASSUMED CENTERED OVER X0/Y0 AT THE START OF PROGRAM SEQUENCE.

FIGURE 5–2

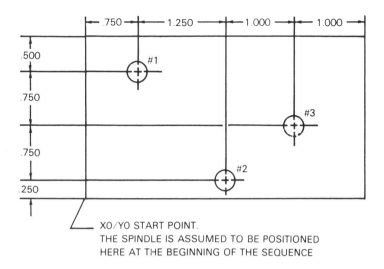

FIGURE 5–3

Notice that the sign of Y was negative when moving to hole #2. Since incremental positioning was being used, hole #1 became the X0/Y0 point for the movement to hole #2. With incremental drawings, it is necessary to add and subtract dimensions in order to correctly program the part, even when using delta dimensioned drawings.

Referring again to Figure 5–2, assume that an incremental positioning machine is to be used. Determine the coordinates necessary to drill the part. The correct coordinates are:

#1	X1.0000, Y0.5000	#5	X0.0000, Y − 1.0000
#2	X − 0.5000, Y0.5000	#6	X0.5000, Y − 0.5000
#3	X − 1.0000, Y0.0000	#7	X1.0000, Y0.0000
#4	X − 0.5000, Y − 0.5000	#8	X0.5000, Y0.5000

Even though this is a datum dimensioned drawing, it is often possible to program incrementally from it.

MILLING OPERATIONS

The system of coordinates presented thus far is used for centering a spindle over a particular location specified on a drawing. This means that when a coordinate location is given to the machine, the center of the spindle is sent to that location. In the case of milling cutters, this technique would cause a problem in that more than the correct amount of stock would be removed from the part (an amount equal to the radius of the cutter). When positioning the spindle for a milling operation, an allowance must be made for the radius of the cutter.

A .500-inch-diameter end mill is to be used to mill the part in Figure 5–4, and an absolute positioning mill will be used. Sending the cutter to X0/Y0 to begin a milling pass from location #1 to location #2 will remove an additional .250 inch of metal from the part that is called out in the drawing. To allow for the radius of the cutter, calculate the cutter coordinate by subtracting half the diameter of the cutter from the coordinate location in each axis. For location #1 the coordinates are X – 0.2500, Y – 0.2500. The coordinates for all four locations are as follows:

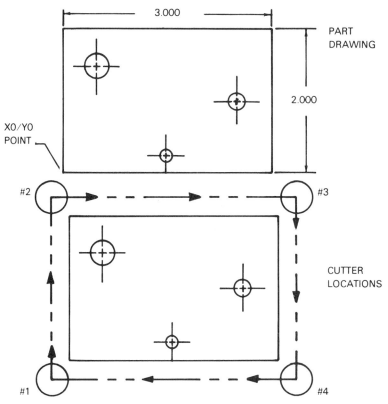

NOTES: CUTTER TRAVEL CLOCKWISE BEGINNING AT LOCATION #1

FOR INCREMENTAL MOVES THE SPINDLE IS ASSUMED TO BE LOCATED AT X0/Y0 WHEN THE SEQUENCE STARTS

FIGURE 5–4

#1   X − 0.2500, Y − 0.2500
#2   X − 0.2500, Y2.2500
#3   X3.2500, Y2.2500
#4   X3.2500, Y − 0.2500

Assume that an absolute positioning machine is to be used to mill points indicated on the part drawing in Figure 5−5. The coordinates for this part are:

#1   X0.7500, Y0.7500
#2   X0.7500, Y1.2500
#3   X2.2500, Y1.2500
#4   X2.2500, Y0.7500

Try to determine the coordinates required to mill the parts in Figures 5−4 and 5−5, using an incremental positioning machine. The correct coordinates for Figure 5−4 are as follows:

#1   X − 0.2500, Y − 0.2500
#2   X0.0000, Y2.5000
#3   X3.5000, Y0.0000
#4   X0.0000, Y − 2.5000

The correct coordinates for Figure 5−5 are as follows:

#1   X.07500, Y0.7500
#2   X0.0000, Y0.5000
#3   X1.5000, Y0.0000
#4   X0.0000, Y − 0.5000

Movement of the Z axis is easier than that of the X or Y axis. To drill any of the parts examined in this chapter, all that is required is to give the Z axis a co-ordinate that would place the end of the tool thru the part.

Assume that the zero point for the Z axis is the top of a .250-inch-thick part. A ¼-inch-diameter hole is to be used to drill a hole in the part. The coordinate for the Z axis would be the thickness of the part plus the length of the drill point. For a ¼-inch drill the coordinate would be Z − 0.3250. The length of a drill point is calculated by multiplying the diameter of the drill by .3. In this case, the drill point is .075 inch long. This length added to the part depth (.250 inch) results in the .3250 length. In practice, it is wise to allow a small amount of additional movement to compensate for differences in drill point and part thickness tolerances. A movement toward the machine table would be a − Z movement. Movement toward the head of the machine would be a + Z movement. Chapter 7 covers three-axis milling and use of the Z axis in more detail. At this point, an understanding of the X and Y movements necessary to program coordinates will suffice.

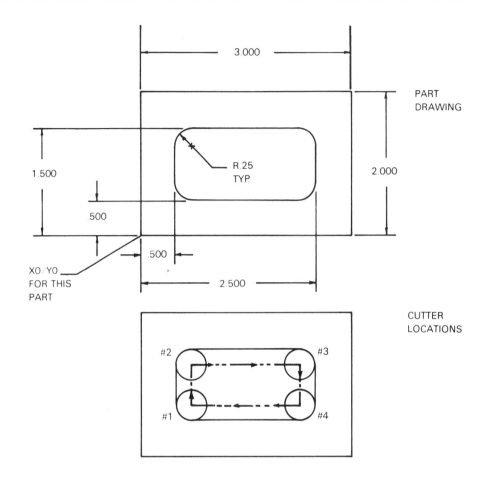

PART
DRAWING

CUTTER
LOCATIONS

NOTES: CUTTER TRAVEL - CLOCKWISE
BEGINNING AT LOCATION #1

FOR INCREMENTAL MOVES, THE SPINDLE IS
ASSUMED TO BE LOCATED AT X0/Y0
WHEN THE SEQUENCE STARTS

FIGURE 5-5

MIXING ABSOLUTE AND INCREMENTAL POSITIONING

CNC machines are capable of both incremental and absolute positioning. This gives the programmer a great deal of flexibility in programming parts. Assume that the part in Figure 5–6 is to be drilled using both absolute and incremental positioning. Hole #1 is to be drilled first, using absolute positioning; holes #2, #3, and #4 are to be drilled next, using incremental positioning; hole #5 is to be programmed next, using absolute positioning; and holes #6, #7, and #8 will be drilled using incremental positioning. Notice that the method of programming these coordinates is similar to the dimensioning used on the part print. Determine the coordinates to program the hole locations before looking at the following correct coordinates.

#1	X0.5000, Y − 0.5000	#5	X2.7500, Y − 2.0000
#2	X0.0000, Y − 0.7500	#6	X0.0000, Y − 0.7500
#3	X1.0000, Y0.0000	#7	X0.7500, Y0.0000
#4	X0.0000, Y0.7500	#8	X0.0000, Y0.7500

METRIC COORDINATES

Some industries have converted all or part of their operations to metric units of measure. Most countries outside of the United States use metric measurement. It is advantageous, therefore, for companies with worldwide markets to use this system in manufacturing their products. Automobile manufacturers are but one example of a number of industries now converting to the metric system. It appears that both the inch and metric systems will be used for quite some time in the United States. Many experts agree that the United States will never fully convert to the metric system, but the numerical control programmer will have to deal with metric measures and should become familiar with their use in the shop.

The metric system in use today is called the *Système International d'Unites,* or the *SI* metric system. There are seven base units used in the metric system. Length is based on the *meter* (m), mass on the *kilogram* (kg), time on the *second* (s), electric current on the *ampere* (A), temperature on the *kelvin* (K), amount of substance on the *mole* (mol), and luminous intensity on the *candela* (cd). All metric units are built on a base-ten system. In the machine shop, measurement is based on the meter, which can be broken down into smaller units. A decimeter is 0.1 meter; a centimeter is 0.01 meter (0.1 decimeter); a millimeter is 0.001 meter (0.1 centimeter).

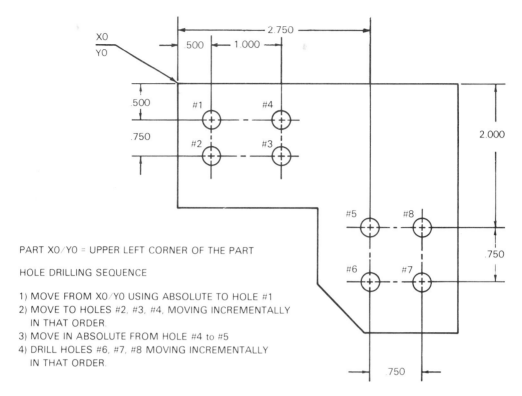

PART XO/YO = UPPER LEFT CORNER OF THE PART

HOLE DRILLING SEQUENCE

1) MOVE FROM XO/YO USING ABSOLUTE TO HOLE #1
2) MOVE TO HOLES #2, #3, #4, MOVING INCREMENTALLY
   IN THAT ORDER.
3) MOVE IN ABSOLUTE FROM HOLE #4 to #5
4) DRILL HOLES #6, #7, #8 MOVING INCREMENTALLY
   IN THAT ORDER.

FIGURE 5–6

    In the inch system, length measurement is based on the yard. Units
smaller than a yard are built on fractions of a yard, foot, or inch, whichever is
most compatible (one-half of a yard = 1½ feet or 18 inches). In the machine
shop, however, measurement is referenced to thousandths of inches. In the
shop, 1″ would be one inch; .500″ however, is not thought of as five-tenths of
an inch but, rather, as five-hundred thousandths of an inch. Therefore, .0005″
is not usually called five ten-thousands of an inch, but five tenths, meaning five
tenths of one-thousandth of an inch. When dealing with metric measurement
in the shop, measurement is referenced to millimeters. One centimeter (1 cm)
is not spoken of as one centimeter but is called ten millimeters (10 mm); one
millimeter is approximately .0394 inch. Units smaller than one millimeter are
also referenced in terms of millimeters; 0.01 is one-hundredth of a millimeter;
0.001 is one-thousandth of a millimeter; 0.001 inch is approximately .0254
millimeter; and .0001 inch is approximately 0.00254 millimeter. Many times a

metric print tolerance will call for a two-place decimal to be held to + or − 0.02 mm. This roughly corresponds to holding an inch dimension to ± .001 inch.

Metric units are easy to work with as long as a company's commitment to metric conversion is carried all the way through from drafting room to tool crib. If metric cutters are available, working with metric dimensions is no problem. Modern CNC machinery has the capability to accept either metric or inch dimensions. The only difference in writing a program in metric versus inch measurements is that the coordinates are expressed differently. If inch tooling is used, it is necessary to convert the cutter sizes to metric units, so that proper milling coordinates can be programmed. To convert an inch dimension to a metric one, multiply the inch dimension by 25.4. To convert a metric dimension to one of inches, multiply the metric dimension by .03937 (or divide the metric dimension by 25.4).

Having learned the use of absolute and incremental positioning, and understanding how the Cartesian coordinate system works, a numerical control program may now be written.

SUMMARY

The important concepts presented in this chapter are:

- To program a hole location coordinate, the center line for the hole is used.
- To program a coordinate for milling operations, the coordinate for the location must include an appropriate allowance for the radius of the cutter.
- For absolute positioning, the datum reference plane remains the X0, Y0 point for all programmed moves.
- For incremental positioning, the current coordinate location is the X0, Y0 point for the next move.
- CNC machines are capable of mixing absolute and incremental positioning. This allows for flexibility in programming.
- Metric measurement in the machine shop is based on the millimeter, where .02 mm is roughly equivalent to .001 inch.
- To convert an inch dimension to millimeters, multiply the inch dimension by 25.4. To convert a metric dimension to inches, multiply the metric dimension by .03937, or divide the metric dimension by 25.4.

REVIEW
QUESTIONS

1. What is a hole operation?
2. Where does the spindle centerline have to be programmed for a hole operation? For a milling operation?
   (Questions #3, #4 and #5 refer to Figure 5–7.)
3. What would the absolute coordinates for holes #2, #3, and #4 be?
4. What would the incremental coordinates be for holes #1, #3, and #4, moving to the holes in that order and starting at the lower left corner of the part?
5. Assume the spindle is positioned at hole #3. What would the incremental coordinates be to move from there to holes #2, #1, and #4, in that order? What would the absolute coordinates be?
6. Using a .625-inch-diameter end mill, what would the four absolute coordinates necessary to mill the part periphery be? What would the incremental coordinates be?
7. Assume that the hole patterns in Figure 5–8 are to be drilled using a CNC machine capable of both incremental and absolute positioning. Give the absolute coordinates to drill hole #1. Give the incremental coordinates to then drill holes a, b, c, and d, respectively. Give the absolute coordinate to drill hole #2, and the incremental coordinates to then drill holes e, f, g, and h, respectively.
8. Convert the following inch measurements to metric measurements.
   a.   .500
   b.   .4375
   c.   .3125
   d.   .125
9. Convert the following metric measurements to inch measurements.
   a.   0.02
   b.   0.005
   c.   2.5
   d.   8.0

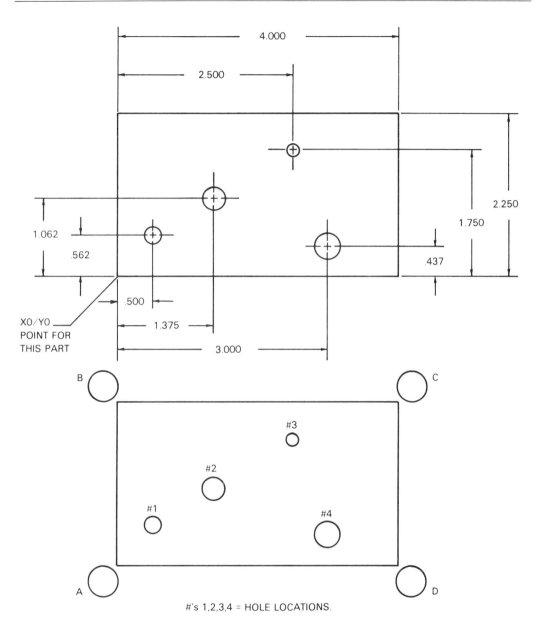

#'s 1,2,3,4 = HOLE LOCATIONS.

A,B,C,D = CUTTER LOCATIONS

FIGURE 5-7

PART DRAWING

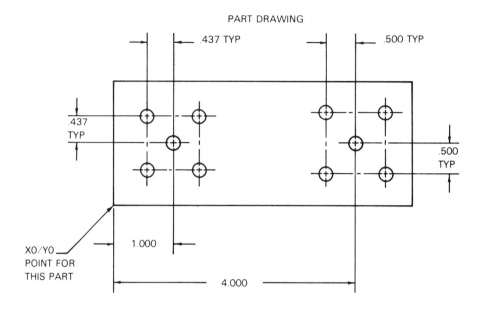

XO/YO
POINT FOR
THIS PART

HOLE LOCATION DRAWING

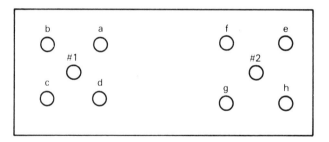

FIGURE 5-8

CHAPTER 6

Two-Axis Programming

OBJECTIVES   Upon completion of this chapter, you will be able to:

*   Write simple two-axis programs in Machinist Shop Language format to perform hole operations.
*   Write simple two-axis programs in word address format to perform hole operations.
*   Write simple two-axis milling programs using Machinist Shop Language.
*   Write simple two-axis milling programs using word address programming format.

This text is concerned primarily with manual programming of CNC machinery. Each successive chapter will introduce a more advanced level of numerical control programming. For purposes of continuity, two basic machines will be used for the next several chapters. The following point cannot be overemphasised: *no two CNC machines program exactly alike.* There are, however, similarities between them. By learning to write programs for these machines, only minimal effort will be required to program other CNC machines.

Programming in this text is done primarily in two formats throughout: *Machinist Shop Language* and *word address format.* The instructional examples used in the next several chapters are milling and drilling examples. Chapters 13 and 14 deal with CNC lathes. The first machine programmed in this chapter is a CNC mill equipped with an Analam Crusader II controller. This machine uses a conversational programming language called *Machinist Shop Language.* The second machine programmed is a vertical machining center equipped with a General Numerics controller; it is programmed using word address format. This chapter deals with two-axis programming. Chapter 7 will introduce three-axis programming.

MACHINIST SHOP LANGUAGE

Machinist Shop Language, as mentioned above, is a conversational language. The commands used in programming with this language are common shop words rather than NC codes. Conversational languages are easy to learn since they use English commands. The MCU on a conversational language machine converts the English commands into the codes required for the program.

In Machinist Shop Language, the machine is given its instructions by means of commands placed in program lines called *events*. There are two types of events used in a Machinist Shop Language program: events *with motion* and events *without motion*. Events with motion position the machine to a desired coordinate location; events without motion perform such tasks as assigning feedrates or initiating planned program dwell. Following are some Machinist Shop Language commands.

Commands

X— When used before a number, X designates an X-axis coordinate. An X-axis coordinate is entered in the format XXX.XXXX for an inch coordinate. The format is XX.XXX for metric coordinates.

Y— When used before a number, Y designates a Y-axis coordinate. A Y-axis coordinate is entered in the format YYY.YYYY for an inch coordinate. The format is YY.YYY for metric coordinates.

Z— When used before a number, Z designates a Z-axis coordinate. A Z-axis coordinate is entered in the format ZZZ.ZZZZ for an inch coordinate. The format is ZZ.ZZZ for metric coordinates.

F— Initiates a move at the programmed feedrate.

A— Specifies that absolute positioning mode is to be used.

I— Specifies that incremental positioning mode is to be used.

FEED—Assigns a feedrate to be used as needed in the program. The format for the FEED command in inch mode is FEED ##.##. For metric mode the format is FEED ###.#.

DWELL—Causes the machine to halt execution of the program. A dwell may be entered with a time element specified, so that the program will recommence after the specified time interval. If no time element is specified, the machine will halt execution of the program until the start button is pushed. When entering a timed DWELL, the time interval is entered as an X-axis value.

TOOL—Acts like an untimed dwell, halting program execution until the start button is depressed. TOOL also assigns length and cutter diameter values to a particular tool. The format for a TOOL command is TOOL # where # is the tool desired (for example, TOOL 1, TOOL 2, etc.). The format for assigning tool lengths and diameters is TOOL 10## where ## is the number of the tool assigned (for example, TOOL 1001, TOOL 1025, etc.). The use of TOOL 10## will be further discussed in Chapter 7.

END—Signals the end of a program. It is also used to mark the end of a do loop or a subroutine. Do loops and subroutines will be covered in Chapter 11.

This is not a complete listing of all Machinist Shop Language commands (see Appendix 2). Other commands will be introduced as more operations are presented.

DRILLING IN MACHINIST SHOP LANGUAGE

The CNC mill to be programmed using Machinist Shop Language is a vertical turret milling machine with a computer numerical control unit. It is a continuous-path machine, with the ability to use both absolute and incremental positioning. The milling machine uses manual tool change. This means that the operator must change the tools. Since the machine has no *set* tool change location, the location must be specified in the program.

The part in Figure 6–1 is to be drilled on the CNC mill. Notice the X0/Y0 point for the part in the lower left corner. A tool change position has been selected at X – 2.0, Y – 1.5. A separate tool change location positions the spindle out of the way of the workpiece during tool changes. It also aids in removing the part from the vise or fixture. In Figure 6–2, a metric part is shown.

The first program for this part is shown in Figure 6–3. This program was written using absolute positioning. Figure 6–4 is the same program but written for the metric part pictured in Figure 6–2. The programming logic for the two parts is identical. The only difference between the two programs is the coordinates. When entering a program in the MCU of this machine, a button is first pushed to select inch or metric input.

Leading and Trailing Zeros

Before presenting program explanations, a discussion of leading and trailing zeros is in order. In the shop and on part drawings, dimensions often contain trailing zeros. For example, the dimension .500 contains two trailing zeros. On part dimensions, the trailing zeros are necessary to communicate the significance of a particular dimension (that is, three-place versus two-place decimals). Sometimes, for the sake of clarity, a leading zero is used, as in 0.500. Early NC equipment required the use of leading and/or trailing zeros in specifying any coordinate. Many CNC controllers do not require the use of either leading or trailing zeros; thus, .500 may be entered as .5 on these machines. The CNC machine will locate all programmed coordinates within the resolution of the machine. The programs in Figures 6–3 and 6–4 reflect this practice of omitting leading and trailing zeros. Not all controllers allow this practice, but for purposes of standardization in teaching, it is assumed that all controllers in this text do.

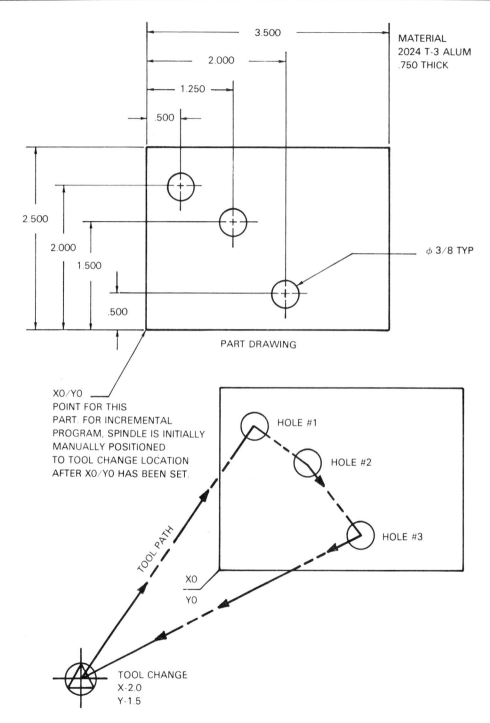

3.500

2.000

1.250

.500

MATERIAL
2024 T-3 ALUM
.750 THICK

2.500

2.000

1.500

.500

φ 3/8 TYP

PART DRAWING

X0/Y0
POINT FOR THIS
PART. FOR INCREMENTAL
PROGRAM, SPINDLE IS INITIALLY
MANUALLY POSITIONED
TO TOOL CHANGE LOCATION
AFTER X0/Y0 HAS BEEN SET.

HOLE #1

HOLE #2

HOLE #3

TOOL PATH

X0
Y0

TOOL CHANGE
X-2.0
Y-1.5

FIGURE 6-1
Hole operations part drawing, nonmetric

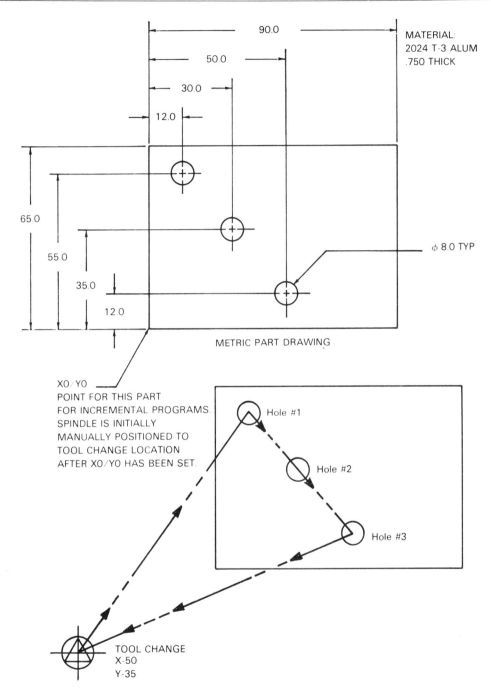

FIGURE 6-2
Hole operations part drawing, metric

```
X0/Y0 = LOWER LEFT CORNER OF PART
TOOL CHANGE = X-2 Y-1.5
SPINDLE SPEED = 2500 RPM

1    X-2 Y-1.5   R A
2    TOOL 1                  REM: 3/8 DRILL
3    X.5 Y2      R A
4    DWELL                   REM: DRILL HOLE
5    X1.25 Y1.5 R A
6    DWELL                   REM: DRILL HOLE
7    X2 Y.5      R A
8    DWELL                   REM: DRILL HOLE
9    X-2 Y-1.5   R A
10 END
```

FIGURE 6-3
Machinist Shop Language drilling program, nonmetric absolute positioning, for the part in Figure 6-1

```
X0/Y0 = LOWER LEFT CORNER OF PART
TOOL CHANGE = X-50 Y-35
SPINDLE SPEED = 2500 RPM

1    X-50 Y-35   R A
2    TOOL 1                  REM: 8mm DRILL
3    X12 Y55     R A
4    DWELL                   REM: DRILL HOLE
5    X30 Y35     R A
6    DWELL                   REM: DRILL HOLE
7    X50 Y12     R A
8    DWELL                   REM: DRILL HOLE
9    X-50 Y-35   R A
10 END
```

FIGURE 6-4
Machinist Shop Language drilling program, metric absolute positioning, for the part in Figure 6-2

PROGRAM EXPLANATION—ABSOLUTE POSITIONING

(Refer to Figures 6-3 and 6-4.)

Notice the use of the term "REM" in the program. "REM" is used in this case for the word "remark." Remark statements are usually provided for by the controller manufacturer. They are ignored by the controller; some symbol or G code is used to precede the statement (in this case the term "REM"). It is good practice to use remark statements in a program manuscript. They help not only the operator to determine what is happening in the program but also aid in debugging the program prior to running the first part.

EVENT 1

- X/Y tool change coordinates. These coordinates position the spindle to the tool change location. The operator will install a 3/8-inch self-centering drill in the spindle (8 mm drill in the metric program).
- R—Instructs the machine to make the move at rapid traverse, which on this machine is 100 in./min.
- A—Tells the machine to use the absolute positioning mode.

A and R are programmed in, using a button on the MCU console. Once activated, rapid/feedrate movement and absolute/incremental positioning remain in force until cancelled by its complementary command. When writing a program manuscript, it is good practice to enter a rapid/feedrate mode command and a positioning system command on every event with motion. If the program is entered in the MCU via the MCU keyboard, the programmer will be reminded to double check the positioning and feed modes for the correct setting at each affected event.

EVENT 2

- TOOL 1—Causes the machine to stop executing the program until the operator depresses the start button. This gives the operator time to safely install the tool without danger of machine movement. The remark (REM) notes that the operator will install a 3/8-inch self-centering drill in the spindle (8-mm drill in the metric program). This machine has no computer control of the spindle motor except the panic stop button, which kills power to the drive motors and the spindle. The operator would turn on the spindle motor to the correct speed at this time.

EVENT 3

- X/Y coordinates—To move from tool change to hole #1.
- R—Specifies rapid traverse.
- A—Specifies that absolute positioning is being used.

EVENT 4

- DWELL—Halts execution of the program until the start button is pushed. The operator then drills hole #1.

EVENT 5

- X/Y coordinates—To move from hole #1 to hole #2.
- R—Specifies rapid traverse.
- A—Specifies that absolute positioning is being used.

EVENT 6

- DWELL—Halts program execution. Hole #2 is drilled.

EVENT 7

- X/Y coordinates—To move from hole #2 to hole #3.
- R—Specifies rapid traverse.
- A—Specifies absolute positioning.

EVENT 8

- DWELL—As in events 4 and 6, the program is halted and hole #3 is drilled.

EVENT 9

- X/Y coordinates—To move from hole #3 to tool change. It is good practice to send the spindle back to tool change at the end of a program, even if machining only one part. Aside from forming good habits for multipiece programming, this practice safely positions the tool out of the way of the part.
- R—Specifies rapid traverse.
- A—Specifies absolute positioning.

EVENT 10

- END—This instructs the machine that this is the end of the program.

All the moves made in the absolute and incremental programs are in the rapid traverse mode. When drilling holes, the faster the speed of the axis movements between holes, the quicker and more economical is the machining. No feedrate need be entered in the program for movement to hole locations as rapid movement is the most efficient mode. Had this been a three-axis machine, it would have been necessary to assign a feedrate to control the rate at which the drills went thru the part.

PROGRAM EXPLANATION—INCREMENTAL POSITIONING

(Refer to Figures 6–5 and 6–6.)

Figure 6–5 is an incremental program for the part in Figure 6–1. Figure 6–6 is an incremental program for the part in Figure 6–2. The program sequence is identical to the two absolute positioning programs just discussed. Only the coordinates for the three hole locations and tool change location differ. Note that it is mandatory for the spindle to begin and finish at the same location for the program to cycle correctly on to the second part. For the programs in Figures 6–5 and 6–6, the corner of the part was established as the zero point at setup. The machine was then manually positioned to the tool change location. No move to tool change is possible in the first event because the spindle is already in position.

```
PROGRAM STARTING X0/Y0 = TOOL CHANGE
SET TO TOOL CHANGE = X-2 Y-1.5  PRIOR  TO
RUNNING FIRST CYCLE.
TOOL = 3/8 DRILL
SPINDLE SPEED = 2500 RPM

1   TOOL 1              REM: 3/8 DRILL
2   X2.5 Y3.5  R I
3   DWELL              REM: DRILL HOLE
4   X.75 Y-.5  R I
5   DWELL              REM: DRILL HOLE
6   X.75 Y-1   R I
7   DWELL              REM: DRILL HOLE
8   X-4 Y-2    R I
9   END
```

FIGURE 6-5
Machinist Shop Language drilling program, nonmetric incremental positioning, for the part in
Figure 6-1

```
PROGRAM STARTING X0/Y0 = TOOL CHANGE
SET TO TOOL CHANGE = X-50  Y-35 PRIOR  TO
RUNNING FIRST CYCLE.
TOOL= 8mm DRILL

1   TOOL 1              REM: 8mm DRILL
2   X62  Y90   R I
3   DWELL              REM: DRILL HOLE
4   X18  Y-20  R I
5   DWELL              REM: DRILL HOLE
6   X20  Y-23  R I
7   DWELL              REM: DRILL HOLE
8   X-100 Y-47  R I
9   END
```

FIGURE 6-6
Machinist Shop Language drilling program, metric incremental positioning, for the part in
Figure 6-2

EVENT 1

- TOOL 1—Causes the machine to stop executing the program until the operator depresses the start button. This gives the operator time to safely install the tool without danger of machine movement. Since the spindle was manually positioned to tool change at setup, no movement to tool change is needed.

EVENT 2

- X/Y coordinates—To move from tool change to hole #1.
- R—Specifies rapid traverse.
- I—Specifies that incremental positioning is being used.

EVENT 3

- DWELL—Halts execution of the program until the start button is pushed. The operator then drills hole #1.

EVENT 4

- X/Y coordinates—To move from hole #1 to hole #2.
- R—Specifies rapid traverse.
- I—Specifies that incremental positioning is being used.

EVENT 5

- DWELL—Halts program execution. Hole #2 is drilled.

EVENT 6

- X/Y coordinates—To move from hole #2 to hole #3.
- R—Specifies rapid traverse.
- I—Specifies incremental positioning.

EVENT 7

- DWELL—As in events 3 and 5, the program is halted and hole #3 is drilled.

EVENT 8

- X/Y incremental coordinates—To move from hole #3 to tool change.
- R—Specifies rapid traverse.
- I—Specifies incremental positioning.

EVENT 9

- END—Instructs the machine that this is the end of the program.

WORD ADDRESS FORMAT

The next machine to be programmed is a CNC mill using a General Numerics controller. This machine is a continuous-path machine that uses a programming format called *word address*. Word address was developed as a tape programming format. Another name for word address is *variable block* format, so named because the program lines (blocks) may vary in length according to the information contained in them. Earlier tape formats required an entry for all possible machine registers. In these earlier formats, a zero was programmed as a null input if the register values were to be unaffected, but in word address, the blocks need only contain necessary information. Although developed as a tape format, word address is used as the format for manual data input on many CNC machines.

Addresses

The block format for word address is as follows:

N...G..X....Y....Z....I....J....K....F....S....T..M..

Only the information needed on a line need be given. Each of the letters is called an address (or word). The various words are as follows:

N— Designates the start of a block. Program lines or blocks are sometimes also called *sequence lines.* On some machinery the address "O" may also be used to start a block of information.

G— Initiates a preparatory function. Preparatory functions change the control mode of the machine. Examples of preparatory functions are rapid/feedrate mode, drilling mode, tapping mode, boring mode, and circular interpolation. Preparatory functions are called *prep functions,* or more commonly, *G codes.*

X— Designates an X-axis coordinate. X is also used to enter a time interval for a timed dwell.

Y— Designates a Y-axis coordinate.

Z— Designates a Z-axis coordinate.

I— Identifies the X-axis location of an arc centerpoint.

J— Identifies the Y-axis location of an arc centerpoint.

K— Identifies the Z-axis location of an arc centerpoint.

S— Sets the spindle RPM.

F— Assigns a feedrate.

T— Specifies the tool to be used in a tool change.

M— Initiates miscellaneous functions (*M functions*). M functions control auxilliary functions such as the turning on and off of the spindle and coolant, initiating tool changes, and signaling the end of a program.

Other words used in word address will be explained as they are used. A list of EIA codes for word address is contained in Appendix 1.

DRILLING IN WORD ADDRESS FORMAT

Figure 6–7 contains a program written in the word address format to drill the part in Figure 6–1. Figure 6–8 contains the program to drill the part in Figure 6–2. The program sequence is identical to that used in the Machinist Shop Language example. The specific codes used in the programs are:

G00—Puts the machine in rapid traverse mode. All moves made with G00 active are made in rapid traverse.

G01—Linear interpolation; puts the machine in feedrate mode. All moves made with G01 active are made in a straight line at the programmed feedrate.

```
XO/YO = LOWER LEFT CORNER OF PART
TOOL CHANGE = X-2 Y-1.5
TOOL 3/8 DRILL SPINDLE SPEED 2500 RPM

N010 G00 G70 G90 X-2 Y-1.5  M06   REM:3/8
DRILL
N020 X.5 Y2
N030 G04
N040 X1.25 Y1.5
N050 G04
N060 X2 Y.5
N070 G04
N080 X-2 Y-1.5
N090  M30
```

FIGURE 6-7
Word address format drilling program, nonmetric absolute positioning, for the part in Figure 6-1

```
XO/YO = LOWER LEFT CORNER OF PART
TOOL CHANGE = X-50 Y-35
TOOL 8mm DRILL
SPINDLE SPEED 2500 RPM

N010 G00 G71 G90 X-50 Y-35 M06   REM:8   mm
DRILL
N020 X12 Y55
N030 G04
N040 X30 Y35
N050 G04
N060 X50 Y12
N070 G04
N080 X-50 Y-35
N090  M30
```

FIGURE 6-8
Word address format drilling program, metric absolute positioning, for the part in Figure 6-2

G04—Dwell command. Causes a halt in the program execution until the cycle start button is depressed. Some controllers require the use of M00 rather than G04. However, G04 will be used throughout this particular text as the dwell code.

G70—Selects inch input.

G71—Selects metric input.

G90—Selects absolute positioning.

G91—Selects incremental positioning.

M06—Institutes a tool change. In two-axis operation, this command functions as a dwell.

M30—Signals the end of the program and resets the computer to the start of the program.

PROGRAM EXPLANATION — ABSOLUTE POSITIONING

(Refer to Figures 6–7 and 6–8.)

N010

- N010—The sequence number. The word is ignored by the controller. It is used only to identify a block.
- G00—Puts the machine in rapid traverse mode. Moves will be made at rapid traverse speed until the mode is cancelled with a G01, G02, or G03.
- G70—Selects inch input. All numbers entered will be inch coordinates.
- G71—Selects metric input for the metric program.
- G90—Selects absolute positioning.
- X/Y coordinates of the tool change location—These coordinates are absolute dimensions.
- M06—Tool change command. In a two-axis program, this command acts like a dwell. The machine moves to the X/Y tool change coordinates and halts for a tool change. Note that on some controllers, the M06 command may have to be placed on a program line by itself in order to function as explained here. Also note that some controllers using word address format will not recognize an M06 command when two axes are supplied on a machine. In these cases a G04 (or M00) dwell command will be used.

N020

- N020—The sequence number.
- X/Y coordinates—To move from tool change to hole #1.

N030

- N030—The sequence number.
- G04—The dwell command. The program halts its execution, allowing the operator to drill the holes.

N040

- N040—The sequence number.
- X/Y coordinates—To move from hole #1 to hole #2.

N050

- N050—The sequence number.
- G04—The dwell command. This halts the program so that hole #2 can be drilled.

N060

- N060—The sequence number.
- X/Y coordinates—To move from hole #2 to hole #3.

N070

- N070—The sequence number.
- G04—The dwell command. Hole #3 is drilled.

N080

- N080—The sequence number.
- X/Y coordinates—To move from hole #3 to tool change.

N090

- N090—The sequence number.
- M30—Signals that the program has ended and resets the computer's memory to the start of the sequence.

PROGRAM EXPLANATION—INCREMENTAL POSITIONING

(Refer to Figures 6–9 and 6–10.)

The program sequence used in these examples is identical to that used in the Machinist Shop Language programs in Figures 6–5 and 6–6.)

N010

- N010—The sequence number. The word is ignored by the controller as it is used only to identify a block.
- G00—Puts the machine in rapid traverse mode. Moves will be made at rapid traverse speed until the mode is cancelled with a G01, G02, or G03.
- G70/G71—Selects inch or metric input.
- G91—Specifies that incremental positioning is to be used.
- M06—Tool change command. This halts the program to allow insertion of the drill. Note that the spindle had been manually positioned at the tool change location prior to the start of the first program cycle, just as was done with the Machinist Shop Language programs.

N020

- N020—The sequence number.
- X/Y—The incremental coordinates required to move from tool change to hole #1.

N030

- N030—The sequence number.
- G04—The dwell command. The program halts its execution, allowing the operator to drill hole #1.

```
PROGRAM XO/YO = TOOL CHANGE
SET TO TOOL CHANGE = X-2 Y-1.5  PRIOR  TO
STARTING FIRST CYCLE
TOOL 3/8 DRILL SPINDLE SPEED 2500 RPM

NO10 GOO G70 G91 MO6  REM:3/8 DRILL
NO20 X2.5 Y3.5
NO30 G04
NO40 X.75 Y-.5
NO50 G04
NO60 X.75 Y-1
NO70 G04
NO80 X-4 Y-2
NO90  M30
```

FIGURE 6-9
Word address format drilling program, nonmetric incremental positioning, for the part in
Figure 6-1

```
PROGRAM START XO/YO = TOOL CHANGE
SET TO TOOL CHANGE = X-50 Y-35  PRIOR  TO
STARTING FIRST CYCLE
TOOL 8mm DRILL SPINDLE SPEED 2500 RPM

NO10 GOO G71 G91 MO6  REM:8mm DRILL
NO20 X62 Y90
NO30 G04               REM:DRILL HOLE
NO40 X18 Y-20
NO50 G04               REM:DRILL HOLE
NO60 X20 Y-23
NO70 G04               REM:DRILL HOLE
NO80 X-100 Y-47
NO90  M30
```

FIGURE 6-10
Word address format drilling program, metric incremental positioning, for the part in Figure 6-2

N040

- ■ N040—The sequence number.
- ■ X/Y—The incremental coordinates required to move from hole #1 to hole #2.

N050

- ■ N050—The sequence number.
- ■ G04—The dwell command. This halts the program so that hole #2 can be drilled.

N060

- ■ N060—The sequence number.
- ■ X/Y—The incremental coordinates required to move from hole #2 to hole #3.

N070

- ■ N070—The sequence number.
- ■ G04—The dwell command. Hole #3 is drilled.

N080

- ■ N080—The sequence number.
- ■ X/Y—The incremental coordinates required to move from hole #3 to tool change.

N090

- ■ N090—The sequence number.
- ■ M30—Signals that the program has ended and resets the computer's memory to the start of the sequence.

MILLING IN MACHINIST SHOP LANGUAGE

Assume that the part in Figure 6–11 is to be milled. The part is an aluminum casting which requires that only the length and width be machined. Figure 6–12 is a metric part. The part setup drawing is Figure 6–13. Clamping will be done through the center hole. Two passes around the part will be made, a roughing pass and a finishing pass. Left for the finish pass will be .010 inch of stock (0.25 mm metric version).

Two programs, one nonmetric and one metric, written using absolute positioning will be presented first. (The programs are contained in Figures 6–15 and 6–16.) A .500-inch-diameter end mill will be used in the nonmetric program and a 5-mm-diameter end mill in the metric version. Notice that only an X or Y coordinate, rather than an X/Y pair of coordinates, is used in some lines. If the machine is already positioned in one of its axes, a coordinate for that axis need not be given. No movement is to take place; therefore the second coordinate is not required.

Up Milling and Down Milling

When milling cuts are programmed, it is important to understand the difference between *up* and *down milling*. Figure 6–14 illustrates these two machining practices. Notice that in up milling (also called *conventional milling*), the cutter forces acting on the part try to lift the part up off of the table, hence the name

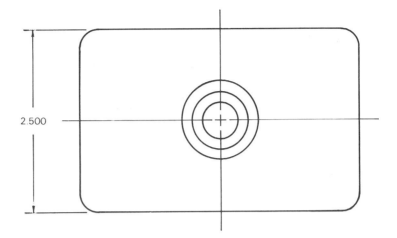

MATERIAL: ALUMINUM CASTING
NOTE: ONLY PERTINENT DIMENSIONS GIVEN

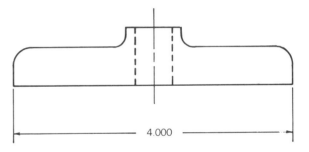

FIGURE 6-11
Milling part drawing, nonmetric

up milling. In down milling the force of the cutter tries to push the part downward onto the table, thus the name down milling. Down milling is also referred to as *climb cutting,* because the cutter is trying to "climb up" on top of the part.

Up milling is used for cutting most ferrous materials, brass and bronze, and for roughing cuts on aluminum and aluminum alloys. Down milling is used for finishing cuts on aluminum and aluminum alloys. It is also occasionally used for finishing cuts on other metals, if conditions warrant it. Down milling requires less power for a particular cut but places more stress on the machine slides and ball screws than does up milling. Exactly when to use up and down milling is something that must be learned through experience as it depends not only on the machine available but also on the cutting tools, coolant, and workpiece materials.

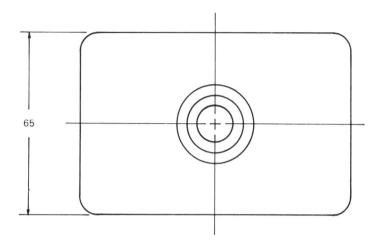

MATERIAL: ALUMINUM CASTING
NOTE: ONLY PERTINENT DIMENSIONS GIVEN

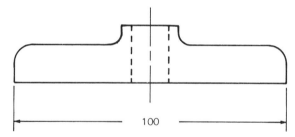

FIGURE 6–12
Milling part drawing, metric

PROGRAM EXPLANATION—ABSOLUTE POSITIONING

(Refer to Figures 6–15 and 6–16.)

EVENT 1
- X/Y coordinates of the tool change location—This move to tool change allows the end mill to be inserted at a safe location.
- R—Specifies a move at rapid traverse.
- A—Specifies absolute positioning.

EVENT 2
- TOOL 1—Halts the program execution to allow the operator to install the end mill in the spindle.

EVENT 3

■ FEED 20—Assigns a feedrate of 20 in./min (500 mm/min metric) to be used when needed. A feedrate may be assigned at any point before it is needed.

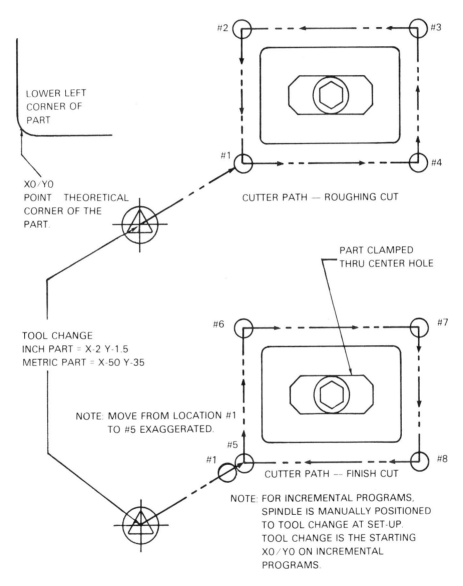

FIGURE 6–13
Setup drawing for the part in Figures 6–11 and 6–12

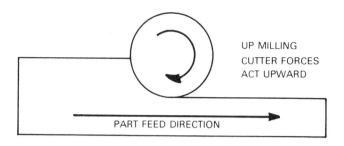

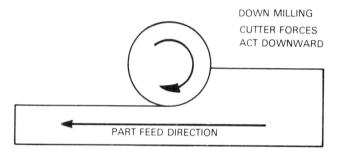

FIGURE 6–14
Up milling and down milling

EVENT 4

- X/Y coordinates—To move from tool change to location #1 as indicated on the cutter path diagram in Figure 6–13. As explained in Chapter 5, half the diameter of the cutter is allowed for in both axes to compensate for the cutter radius.
- R—Specifies rapid traverse.
- A—Specifies absolute positioning.

EVENT 5

- DWELL—Halts the program execution. The operator is instructed to lower the spindle.

EVENT 6

- X coordinate—To move from location #1 to location #2.
- F—Specifies a feedrate move. This means that a milling cut feedrate movement is required.
- A—Specifies absolute positioning.

```
X0/Y0 = LOWER LEFT CORNER OF PART
TOOL CHANGE = X-2 Y-1.5
TOOL = .500 END MILL

 1 X-2 Y-1.5        R A
 2 TOOL 1                      REM:2500 RPM
 3 FEED 20
 4 X-.26 Y-.26      R A
 5 DWELL                       REM:LOWER SPNDL
 6 X4.26            F A
 7 Y2.76            F A
 8 X-.26            F A
 9 Y-.26            F A
10 X-.25 Y-.25      F A
11 Y2.75            F A
12 X4.25            F A
13 Y-.25            F A
14 X-.25            F A
15 DWELL                       REM:RAISE SPNDL
16 X-2 Y-1.5        R A
17 END
```

FIGURE 6–15

Machinist Shop Language milling program, nonmetric absolute positioning, for the part in Figure 6–11

```
X0/Y0 = LOWER LEFT CORNER OF PART
TOOL CHANGE = X-50 Y-35
TOOL = 5mm END MILL

 1 X-50 Y-35        R A
 2 TOOL 1                      REM:2500 RPM
 3 FEED 500
 4 X-2.75 Y-2.75    R A
 5 DWELL                       REM:LOWER SPNDL
 6 X102.75          F A
 7 Y67.75           F A
 8 X-2.75           F A
 9 Y-2.75           F A
10 X-2.5 Y-2.5      F A
11 Y67.5            F A
12 X102.5           F A
13 Y-2.5            F A
14 X-2.5            F A
15 DWELL                       REM:RAISE SPNDL
16 X-50 Y-35        R A
17 END
```

FIGURE 6–16

Machinist Shop Language milling program, metric absolute positioning, for the part in Figure 6–12

EVENT 7

- Y coordinate—To move from location #2 to location #3.
- F—Specifies a feedrate move.
- A—Specifies absolute positioning.

EVENT 8

- X coordinate—To move from location #3 to location #4.
- F—Specifies a feedrate move.
- A—Specifies absolute positioning.

EVENT 9

- Y coordinate—To move the machine from location #4 to location #1, to complete the milling of the part.
- F—Specifies a feedrate move.
- A—Specifies absolute positioning.

EVENT 10

- X/Y coordinates—To move from location #1 to location #5. This move positions the cutter for the finish pass.
- F—Specifies a feedrate move.
- A—Specifies absolute positioning.

EVENT 11

- Y coordinate—To move from location #5 to location #6. Notice that .010 inch of stock (0.25 mm metric) material is removed from the side of the part during this move.
- F—Specifies a feedrate move.
- A—Specifies absolute positioning.

EVENT 12

- X coordinate—To move from location #6 to location #7.
- F—Specifies a feedrate move.
- A—Specifies absolute positioning.

EVENT 13

- Y coordinate—To move from location #7 to #8.
- F—Specifies a feedrate move.
- A—Specifies absolute positioning.

EVENT 14

- X coordinate—To move from location #8 to location #5, thus completing the finish milling cut.

EVENT 15

- DWELL—Halts the program execution. The operator is instructed to raise the spindle.

EVENT 16
- X/Y coordinates—To move from location #5 to tool change.
- R—Specifies rapid traverse. The milling cut is complete. Therefore, a feedrate move is no longer required.
- A—Specifies absolute positioning.

EVENT 17
- END—Signals the end of program.

PROGRAM EXPLANATION—INCREMENTAL POSITIONING

Incremental versions of these programs are featured in Figures 6–17 and 6–18. As with the drilling programs, the machine is positioned manually to the tool change position after the edge of the part has been located. The programs begin and end at the tool change location.

EVENT 1
- TOOL 1—Halts the program execution to allow the operator to install the end mill in the spindle if not done previously.

EVENT 2
- FEED 20—Assigns a feedrate of 20 in./min (500 mm/min metric) to be used when needed.

EVENT 3
- X/Y—Incremental coordinates required to move from tool change to location #1, as indicated on the cutter path diagram in Figure 6–13.
- R—Specifies rapid traverse.
- I—Specifies incremental positioning.

EVENT 4
- DWELL—Halts the program execution. The operator is instructed to lower the spindle.

EVENT 5
- X—Incremental coordinate required to move from location #1 to location #2.
- F—Specifies a feedrate move. This means that a milling cut feedrate movement is required.
- I—Specifies incremental positioning.

```
PROGRAM START X0/Y0 = TOOL CHANGE
SET TO TOOL CHANGE = X-2 Y-1.5 PRIOR TO
STARTING FIRST CYCLE
TOOL = .500 END MILL

 1 TOOL 1                      REM:2500 RPM
 2 FEED 20
 3 X1.74 Y1.24      R A
 4 DWELL                       REM:LOWER SPNDL
 5 X4.52            F A
 6 Y3.02            F A
 7 X-4.52           F A
 8 Y-3.02           F A
 9 X.01 Y.01        F A
10 Y3.0             F A
11 X4.5             F A
12 Y-3              F A
13 X-4.5            F A
14 DWELL                       REM:RAISE SPNDL
15 X-1.75 Y-1.25    R A
16 END
```

FIGURE 6–17
Machinist Shop Language milling program, nonmetric incremental positioning, for the part in
Figure 6–11

```
PROGRAM START X0/Y0 = TOOL CHANGE
SET TO TOOL CHANGE = X-50 Y-35 PRIOR TO
STARTING FIRST CYCLE
TOOL = 5mm END MILL

 1 TOOL 1                      REM:2500 RPM
 2 FEED 500
 3 X47.75 Y32.25    R A
 4 DWELL                       REM:LOWER SPNDL
 5 X105.5           F A
 6 Y70.5            F A
 7 X-105.5          F A
 8 Y-70.5           F A
 9 X.25 Y.25        F A
10 Y70              F A
11 X105             F A
12 Y-70             F A
13 X-105            F A
14 DWELL                       REM:RAISE SPNDL
15 X-47.5 Y-32.5    R A
16 END
```

FIGURE 6–18
Machinist Shop Language milling program, metric incremental positioning, for the part in
Figure 6–12

EVENT 6

- Y—Incremental coordinate required to move from location #2 to location #3.
- F—Specifies a feedrate move.
- I—Specifies incremental positioning.

EVENT 7

- X—Incremental coordinate required to move from location #3 to location #4.
- F—Specifies a feedrate move.
- I—Specifies incremental positioning.

EVENT 8

- Y—Incremental coordinate required to move the machine from location #4 to location #1 to complete the roughing cut.
- F—Specifies a feedrate move.
- I—Specifies incremental positioning.

EVENT 9

- X/Y—Incremental coordinates required to move from location #1 to location #5. This move positions the spindle for the finish pass.

EVENT 10

- Y—Incremental coordinate required to move from location #5 to location #6. Left for finish cut is .010 inch of stock (0.25 mm metric).
- F—Specifies a feedrate move.
- I—Specifies incremental positioning.

EVENT 11

- X—Incremental coordinate required to move from location #6 to location #7.
- F—Specifies a feedrate move.
- I—Specifies incremental positioning.

EVENT 12

- Y—Incremental coordinate required to move from location #7 to location #8.
- F—Specifies a feedrate move.
- I—Specifies incremental positioning.

EVENT 13

- X—Incremental coordinate required to move from location #8 to location #5 to complete the finish pass.
- F—Specifies a feedrate move.
- I—Specifies incremental positioning.

EVENT 14

- DWELL—Halts the program execution. The operator is instructed to raise the spindle.

EVENT 15

- X/Y—Incremental coordinates required to move from location #5 to tool change.
- R—Specifies rapid traverse (the milling cut is complete).
- I—Specifies incremental positioning.

EVENT 16

- END—Signals that the program has ended.

MILLING IN WORD ADDRESS FORMAT

The part pictured in Figure 6–11 will now be milled using word address format. Figure 6–19 is the word address program in absolute positioning. Figure 6–20 is the metric version to mill the part in Figure 6–12. Figure 6–21 is an incremental program for the part in Figure 6–11, and Figure 6–22 is the metric version to mill the part in Figure 6–12. The programming logic for these programs is identical to that of the Machinist Shop Language programs.

```
X0/Y0 = LOWER LEFT CORNER OF PART
TOOL CHANGE = X-2 Y-1.5
TOOL = .500 END MILL

N010 G00 G70 G90 X-2 Y-1.5 M06     REM:2500 RPM.
N020 X-.26 Y-.26
N030 G04                 REM:LOWER SPNDL
N040 G01 X4.26 F20
N050 Y2.76
N060 X-.26
N070 Y-.26
N080 X-.25 Y-.25
N090 Y2.75
N100 X4.25
N110 Y-.25
N120 X-.25
N130 G04                 REM:RAISE SPNDL
N140 G00 X-2 Y-1.5
N150 M30
```

FIGURE 6–19
Word address format milling program, nonmetric absolute positioning, for the part in Figure 6–11

```
X0/Y0 = LOWER LEFT CORNER OF PART
TOOL CHANGE = X-50 Y-35
TOOL = 5mm END MILL

N010 G00 G71 G90 X-50 Y-35 M06    REM:2500
RPM.
N020 X-2.75 Y-2.75
N030 G04
REM:LOWER SPNDL
N040 G01 X102.75 F500
N050 Y67.75
N060 X-2.75
N070 Y-2.75
N080 X-2.5 Y-2.5
N090 Y67.5
N100 X102.5
N110 Y-2.5
N120 X-2.5
N130 G04
REM:RAISE SPNDL
N140 G00 X-50 Y-35
N150 M30
```

FIGURE 6–20
Word address format milling program, metric absolute positioning, for the part in Figure 6–12

```
PROGRAM START X0/Y0 = TOOL CHANGE
MANUALY SET TO TOOL CHANGE PRIOR TO
RUNNING FIRST CYCLE
TOOL = .500 END MILL

N010 G00 G70 G91 M06      REM:2500 RPM.
N020 X1.74 Y1.24
N030 G04                  REM:LOWER SPNDL
N040 G01 X4.52 F500
N050 Y3.02
N060 X-4.52
N070 Y-3.02
N080 X.01 Y.01
N090 Y3
N100 X4.5
N110 Y-3
N120 X-4.5
N130 G04                  REM:RAISE SPNDL
N140 G00 X-1.75 Y-1.25
N150 M30
```

FIGURE 6–21
Word address format milling program, nonmetric incremental positioning, for the part in Figure 6–11

```
PROGRAM START X0/Y0 = TOOL CHANGE
MANUALY SET TO TOOL CHANGE PRIOR TO
RUNNING FIRST CYCLE
TOOL = 5mm END MILL

N010 G00 G71 G91 M06        REM:2500 RPM.
N020 X47.25 Y32.25
N030 G04                    REM:LOWER SPNDL
N040 G01 X105.5 F500
N050 Y70.5
N060 X-105.5
N070 Y-70.5
N080 X.25 Y.25
N090 Y70
N100 X105
N110 Y-70
N120 X-105
N130 G04                    REM:RAISE SPNDL
N140 G00 X-47.5 Y-32.5
N150 M30
```

FIGURE 6-22
Word address format milling program, metric incremental positioning, for the part in Figure 6-12

PROGRAM EXPLANATION—ABSOLUTE POSITIONING

(Refer to Figures 6-19 and 6-20.)

N010

- N010—The sequence number.
- G00—Puts the machine in rapid traverse mode.
- G70/G71—Specifies inch/metric input.
- G90—Specifies absolute positioning.
- X/Y coordinates—To move to tool change.
- M06—Initiates a tool change. The .500-inch-diameter end mill (5 mm metric) is installed in the spindle. As previously mentioned, some controllers will not use this command on a manual tool change machine. In those cases, a G04 (dwell) or an M00 (program stop) is used for manual tool changes.

N020

- N020—The sequence number.
- X/Y coordinates—To move from tool change to location #1 (see Figure 6-13).

N030

- N030—The sequence number.
- G04—A dwell command. The operator is instructed to lower the spindle.

N040

- N040—The sequence number.
- G01—Puts the machine in feedrate mode.
- X coordinate—Required to move from location #1 to location #2 (see Figure 6–13).
- F20—Assigns a feedrate of 20 in./min (500 mm/min metric).

N050

- N050—The sequence number.
- Y coordinate—Required to move from location #2 to location #3.

N060

- N060—The sequence number.
- X coordinate—Required to move from location #3 to location #4.

N070

- N070—The sequence number.
- Y coordinate—Required to move from location #4 to location #1. This move completes the milling roughing pass.

N080

- N080—The sequence number.
- X/Y coordinates—Required to move from location #1 to location #5. This is a feed move to position the cutter for the finish pass.

N090

- N090—The sequence number.
- Y coordinate—To move from location #5 to location #6. Notice that this is a down milling cut. Down milling gives a nice surface finish on aluminum alloys such as the aluminum casting used for this example.

N100

- N100—The sequence number.
- X coordinate—To move from location #6 to location #7.

N110

- N110—The sequence number.
- Y coordinate—To move from location #7 to location #8.

N120

- N120—The sequence number.
- X coordinate—To move from location #8 to location #5. This completes the finish milling pass.

N130

- ■ N130—The sequence number.
- ■ G04—A dwell command. The operator is instructed to raise the spindle.

N140

- ■ N140—The sequence number.
- ■ G00—Puts the machine in rapid traverse mode.
- ■ X/Y coordinates—To move from location #5 to tool change.

N150

- ■ N150—The sequence number.
- ■ M30—Signals that the program has ended. The computer's memory is reset to the start of the program.

PROGRAM EXPLANATION—INCREMENTAL POSITIONING

(Refer to Figures 6–21 and 6–22.)

N010

- ■ N010—The sequence number.
- ■ G00—Puts the machine in rapid traverse mode.
- ■ G70/G71—Specifies inch/metric input.
- ■ G91—Specifies incremental positioning.
- ■ M06—Initiates a tool change. The .500-inch-diameter end mill (5 mm metric) is installed in the spindle. For controllers that will not use this command on a manual tool change machine, a G04 (dwell) or an M00 (program stop) is used. To insure that the program started and stopped in the same location, the spindle was positioned at the tool change location prior to starting the program.

N020

- ■ N020—The sequence number.
- ■ X/Y incremental coordinates—To move from tool change to location #1 (see Figure 6–13).

N030

- ■ N030—The sequence number.
- ■ G04—A dwell command. The operator is instructed to lower the spindle.

N040

- N040—The sequence number.
- G01—Puts the machine in feedrate mode.
- X incremental coordinate—Required to move from location #1 to location #2 (see Figure 6–13).
- F20—Assigns a feedrate of 20 in./min (500 mm/min metric).

N050

- N050—The sequence number.
- Y incremental coordinate—Required to move from location #2 to location #3.

N060

- N060—The sequence number.
- X incremental coordinate—Required to move from location #3 to location #4.

N070

- N070—The sequence number.
- Y incremental coordinate—Required to move from location #4 to location #1, completing the milling roughing pass.

N080

- N080—The sequence number.
- X/Y incremental coordinates—Required to move from location #1 to location #5.

N090

- N090—The sequence number.
- Y incremental coordinate—To move from location #5 to location #6.

N100

- N100—The sequence number.
- X incremental coordinate—To move from location #6 to location #7.

N110

- N110—The sequence number.
- Y incremental coordinate—To move from location #7 to location #8.

N120

- N120—The sequence number.
- X incremental coordinate—Required to move from location #8 to location #5, completing the finish milling pass.

N130

- N130—The sequence number.
- G04—A dwell command. The operator is instructed to raise the spindle.

N140

- N140—The sequence number.
- G00—Puts the machine in rapid traverse mode.
- X/Y incremental coordinates — To move from location #5 to tool change.

N150

- N150—The sequence number.
- M30—Signals that the program has ended. The computer's memory is reset to the start of the program.

Note that the programming logic was identical for the absolute and incremental positioning programs. Incremental positioning is primarily used within an absolute program rather than as a program in and of itself. From this point on, incremental coordinates will be used within the body of a general absolute program.

FANUC CONTROLLER APPLICATIONS

Fanuc CNC controllers have become a common choice of CNC manufacturers. In some areas of the country Fanuc controls dominate. This chapter is one of several in which a special section containing solutions to the chapter examples in Fanuc format is presented to assist those schools and industries utilizing Fanuc controllers. There are a few important differences in Fanuc format from those of the generic examples presented in this chapter.

Each part program begins with an O number (the letter "O"). This number identifies the program to the controller. Through the use of "O" numbers, multiple programs can be stored in the MCU memory, and called up as required.

The program manuscript does not contain spaces between command words. While some Fanuc models would ignore the spaces, in some older models, they would cause the controller to go into alarm and halt the reading of the tape. Third, comment lines are placed between parentheses. The open parenthesis "(" is known as a control-out character. When the MCU encounters this character, it interprets all information following as a comment until a closed parenthesis ")" is encountered. The closed parenthesis is called a control-in character.

The program stop command used by Fanuc is M00, as opposed to the G04 command used in the text examples. Also, Fanuc controls do not require the use of trailing zeros. If a whole number is programmed (i.e., 1 inch, 3 inches, etc.), the decimal point must be programmed, but the trailing zeros are omitted. 1.000 would be written 1., 3.000 would be 3..

Figure 6–23 is a Fanuc program written in to drill the part in Figure 6–1 using absolute positioning. Figure 6–24 is a program for the same part using incremental positioning. Figures 6–25 (absolute positioning) and Figure 6–26 (incremental positioning) are mill programs for the part in Figure 6–11. Figure 6–27 shows a Fanuc series 15–T control.

```
%
O0601
(-------------------------------------------------------)
(THIS PROGRAM USES ABSOLUTE POSITIONING)
(X/Y ORIGIN IS LOWER LEFT CORNER OF PART)
(PLACE 3/8 DRILL IN SPINDLE PRIOR TO START OF CYCLE)
(-------------------------------------------------------)
(SET PARAMETERS TO RAPID - INCH INPUT - ABS. POS.)
N010G00G70G90
(MOVE TO 1ST HOLE AND PROG. STOP TO DRILL)
N020X.5Y2.
N030M00
(MOVE TO 2ND HOLE AND PROG. STOP TO DRILL)
N040X1.25Y1.5
N050M00
(MOVE TO 3RD HOLE AND PROG. STOP TO DRILL)
N060X2.Y.5
N070M00
(RETURN TO TOOL CHANGE AND END CYCLE)
N080X-2.Y-1.5
N090M30
%
```

FIGURE 6–23

```
%
O0601
(-------------------------------------------------------)
(THIS PROGRAM USES INCREMENTEAL POSITIONING)
(CYCLE STARTS THE TOOL CHANGE LOCATION)
(PLACE 3/8 DRILL IN SPINDLE PRIOR TO START OF CYCLE)
(-------------------------------------------------------)
(SET PARAMETERS TO RAPID - INCH INPUT - INCR. POS.)
N010G00G70G91
(MOVE TO 1ST HOLE AND PROG. STOP TO DRILL)
N020X2.5Y3.5
N030M00
(MOVE TO 2ND HOLE AND PROG. STOP TO DRILL)
N040X.75Y-.5
N050M00
(MOVE TO 3RD HOLE AND PROG. STOP TO DRILL)
N060X.75Y-1.
N070M00
(RETURN TO TOOL CHANGE AND END CYCLE)
N080X-4.Y-2.
N090M30
%
```

FIGURE 6–24

```
%
O0611
(-------------------------------------------------------)
(THIS PROGRAM USES ABSOLUTE POSITIONING)
(X/Y ORIGIN IS LOWER LEFT CORNER OF PART)
(PLACE 1/2 END MILL IN SPINDLE PRIOR TO START OF CYCLE)
(-------------------------------------------------------)
(SET PARAMETERS TO RAPID - INCH INPUT - ABS. POS.)
N010G00G70G90
(AT PROG. STOP - LOWER SPINDLE AND CLAMP)
N020X-.26Y-.26
N030M00
(BEGIN ROUGH MILL CUT AT FEEDRATE)
N040G01X4.26F20.0
N050Y2.76
N060X-.26
N070Y-.26
(BEGIN FINISH MILL CUT)
N080X-.25Y-.25
N090Y2.75
N100X4.25
N110Y-.25
N120X-.25
(AT PROG. STOP UNCLAMP AND RAISE SPINDLE)
N130M00
(RETURN TO TOOL CHANGE LOCATION AND END CYCLE)
N140G00X-2.Y-1.5
N150M30
%
```

FIGURE 6-25


```
%
O0611
(-------------------------------------------------------)
(THIS PROGRAM USES INCREMENTAL POSITIONING)
(CYCLE STARTS FROM TOOL CHANGE LOCATION)
(PLACE 1/2 END MILL IN SPINDLE PRIOR TO START OF CYCLE)
(-------------------------------------------------------)
(SET PARAMETERS TO RAPID - INCH INPUT - INCR. POS.)
N010G00G70G91
(AT PROG. STOP - LOWER SPINDLE AND CLAMP)
N020X1.74Y1.24
N030M00
(BEGIN ROUGH MILL CUT AT FEEDRATE)
N040G01X4.52F20.0
N050Y3.02
N060X-4.52
N070Y-3.02
(BEGIN FINISH MILL CUT)
N080X.01Y.01
N090Y3.
N100X4.5
N110Y-3.
N120X-4.5
(AT PROG. STOP UNCLAMP AND RAISE SPINDLE)
N130M00
(RETURN TO TOOL CHANGE LOCATION AND END CYCLE)
N140G00X-1.75Y1.25
N150M30
%
```

FIGURE 6-26

FIGURE 6-27
Fanuc series 15-T control

SUMMARY

The important concepts presented in this chapter are:

- Some procedure for tool change must be included in a program. For a manual tool change mill, a tool change location is used to safely position the spindle away from the part. The program must then be halted to allow the safe insertion of the tool. In Machinist Shop Language, the TOOL command is used to perform this function at the hole location; in word address format, an M06 is used.
- The spindle must be positioned safely out of the way at the end of the program to allow safe loading and unloading of the workpiece. This is accomplished in both the milling and drilling examples by sending the spindle back to its tool change location at the end of the program.

- Incremental programs differ from absolute programs only in the coordinates used. Programs in absolute and incremental positioning use the same programming logic. In incremental positioning, it is imperative that the machine start and stop in the same location. Failure to program for this will result in incorrect positioning for the second cycle.
- To perform hole operations, it is necessary to position the spindle over the centerline of the hole.
- A dwell command is used at hole locations to halt the program and enable the operator to drill the hole.
- When programming coordinates for milling, an allowance must be made for the size of the cutter.
- F is used to specify a feedrate move in Machinist Shop Language.
- R is used to specify a rapid move in Machinist Shop Language.
- G00 is used in word address format to specify a rapid move.
- G01 is used in word address format to specify a feedrate move.

REVIEW
QUESTIONS

1. What do each of the following Machinist Shop Language commands mean: X, Y, Z, R, F, A, FEED, DWELL, END?
2. What do the following addresses stand for in word address format: X, Y, G, M, S, F, N?
3. What is a preparatory function?
4. What are miscellaneous functions?
   (Questions #5 – #8 refer to the part in Figure 6 – 28. The cutter path drawing is given in Figure 6–29.)
5. Write a program in Machinist Shop Language to mill and drill the part using absolute positioning.
6. Write a program in Machinist Shop Language to mill and drill the part using incremental positioning.
7. Write a program in word address format to mill and drill the part using absolute positioning.
8. Write a program in word address format to mill and drill the part using incremental positioning.

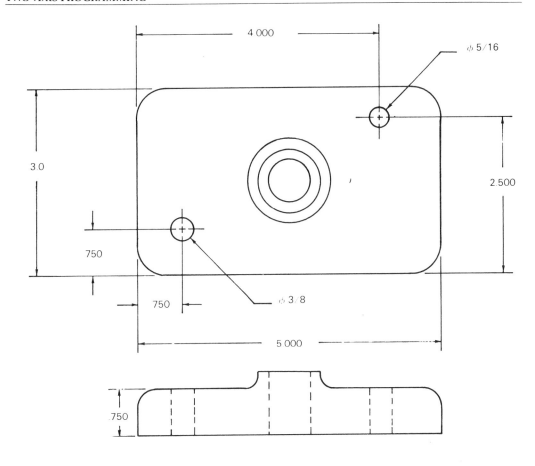

INSTRUCTIONS 1) MILL AND DRILL PART
2) USE LOWER LEFT CORNER FOR XO YO

FIGURE 6–28
Part drawing for review questions #5–8

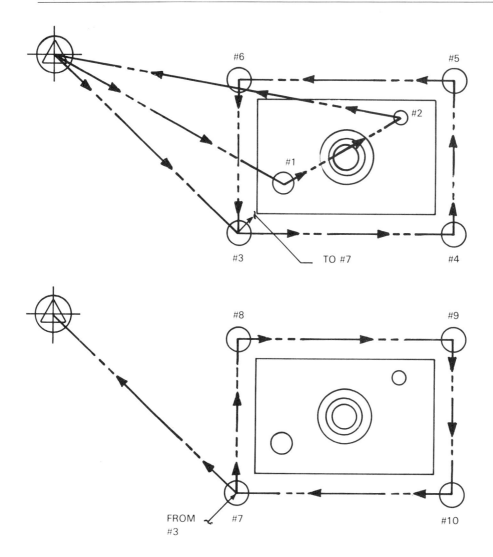

TOOL CHANGE = X-6 Y1
XO/YO = LOWER LEFT CORNER

FIGURE 6-29
Cutter path for Figure 6-28

CHAPTER 7

Three-Axis Programming

OBJECTIVES Upon completion of this chapter, you will be able to:

- Write simple programs to perform three-axis hole operations and simple milling cuts using Machinist Shop Language.
- Write simple programs to perform three-axis hole operations and simple milling cuts using word address format.
- Explain the difference between initial level and reference level on CNC machinery.
- Explain the difference between a modal and nonmodal command.

In this chapter, drilling and milling operations are programmed using all three machine axes. The first program is written using Machinist Shop Language; the second using word address format. When writing three-axis programs using the CNC, one must either know the lengths of the tools in their toolholders, or leave empty lines in the program to allow the operator to enter the tool lengths. The concept of tool length offset was discussed in Chapter 4. The programs in this chapter will put the concept to use.

A THREE-AXIS PROGRAMMING TASK

The part in Figure 7–1 is to be milled. The program is to be written in Machinist Shop Language, using incremental positioning. Figure 7–2 depicts the cutter paths necessary to machine the part. In accordance with the cutter path drawing, the following sequence of events is to be performed:

1. At the tool change location, place a drill into the spindle. Move to location #1 and turn the spindle and coolant on.
2. Drill hole #1.
3. Drill hole #2.
4. Drill hole #3.
5. Drill hole #4.
6. Drill hole #5.
7. Turn off the spindle and coolant and return to tool change for a 1-inch-diameter end mill.

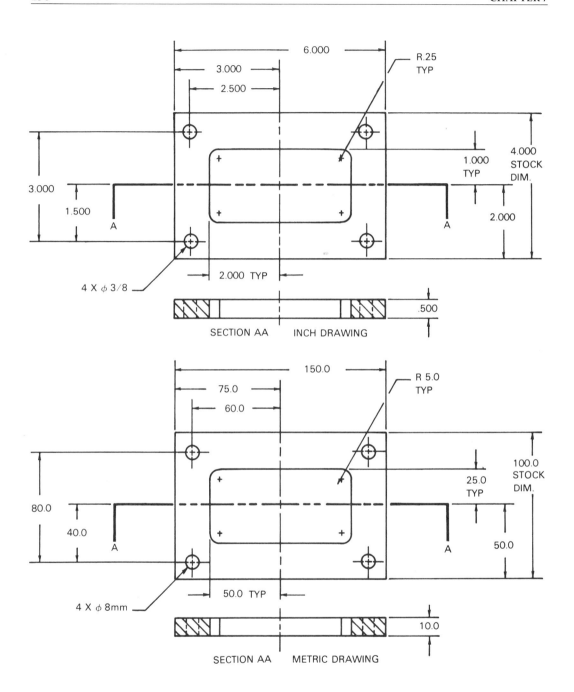

FIGURE 7–1
Part drawing for three-axis programming task

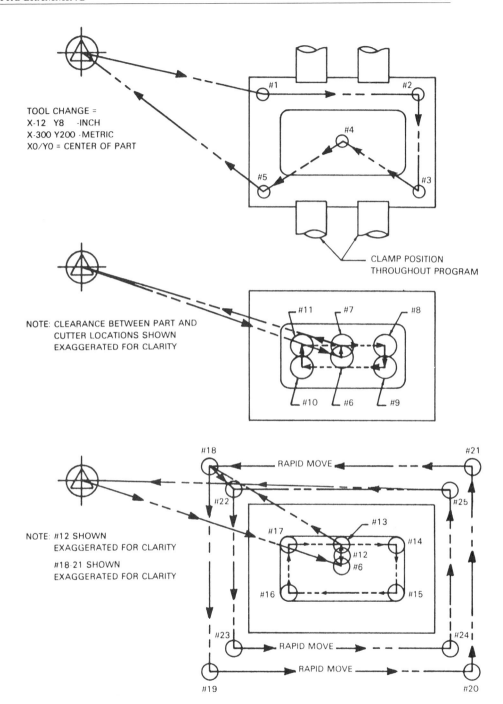

TOOL CHANGE =
X-12 Y8 -INCH
X-300 Y200 -METRIC
X0/Y0 = CENTER OF PART

CLAMP POSITION
THROUGHOUT PROGRAM

NOTE: CLEARANCE BETWEEN PART AND
CUTTER LOCATIONS SHOWN
EXAGGERATED FOR CLARITY

NOTE: #12 SHOWN
EXAGGERATED FOR CLARITY

#18-21 SHOWN
EXAGGERATED FOR CLARITY

RAPID MOVE

FIGURE 7-2
Cutter paths for part drawing in Figure 7-1

8. Move to location #6 at rapid traverse, turn the spindle and coolant on, and plunge cut a hole thru the part.
9. Feed from #6 to #7.
10. Feed from #7 to #8.
11. Feed from #8 to #9.
12. Feed from #9 to #10.
13. Feed from #10 to #11.
14. Feed from #11 to #7.
15. Retract the spindle.
16. Turn the spindle and coolant off and return to tool change for a .500-inch-diameter end mill.
17. Rapid traverse to location #12, and turn the spindle and coolant on.
18. Lower the spindle to depth.
19. Feed from #12 to #13.
20. Feed from #13 to #14.
21. Feed from #14 to #15.
22. Feed from #15 to #16.
23. Feed from #16 to #17.
24. Feed from #17 to #13.
25. Feed from #13 to #12.
26. Retract the spindle.
27. Rapid traverse from #12 to #18.
28. Lower the spindle to depth.
29. Feed from #18 to #19.
30. Retract the spindle.
31. Rapid traverse from #19 to #20, jumping over the clamps.
32. Lower the spindle to depth.
33. Feed from #20 to #21.
34. Retract the spindle.
35. Rapid move from #21 to #18.
36. Lower the spindle to depth.
37. Feed from #18 to #22.
38. Feed from #22 to #23.
39. Retract the spindle.
40. Rapid traverse from #23 to #24, jumping over the clamps.
41. Lower the spindle to depth.
42. Feed from #24 to #25.
43. Retract the spindle.
44. Turn the spindle and coolant off and rapid traverse to the spindle park position, ending the task.

Figure 7–3 is a Machinist Shop Language program written for inch specifications; Figure 7–4 is an identical program written for metric. Both a roughing and finishing milling cut are taken on the surfaces to be milled. The part is clamped to the table along the surfaces that do not require machining.

```
XO/YO = CENTER OF PART
TOOL CHANGE = X-12 Y8
TOOLS LIST:
TOOL 1 = 3/8 COMB. DRILL
TOOL 2 = 1.000 4 FLUTE END MILL
TOOL 3 = .500 4 FLUTE END MILL
BUFFER HEIGHT:  TOP OF PART FOR TOOL 1,  .100 INCH
FOR TOOLS 2 AND 3
CLEARANCE OVER CLAMPS:  3.000 IN.

 1 TOOL 1001
 2 Z4   A
 3 TOOL 1002
 4 Z3   A
 5 TOOL 1003
 6 Z3.5  A
 7 X-12 Y8 RA
 8 TOOL 1                       REM:3/8 DRILL 1066 RPM
 9 V20 12.8
10 V21 .1
11 G81
12 X-2.5 Y1.5 Z-.62 RA         REM:DRILL #1
13 Z3 RA                        REM:RETRACT SPNDL
14 X2.5 RA                      REM:DRILL #2
15 Y-1.5 RA                     REM:DRILL #3
16 XO YO RA                     REM:DRILL #4
17 X-2.5 Y-1.5 RA               REM:DRILL #5
18 G80
19 TOOL 0                       REM:CANCEL OFFSET
20 ZO RA                        REM:RETRACT SPNDL
21 X-12 Y8 RA                   REM:TCH
22 TOOL 2                       REM:1.000 E/M 425 RPM
23 FEED 6.8
24 XO YO RA                     REM:RAPID TO #6
25 ZO RA
26 Z-.62 FA                     REM:FEED TO DEPTH
27 Y.48 FA                      REM:FEED TO #7
28 X1.48 FA                     REM:FEED TO #8
29 Y-.48 FA                     REM:FEED TO #9
30 X-1.48 FA                    REM:FEED TO #10
31 Y.48 FA                      REM:FEED TO #11
32 XO FA                        REM:FEED TO #7
33 TOOL 0                       REM:CANCEL OFFSET
34 ZO RA                        REM:RETRACT SPNDL
35 X-12 Y8 RA                   REM:TCH
36 TOOL 3                       REM:.500 E/M 800 RPM
37 FEED 12,8
38 XO YO RA                     REM:RAPID TO #6
39 ZO RA                        REM:RAPID TO BUFFER
40 Z-.62 FA                     REM:FEED TO DEPTH
41 Y.75 FA                      REM:FEED TO #13
42 X1.75 FA                     REM:FEED TO #14
43 Y-.75 FA                     REM:FEED TO #15
44 X-1.75 FA                    REM:FEED TO #16
45 Y.75 FA                      REM:FEED TO #17
46 XO FA                        REM:FEED TO #13
47 Y.74 FA
```

```
48  Z3 RA
49  X-3.26 Y2.26 RA              REM:RAPID TO #18
50  Z0 RA                        REM:RAPID TO BUFFER
51  Z-.62 FA                     REM:FEED TO DEPTH
52  Y-2.26 FA                    REM:FEED TO #19
53  Z3 RA
54  X3.26 RA                     REM:RAPID TO #20
55  Z0 RA                        REM:RAPID TO BUFFER
56  Z-.62 FA                     REM:FEED TO DEPTH
57  Y2.26 FA                     REM:FEED TO #21
58  Z3 RA
59  X-3.26 RA                    REM:RAPID TO #18
60  Z0 RA                        REM:RAPID TO BUFFER
61  Z-.62 FA                     REM:FEED TO DEPTH
62  X-3.25 Y2.25 FA              REM:FEED TO #22
63  Y-2.25 FA                    REM:FEED TO #23
64  Z3 RA
65  X3.25 RA                     REM:RAPID TO #24
66  Z0 RA                        REM:RAPID TO BUFFER
67  Z-.62 FA                     REM:FEED TO DEPTH
68  Y2.25 FA                     REM:FEED TO #25
69  TOOL 0                       REM:CANCEL OFFSET
70  Z0 RA                        REM:RETRACT QUILL
71  X-12 Y8 RA                   REM:TCH
72  END
```

FIGURE 7-3
Machinist Shop Language three-axis program, nonmetric, for the part in Figure 7-1

MACHINIST SHOP LANGUAGE

To use all three axes on the CNC machine, it is necessary to introduce some new commands.

TOOL—Used in Chapter 6 to call up a tool at tool change, TOOL is also used to assign the length of a tool into an offset register.

G—This is a preparatory function (G code). With three-axis operation, many different cycles may be used, each called up by a G code in both Machinist Shop Language and word address format. The G code G04 was used in the last chapter to cause a dwell to occur in word address format.

V—This means *variable,* and allows the programmer to assign values to specific things. V20, for example, specifies the feedrate for the Z axis when using G codes. V21 specifies the height off the surface of the workpiece at which the tool begins and ends its feedrate movements when using G codes. A listing of Machinist Shop Language G code and V code commands is contained in Appendix 2.

```
X0/Y0 = CENTER OF PART
TOOL CHANGE = X-300 Y200
TOOLS LIST:
TOOL 1 = 8mm COMB. DRILL
TOOL 2 = 25mm 4 FLUTE END MILL
TOOL 3 = 10mm 4 FLUTE END MILL
BUFFER HEIGHT:  TOP OF PART FOR TOOL 1,2.54mm FOR
TOOLS 2 AND 3
CLEARANCE OVER CLAMPS:  75mm


 1 TOOL 1001
 2 Z100 A
 3 TOOL 1002
 4 Z75  A
 5 TOOL 1003
 6 Z87.5 A
 7 X-300 Y200 RA
 8 TOOL 1                       REM:8mm DRILL 1066 RPM
 9 V20 216
10 V21 2.54
11 G81
12 X-60 Y40 Z-13 RA            REM:DRILL #1
13 Z75 RA                      REM:RETRACT SPNDL
14 X60 RA                      REM:DRILL #2
15 Y-40 RA                     REM:DRILL #3
16 X0 Y0 RA                    REM:DRILL #4
17 X-60 Y-40 RA               REM:DRILL #5
18 G80
19 TOOL 0                      REM:CANCEL OFFSET
20 Z0 RA                       REM:RETRACT SPNDL
21 X-300 Y200 RA              REM:TCH
22 TOOL 2                      REM:25mm E/M 425 RPM
23 FEED 172.5
24 X0 Y0 RA                    REM:RAPID TO #6
25 Z0 RA
26 Z-13 FA                     REM:FEED TO DEPTH
27 Y12.25 FA                   REM:FEED TO #7
28 X37.25 FA                   REM:FEED TO #8
29 Y-12.25 FA                  REM:FEED TO #9
30 X-37.25 FA                  REM:FEED TO #10
31 Y12.25 FA                   REM:FEED TO #11
32 X0 FA                       REM:FEED TO #7
33 TOOL 0                      REM:CANCEL OFFSET
34 Z0 RA                       REM:RETRACT SPNDL
35 X-300 Y200 RA              REM:TCH
36 TOOL 3                      REM:10.0 E/M 800 RPM
37 FEED 325.1
38 X0 Y0 RA                    REM:RAPID TO #6
39 Z0 RA                       REM:RAPID TO BUFFER
40 Z-13 FA                     REM:FEED TO DEPTH
41 Y20 FA                      REM:FEED TO #13
42 X45 FA                      REM:FEED TO #14
43 Y-20 FA                     REM:FEED TO #15
44 X-45 FA                     REM:FEED TO #16
45 Y20 FA                      REM:FEED TO #17
46 X0 FA                       REM:FEED TO #13
47 Y44.75 FA
48 Z75 RA
```

```
49 X-80.25 Y55.25 RA        REM:RAPID TO #18
50 Z0 RA                    REM:RAPID TO BUFFER
51 Z-13 FA                  REM:FEED TO DEPTH
52 Y-55.25 FA               REM:FEED TO #19
53 Z75 RA
54 X80.25 RA                REM:RAPID TO #20
55 Z0 RA                    REM:RAPID TO BUFFER
56 Z-13 FA                  REM:FEED TO DEPTH
57 Y55.25 FA                REM:FEED TO #21
58 Z75 RA
59 X-80.25 RA               REM:RAPID TO #18
60 Z0 RA                    REM:RAPID TO BUFFER
61 Z-13 FA                  REM:FEED TO DEPTH
62 X-80 Y55 FA              REM:FEED TO #22
63 Y-55 FA                  REM:FEED TO #23
64 Z75 RA
65 X80 RA                   REM:RAPID TO #24
66 Z0 RA                    REM:RAPID TO BUFFER
67 Z-13 FA                  REM:FEED TO DEPTH
68 Y55 FA                   REM:FEED TO #25
69 TOOL 0                   REM:CANCEL OFFSET
70 Z0 RA                    REM:RETRACT QUILL
71 X-300 Y200 RA            REM:TCH
72 END
```

FIGURE 7–4
Machinist Shop Language three-axis program, metric, for the part in Figure 7–2

Tool Length Offsets

As noted, tool length offsets are assigned using the TOOL command. If the tool lengths are not known by the programmer (and they usually are not), a blank event is created in the program at the proper spot to allow the setup man to enter tool length values. In this chapter it is assumed that the tools have been measured and the lengths are known: a 3/8-inch combination center drill/drill; 1-inch-diameter, four-flute end mill; and 1/2-inch-diameter, four-flute end mill. Their lengths are as follows:

- Drill—1.000 inch long
- 1-inch-diameter end mill—2.000 inches long
- 1/2-inch-diameter end mill—1.500 inches long

A buffer area of .100 inch is to be used between the top of the part and the tool when the tool offset is active for the two end mills. Because of the way a Machinist Shop Language preparatory function cycle operates, the buffer for the drill will be programmed within the drilling cycle. The buffer is established by setting the tool height using a .100-inch gaging block on top of the part as explained in Chapter 4. A total of 3 inches of clearance is desired between the start of the buffer level and the longest tool when the spindle is retracted.

The format for assigning tool lengths is

TOOL 10##

Where the first two digits, "10" (which always remain the same) tell the controller that tool information will be defined in the following event, and the second two digits (##) are the tool number of the tool being defined. Although they may be given at any time before they are used, tool statements are generally placed first in the program for the convenience of the setup man.

The tool statements for the tools used in the following part program are:

1. TOOL 1001   3. TOOL 1002   5. TOOL 1004
2. Z4 A   4. Z3 A   6. Z3.5 A

For the first tool, the tool command specifies that tool information for tool 01 is being assigned. The second event sets the first tool offset at 4 inches. With the longest tool (2 inches) in the spindle, the clearance zone is 3 inches. With an inch-long tool, the clearance zone increases to 4 inches. That is, 4 inches is the distance necessary to move the end of the tool from the spindle-retracted position to the start of the buffer zone. The offset is entered as an absolute Z coordinate. The remaining tools are entered in like manner.

Remark Statements

Notice the use of the term "REM" in the program. REM, as used in this case, stands for the word *remark* (or reminder). Remark statements are usually provided for by the controller manufacturer and are ignored by the controller. Their inclusion in the program has no effect on the actual program; their main purpose in the program listing is to remind the operator what is happening in the program or to tell someone else what the program intends to accomplish in each part. They also aid in debugging the program prior to milling the first part.

In word address systems, a comment (remark) is often a statement placed between parentheses on a line of its own. While displayed on the machine MCU monitor, the comment is ignored by the controller. For consistency however, this text will use the term REM throughout the examples to indicate remarks to the student.

PROGRAM EXPLANATION

(Refer to Figures 7–3 and 7–4.)

EVENT 1

- TOOL 1001—Signals the MCU that a tool length is to be assigned for tool #1. The first two digits (10) specify that tool information is contained in the event that follows. The last two digits (01) tell the MCU that this offset is to be assigned to tool #1. The offset will become active when the command TOOL 1 is given.

EVENT 2

- Z coordinate that equals the tool offset value—This coordinate is always entered in absolute mode.

EVENT 3

- TOOL 1002—Assigns the offset in Event 4 to tool #2.

EVENT 4

- Z coordinate—For tool #2 offset.

EVENT 5

- TOOL 1003—Assigns the offset in Event 6 to tool #3.

EVENT 6

- Z coordinate—For tool #3 offset.

EVENT 7

- X/Y coordinates—For the tool change location.
- R—Specifies rapid traverse.
- A—Specifies absolute positioning.

EVENT 8

- TOOL 1—In three-axis operation, this causes two things to happen: dwell is automatically assigned to the controller, thereby allowing the operator to install the first tool in the spindle; and the offset that was entered using TOOL 1001 is activated.

EVENT 9

- V20—A variable register code that is unique to Machinist Shop Language. V20 is used to assign a feedrate to be used by the Z axis whenever an 80 series G code cycle is called up. G code cycles are often called *canned cycles* because they are built into the executive program. In this case, a feedrate of 12.8 in./min (172.5 mm/min) is used.

EVENT 10

- V21—Sets the amount of buffer to be established between the Z0 point (offset active) and the tool. A buffer gives a safety cushion for the tool to begin the feedrate move. Leaving a cushion allows for deceleration of the tool, and insures that the feedrate will be active when the tool cutting edge contacts the metal. V21.1 sets a .100-inch (2.54-mm) buffer zone. With this feature, a buffer need not be built into the drill, as was done with the end mills at setup. Appendix 2 lists V codes used in Machinist Shop Language.

EVENT 11

- G81—Calls up the canned drilling cycle. When a G81 is issued, the spindle rapids to the X/Y coordinate, rapids to the start of the buffer zone, feeds to the indicated Z axis coordinate, and rapids back out to the start of the buffer zone. At the end of this chapter is a brief summary of the more common G codes. Appendix 2 lists all the G codes used in Machinist Shop Language.

EVENT 12

- X/Y coordinates—To move from tool change to hole #1 and drill the hole.
- R—Specifies rapid traverse.
- A—Specifies absolute positioning.

EVENT 13

- Z-axis coordinate—To raise the spindle. Since a 3-inch clearance was allowed with the longest tool, there are at least three inches of upward movement possible for the spindle. With the G81 active, the spindle retracted to the buffer zone height after drilling hole #1. A clamp is in the way of the move to hole #2. The spindle was therefore raised to allow the tool to clear the clamp. Another technique that could be employed here is to cancel the G code, cancel the tool offset, raise the spindle to Z0 (the fully retracted position), call up the tool offset, and reinstitute the G code. By using the practice chosen, the tool offset remains active the entire time the tool is used.
- R—Specifies rapid traverse.
- A—Specifies absolute positioning.

EVENT 14

- X-axis coordinate—To move from hole #1 to #2. G81 is still active; therefore hole #2 is drilled.
- R—Specifies rapid traverse.
- A—Specifies absolute positioning.

EVENT 15

- Y-axis coordinate—To drill hole # 3.
- R—Specifies rapid traverse.
- A—Specifies absolute positioning.

EVENT 16

- X/Y coordinates—To drill hole #4.
- R—Specifies rapid traverse.
- A—Specifies absolute positioning.

EVENT 17
- ▪ X/Y coordinates—To drill hole #5.
- ▪ R—Specifies rapid traverse.
- ▪ A—Specifies absolute positioning.

EVENT 18
- ▪ G80—Cancels the drilling cycle.

EVENT 19
- ▪ TOOL 0—Cancels the tool length offset. Z0 is now the fully retracted spindle position.

EVENT 20
- ▪ Z0—Retracts the spindle.
- ▪ R—Specifies rapid traverse.
- ▪ A—Specifies absolute positioning.

EVENT 21
- ▪ X/Y coordinates—To move from hole #5 to tool change.
- ▪ R—Specifies rapid traverse.
- ▪ A—Specifies absolute positioning.

EVENT 22
- ▪ TOOL 2—Calls up the offset for tool #2 and halts the program so that the operator can install the end mill.

EVENT 23
- ▪ FEED 6.8 (172.5 in the metric version)—Assigns a feedrate to be used for feedrate moves.

EVENT 24
- ▪ X/Y coordinates—To move from tool change to location #6.
- ▪ R—Specifies rapid traverse.
- ▪ A—Specifies absolute positioning.

EVENT 25
- ▪ Z0 coordinate—Rapids the spindle to the start of the .100 buffer zone that was built into the tool offset at setup per the programmer's instruction.
- ▪ R—Specifies rapid traverse.
- ▪ A—Specifies absolute positioning.

EVENT 26
- ▪ Z coordinate—Feeds the end mill to depth. The coordinate is derived by adding the thickness of the part, the height of the buffer zone, and the additional space below the part that the programmer desires the tool to feed.
- ▪ F—Specifies a feedrate move.
- ▪ A—Specifies absolute positioning.

EVENT 27
- Y coordinate—To feed from #6 to #7.
- F—Specifies a feedrate move.
- A—Specifies absolute positioning.

EVENT 28
- X coordinate—To feed from #7 to #8.
- F—Specifies a feedrate move.
- A—Specifies absolute positioning.

EVENT 29
- Y coordinate—To feed from #8 to #9.
- F—Specifies a feedrate move.
- A—Specifies absolute positioning.

EVENT 30
- X coordinate—To feed from #9 to #10.
- F—Specifies a feedrate move.
- A—Specifies absolute positioning.

EVENT 31
- Y coordinate—To feed from #10 to #11.
- F—Specifies a feedrate move.
- A—Specifies absolute positioning.

EVENT 32
- X coordinate—To feed from #11 to #7.
- F—Specifies a feedrate move.
- A—Specifies absolute positioning.

EVENT 33
- TOOL 0—Cancels the active tool offset. Z0 becomes the fully re-tracted spindle position.

EVENT 34
- Z0—Positions the spindle at the fully retracted location.
- R—Specifies rapid traverse.
- A—Specifies absolute positioning.

EVENT 35
- X/Y coordinates—To move from #7 to tool change.
- R—Specifies rapid traverse.
- A—Specifies absolute positioning.

EVENT 36
- TOOL 3—Calls up the offset for tool #3. The .500-inch-diameter (10-mm metric) end mill is installed in the spindle.

EVENT 37

- FEED 12.8 (325.1 in the metric version)—Assigns a feedrate to be used for feedrate moves.

EVENT 38

- X0 Y0 coordinates—To move from tool change to location #6.
- R—Specifies rapid traverse.
- A—Specifies absolute positioning.

EVENT 39

- Z0—Moves the spindle to the start of the buffer zone. Z0 is .100 above the top of the part when the tool offset is active.
- R—Specifies rapid traverse.
- A—Specifies absolute positioning.

EVENT 40

- Z coordinate—To feed the end mill to depth.
- F—Specifies a feedrate move.
- A—Specifies absolute positioning.

EVENT 41

- Y coordinate—To feed from #12 to #13.
- F—Specifies a feedrate move.
- A—Specifies absolute positioning.

EVENT 42

- X coordinate—To feed from #13 to #14.
- F—Specifies a feedrate move.
- A—Specifies absolute positioning.

EVENT 43

- Y coordinate—To feed from #14 to #15.
- F—Specifies a feedrate move.
- A—Specifies absolute positioning.

EVENT 44

- X coordinate—To feed from #15 to #16.
- F—Specifies a feedrate move.
- A—Specifies absolute positioning.

EVENT 45

- Y coordinate—To feed from #16 to #17.
- F—Specifies a feedrate move.
- A—Specifies absolute positioning.

EVENT 46

- X axis coordinate—To feed from #17 to #13.
- F—Specifies a feedrate move.
- A—Specifies absolute positioning.

EVENT 47

- Y coordinate—To feed from #13 to #12.
- F—Specifies a feedrate move.
- A—Specifies absolute positioning.

EVENT 48

- Z coordinate—To retract the spindle to clear the clamps.
- R—Specifies rapid traverse.
- A—Specifies absolute positioning.

EVENT 49

- X/Y coordinates—To move from #12 to #18.
- R—Specifies rapid traverse.
- A—Specifies absolute positioning.

EVENT 50

- Z0—Rapids the spindle to the start of the buffer zone.
- R—Specifies rapid traverse.
- A—Specifies absolute positioning.

EVENT 51

- Z coordinate—To feed the end mill to depth.
- F—Specifies a feedrate move.
- A—Specifies absolute positioning.

EVENT 52

- Y coordinate—To feed from #18 to #19.
- F—Specifies a feedrate move.
- A—Specifies absolute positioning.

EVENT 53

- Z coordinate—To raise the spindle to clear the clamps.
- R—Specifies rapid traverse.
- A—Specifies absolute positioning.

EVENT 54

- X coordinate—To move from #19 to #20.
- R—Specifies rapid traverse.
- A—Specifies absolute positioning.

EVENT 55

- Z0—Positions the spindle at the start of the buffer zone.
- R—Specifies rapid traverse.
- A—Specifies absolute positioning.

EVENT 56

- Z coordinate—To feed the end mill to depth.
- F—Specifies a feedrate move.
- A—Specifies absolute positioning.

EVENT 57

- Y cocrdinate—To feed from #20 to #21.
- F—Specifies a feedrate move.
- A—Specifies absolute positioning.

EVENT 58

- Z coordinate—To raise the spindle to clear the clamps.
- R—Specifies rapid traverse.
- A—Specifies absolute positioning.

EVENT 59

- X coordinate—To move from #21 to #18.
- R—Specifies rapid traverse.
- A—Specifies absolute positioning.

EVENT 60

- Z0—Positions the spindle at the start of the buffer zone.
- R—Specifies rapid traverse.
- A—Specifies absolute positioning.

EVENT 61

- Z coordinate—To feed the end mill to depth.
- F—Specifies a feedrate move.
- A—Specifies absolute positioning.

EVENT 62

- X/Y coordinates—To feed from #18 to #22. This move positions the cutter for the finish cut on the outside surfaces of the part.
- F—Specifies a feedrate move.
- A—Specifies absolute positioning.

EVENT 63

- Y coordinate—To feed from #22 to #23.
- F—Specifies a feedrate move.
- A—Specifies absolute positioning.

EVENT 64

- Z coordinate—To raise the spindle to clear the clamps.
- R—Specifies rapid traverse.
- A—Specifies absolute positioning.

EVENT 65

- X coordinate—To move from #23 to #24.
- R—Specifies rapid traverse.
- A—Specifies absolute positioning.

EVENT 66
- Z0—Positions the spindle at the start of the buffer zone.
- R—Specifies rapid traverse.
- A—Specifies absolute positioning.

EVENT 67
- Z coordinate—To feed the end mill to depth.
- F—Specifies a feedrate move.
- A—Specifies absolute positioning.

EVENT 68
- Y coordinate—To feed from #24 to #25.
- F—Specifies a feedrate move.
- A—Specifies absolute positioning.

EVENT 69
- TOOL 0—Cancels the tool offset, making Z0 the fully retracted spindle position.

EVENT 70
- Z0—Fully retracts the spindle.
- R—Specifies rapid traverse.
- A—Specifies absolute positioning.

EVENT 71
- X/Y coordinates—Specifies the tool change location.
- R—Specifies rapid traverse.
- A—Specifies absolute positioning.

EVENT 72
- END—Signals the end of the program. The computer's memory resets to the start of the program.

Modal/Nonmodal Commands

Notice that in this program, and in those in the previous chapter, certain commands remained active until canceled by another code. Codes that are active for more than the line in which they are issued are called *modal* commands. In this program, rapid traverse, feedrate moves, and the G81 canned cycle were examples of modal commands. A *nonmodal* command is one which is active only in the program line in which it is issued. TOOL and DWELL are examples of nonmodal commands.

WORD ADDRESS FORMAT

The part program using word address format follows the same basic programming logic as that just given in Machinist Shop Language. Although the sequence of operations is the same, the codes used to carry out the operations vary somewhat from the Machinist Shop Language commands. They are:

G00—Selects rapid traverse mode.

G01—Selects feedrate mode.

G70—As in two-axis operation, selects inch input.

G90/G91—As in two-axis programming, selects absolute or incremental positioning.

G10—Used when assigning tool length offsets, G10 fulfills the same function as the first two digits in the Machinist Shop Language command TOOL 1001; namely, to tell the MCU that tool length information is to be assigned.

H—Used to assign a tool register (just as the last two digits of a TOOL 10## in Machinist Shop Language did). H01 would assign the information given to offset register #1. H02 would assign the information to offset register #2. H is used in conjunction with G10. A tool assignment statement would be: G10 H##, where ## is the register number.

G45—Calls up the tool length offset. G45 accomplishes a Z0 shift toward the workpiece. The coding used for the programs in Figures 7–5 and 7–6 is in the General Numerics format. The controller actually uses G43 and G44 for tool length offsets, but these codes conflict with EIA standards; therefore, G45 will be used in this text (G45 is normally an unassigned code). As has been pointed out, codes vary from controller to controller and machine to machine. The coding used in these programs is also similar to that used on Fanuc controllers. The only way to know which code to use is to check the programming manual for a particular machine. Always remember: *when in doubt, check the manual!*

G49—This is the tool length offset cancel code.

G81—As in Machinist Shop Language, G81 is the canned drill cycle. It functions in the same manner explained in the previous example.

G80—As used previously, G80 cancels an 80 series canned cycle.

R—This address stands for reference level. The *reference level* is the spot where the programmer desires the canned cycle to start feeding into the workpiece. The reference level is also called the *rapid* or *gage level*. The reference level is usually the same height as the buffer zone, but it may not be.

G98/G99—G98 is the return to initial level command. G99 is the return to rapid (reference) level command. When an 80 series canned cycle is active in word address format, the spindle may be directed to return to the rapid level with a G99. G99 is modal and will remain active until a G80 cancel code is issued, or until canceled with a G98. If a clamp is in the path of movement, or if the spindle is at the last location in a series, the spindle may be retracted to the initial starting point in the cycle by using a G98 command.

M03—M functions, as briefly explained in Chapter 6, control a number of auxiliary functions. M03 is the code for turning the spindle on in the clockwise direction. In Appendix 3 is a list of common M functions used on numerical control machinery.

M05—Turns the spindle off.

M06—Tool change code.

M08—Turns the coolant on.

M09—Turns the coolant off.

T—Selects the tool to be put in the spindle by the tool changer.

F—Assigns feedrates, as in two-axis programming.

S—Designates the spindle speed.

PROGRAM EXPLANATION

(Refer to Figures 7–5 and 7–6.)

The machine used for the programs in Figures 7–5 and 7–6 is a vertical machining center using automatic tool change such as that shown in Figure 7–7. Figure 7–5 is a word address format program for the inch-dimensioned part in Figure 7–1. Figure 7–6 is the metric version of the program.

N010

- N010—The sequence number.
- G70—Selects inch input.
- G90—Selects absolute positioning.
- G10 H01 Z4—Assigns a 4-inch value to tool register #1 (100 mm in the metric program). G10 tells the controller that tool information is to be assigned. H01 tells the controller to place the information in register #1.

N020

- N020—The sequence number.
- G10 H02 Z3—Assigns a 3-inch value to tool register #2 (in the metric program, a value of 75 mm is assigned).

N030

- N030—The sequence number.
- G10 H03 Z3.5—Assigns a 3.5-inch value to tool register #3 (in the metric program, a value of 87.5 mm is assigned).

```
X0/Y0 = CENTER OF PART
TOOL CHANGE = X-12 Y8
TOOLS LIST:
TOOL 1 = 3/8 COMB. DRILL
TOOL 2 = 1.000 4 FLUTE END MILL
TOOL 3 = .500 4 FLUTE END MILL
BUFFER HEIGHT:
.100 ABOVE PART SURFACE

N010 G70 G90 G10 H01 Z4
N020 G10 H02 Z3
N030 G10 H03 Z3.5
N040 G00 M06 T1                              REM:3/8 DRILL #1
N050 G45 H01 S1066 M03                       REM:SPNDL ON
N060 G81 G98 X-2.5 Y1.5 Z-.62 R0 F12.8 M08   REM:DRILL #1
N070 G99 X2.5                                REM:DRILL #2
N080 Y-1.5                                   REM:DRILL #3
N090 X0 Y0                                   REM:DRILL #4
N100 X-2.5  Y-1.5                            REM:DRILL #5
N110 G80 G49 Z0                              REM:CANCEL DRILL & OFFSET
N120 M06 T2                                  REM:1.000 E/M
N130 X0 Y0 S425 M03                          REM:RAPID TO #6, SPNDL ON
N140 G45 H02 Z0
N150 G01 Z-.62  F6.8 M08                     REM:FEED TO DEPTH
N160 Y.48                                    REM:FEED TO #7
N170 X1.48                                   REM:FEED TO #8
N180 Y-.48                                   REM:FEED TO #9
N190 X-1.48                                  REM:FEED TO #10
N200 Y.48                                    REM:FEED TO #11
N210 X0                                      REM:FEED TO #7
N220 G00 G49 Z0 M09                          REM:RETRACT SPNDL, COOL.
OFF
N230 M06 T3                                  REM:.500 E/M
N240 G45 H03
N250 X0 Y0 S800 F12.8 M03                    REM:RAPID TO #6, SPNDL ON
N260 Z0                                      REM:RAPID TO BUFFER
N270 G01 Z-.62                               REM:FEED TO DEPTH
N280 Y.75                                    REM:FEED TO #13
N290 X1.75                                   REM:FEED TO #14
N300 Y-.75                                   REM:FEED TO #15
N310 X-1.75                                  REM:FEED TO #16
N320 Y.75                                    REM:FEED TO #17
N330 X0                                      REM:FEED TO #13
N340 Y.74                                    REM:FEED TO #12
N350 G00 Z3                                  REM:RETRACT SPNDL
N360 X-3.26 Y2.26                            REM:RAPID TO #18
N370 Z0                                      REM:RAPID TO BUFFER
N380 G01 Z-.62                               REM:FEED TO DEPTH
N390 Y-2.26                                  REM:FEED TO #19
N400 G00 Z3                                  REM:RETRACT SPNDL
N410 X3.26                                   REM:RAPID TO#20
N420 Z0                                      REM:RAPID TO BUFFER
N430 G01 Z-.62                               REM:FEED TO DEPTH
N440 Y2.26                                   REM:FEED TO #21
N450 G00 Z3                                  REM:RETRACT  SPNDL
N460 X-3.26                                  REM:RAPID TO #18
N470 Z0                                      REM:RAPID TO BUFFER
```

```
N480 G01 Z-.62                         REM:FEED TO DEPTH
N490 X-3.25 Y2.25                      REM:FEED TO #22
N500 Y-2.25                            REM:FEED TO #23
N510 G00 Z3                            REM:RETRACT SPNDL
N520 X3.25                             REM:RAPID TO #24
N530 Z0                                REM:RAPID TO BUFFER
N540 G01 Z-.62                         REM:FEED TO DEPTH
N550 Y2.25                             REM:FEED TO #25
N560 G00 G49 Z0 M09                    REM:CANCEL OFFSET, COOL.
OFF
N570 X-12 Y8 M05                       REM:RAPID TO PARK, SPNDL
OFF
N580 M30
```

FIGURE 7-5

Word address format three-axis program, nonmetric, for the part in Figure 7-1

```
X0/Y0 = CENTER OF PART
TOOL CHANGE = X-300 Y200
TOOLS LIST:
TOOL 1 = 8mm COMB. DRILL
TOOL 2 = 25mm 4 FLUTE END MILL
TOOL 3 = 10mm 4 FLUTE END MILL
BUFFER HEIGHT:
.25mm ABOVE PART SURFACE

N010 G71 G90 G10 H01 Z100
N020 G10 H02 Z75
N030 G10 H03 Z87.5
N040 G00 M06 T1                        REM:8mm DRILL #1
N050 G45 H01 S1066 M03                 REM:SPNDL ON
N060 G81 G98 X-60 Y40 Z-13 R0 F216 M08 REM:DRILL #1
N070 G99 X60                           REM:DRILL #2
N080 Y-40                              REM:DRILL #3
N090 X0 Y0                             REM:DRILL #4
N100 X-60 Y-40                         REM:DRILL #5
N110 G80 G49 Z0                        REM:CANCEL DRILL, OFFSET
N120 M06 T2                            REM:25mm E/M
N130 X0 Y0 S425 M03                    REM:RAPID TO #6, SPNDL ON
N140 G45 H02 Z0
N150 G01 Z-13  F172.5 M08              REM:FEED TO DEPTH
N160 Y12.25                            REM:FEED TO #7
N170 X37.25                            REM:FEED TO #8
N180 Y-12.25                           REM:FEED TO #9
N190 X-37.25                           REM:FEED TO #10
N200 Y12.25                            REM:FEED TO #11
N210 X0                                REM:FEED TO #7
N220 G00 G49 Z0 M09                    REM:RETRACT SPNDL, COOL.
OFF
N230 M06 T3                            REM:.500 E/M
N240 G45 H03
N250 X0 Y0 S800 F325.1 M03             REM:RAPID TO #6, SPNDL ON
N260 Z0                                REM:RAPID TO BUFFER
```

```
N270 G01 Z-13                        REM:FEED TO DEPTH
N280 Y20                             REM:FEED TO #13
N290 X45                             REM:FEED TO #14
N300 Y-20                            REM:FEED TO #15
N310 X-45                            REM:FEED TO #16
N320 Y20                             REM:FEED TO #17
N330 X0                              REM:FEED TO #13
N340 Y44.75                          REM:FEED TO #12
N350 G00 Z75                         REM:RETRACT SPNDL
N360 X-80.25 Y55.25                  REM:RAPID TO #18
N370 Z0                              REM:RAPID TO BUFFER
N380 G01 Z-13                        REM:FEED TO DEPTH
N390 Y-55.25                         REM:FEED TO #19
N400 G00 Z75                         REM:RETRACT SPNDL
N410 X80.25                          REM:RAPID TO#20
N420 Z0                              REM:RAPID TO BUFFER
N430 G01 Z-13                        REM:FEED TO DEPTH
N440 Y55.25                          REM:FEED TO #21
N450 G00 Z75                         REM:RETRACT  SPNDL
N460 X-80.25                         REM:RAPID TO #18
N470 Z0                              REM:RAPID TO BUFFER
N480 G01 Z-13                        REM:FEED TO DEPTH
N490 X-80 Y55                        REM:FEED TO #22
N500 Y-55                            REM:FEED TO #23
N510 G00 Z75                         REM:RETRACT SPNDL
N520 X80                             REM:RAPID TO #24
N530 Z0                              REM:RAPID TO BUFFER
N540 G01 Z-13                        REM:FEED TO DEPTH
N550 Y55                             REM:FEED TO #25
N560 G00 G49 Z0 M09                  REM:CANCEL OFFSET, COOL.
OFF
N570 X-300 Y200 M05                  REM:RAPID TO PARK, SPNDL
OFF
N580 M30
```

FIGURE 7-6
Word address format three-axis program, metric, for the part in Figure 7-2

N040

- ▪ N040—The sequence number.
- ▪ G00—Puts the machine in rapid traverse mode.
- ▪ M06—Initiates a tool change.
- ▪ T1—Selects tool #1 to be loaded into the spindle by the tool changer.

N050

- ▪ N050—The sequence number.
- ▪ G45 H01—Calls up an offset for tool #1. Notice that the offset is not automatically assigned to the tool. With this system, any offset can be called up for any tool. (Note: It is wise to use the offset register number that corresponds to the tool number being used to avoid confusion.)
- ▪ S1066—Specifies that a spindle speed of 1066 RPM is to be used.
- ▪ M03—Turns the spindle on in clockwise rotation.

FIGURE 7–7
A light-duty vertical machining center *(Photo courtesy of Bridgeport Machines Division of Textron Inc.)*

N060

- N060—The sequence number.
- G81—Calls up the canned drilling cycle.
- G98—Instructs the MCU to return the spindle to its initial level at the Z-axis position when the 80 series G code was instituted. It is impor-

tant to know beforehand the position of the spindle relative to fixtures and clamps when using G98 to clear parts, fixtures, and clamps (as is done in this program). The tool change that was performed in block N050 left the spindle in its fully retracted position.

- X/Y/Z coordinates—Give the location of hole #1. With an 80 series G code, the machine positions to the X/Y coordinate before moving to the Z axis. The Z-axis coordinate is calculated to feed the drill through the part.
- R0—Sets the rapid level. The tool offset activated in block N050 shifted the Z0 point to the start of the buffer zone. The buffer zone height is also the desired rapid (or reference) level; zero as the Z-axis coordinate specifies this.
- F—Assigns the feedrate to be used for Z-axis feedrate moves with the active G81.
- M08—Turns on the coolant.

N070

- N070—The sequence number.
- G99—Instructs the machine to return the spindle to the rapid (reference) level (the start of the buffer zone).
- X coordinate—Gives the location of hole #2.

N080

- N080—The sequence number.
- Y coordinate—To move from hole #2 to hole #3.

N090

- N090—The sequence number.
- X0/Y0—To move from hole #3 to hole #4. Since the G81 is still active, a hole is drilled at this location.

N100

- N100—The sequence number.
- X/Y coordinates—To move from hole #4 to hole #5.

N110

- N110—The sequence number.
- G80—Cancels the G81.
- G49—Cancels the tool offset, thereby making Z0 the fully retracted position.
- Z0—Retracts the spindle.

N120

- N120—The sequence number.
- M06—Initiates an automatic tool change.
- T2—Specifies that tool #2 is to be used.

N130

- N130—The sequence number.
- X0/Y0—Positions the machine to location #6 in Figure 7–2. G00 is still active from the first block, so this is a rapid move.
- S425—Specifies a spindle speed of 425 RPM.
- M03—Turns the spindle on in clockwise rotation.

N140

- N140—The sequence number.
- G46 H02—Calls up tool offset #2, which is used here with tool #2.
- Z0—Rapids the spindle to the Z0 point when the tool offset is active, which is the start of the buffer zone.

N150

- N150—The sequence number.
- G01—Puts the machine in feedrate mode.
- Z coordinate—Feeds the end mill to proper milling depth.
- F—Assigns a feedrate to be used with feedrate moves, in this case, the Z-axis movement to milling depth.
- M08—Turns on the coolant.

N160

- N160—The sequence number.
- Y coordinate—To feed from #6 to #7.

N170

- N170—The sequence number.
- X coordinate—To feed from #7 to #8.

N180

- N180—The sequence number.
- Y coordinate—To feed from #8 to #9.

N190

- N190—The sequence number.
- X coordinate—To feed from #9 to #10.

N200

- N200—The sequence number.
- Y coordinate—To feed from #10 to #11.

N210

- N210—The sequence number.
- X coordinate—To feed from #11 to #7.

N220

- N220—The sequence number.
- G00—Puts the machine in rapid traverse mode.

- G49—Cancels the tool length offset.
- Z0—Rapids the spindle to the fully retracted position.
- M09—Turns off the coolant.

N230

- N230—The sequence number.
- M06—Initiates an automatic tool change.
- T3—Specifies that tool #3 is to be used.

N240

- N240—The sequence number.
- G46 H03—Calls up tool offset #3, which in this case is to be used with tool #3.

N250

- N250—The sequence number.
- X/Y coordinates—Position the machine to location #6. G00 is active and the move is therefore at rapid traverse.
- S800—Sets the spindle speed to 800 RPM.
- F—Assigns a feedrate to be used with feedrate moves.
- M03—Turns the spindle on in clockwise rotation.

N260

- N260—The sequence number.
- Z0—Rapids the spindle to the start of the buffer zone.

N270

- N270—The sequence number.
- G01—Puts the machine in feedrate mode.
- Z coordinate—Feeds the end mill to depth.

N280

- N280—The sequence number.
- Y coordinate—To feed from #6 to #13.

N290

- N290—The sequence number.
- X coordinate—To feed from #13 to #14.

N300

- N300—The sequence number.
- Y coordinate—To feed from #14 to #15.

N310

- N310—The sequence number.
- X coordinate—To feed from #15 to #16.

N320

- N320—The sequence number.
- Y coordinate—To feed from #16 to #17.

N330

- N330—The sequence number.
- X coordinate—To feed from #17 to #13.

N340

- N340—The sequence number.
- Y coordinate—To feed from #13 to #12. This move is used to pull the tool away from the finished machined part surface. This will prevent tool marks from being left on the part when the spindle is retracted.

N350

- N350—The sequence number.
- G00—Puts the machine in rapid traverse mode.
- Z coordinate—Retracts the spindle a sufficient amount to clear the part and clamps (the tool offset is still active in this case).

N360

- N360—The sequence number.
- X/Y coordinates—To move from #12 to #18.

N370

- N370—The sequence number.
- Z0—Rapids the spindle to the start of the buffer zone.

N380

- N380—The sequence number.
- G01—Puts the machine in feedrate mode.
- Z coordinate—To feed the end mill to depth.

N390

- N390—The sequence number.
- Y coordinate—To feed from #18 to #19.

N400

- N400—The sequence number.
- G00—Puts the machine in rapid traverse mode.
- Z coordinate—Raises the spindle to clear the clamps that will be encountered on the next move.

N410

- N410—The sequence number.
- X coordinate—To move from #19 to #20. The move is in rapid traverse mode.

N420

- N420—The sequence number.
- Z0—Rapids the spindle to the start of the buffer zone.

N430

- N430—The sequence number.
- G01—Puts the machine in feedrate mode.
- Z coordinate—Feeds the end mill to depth.

N440

- N440—The sequence number.
- Y coordinate—To feed from #20 to #21.

N450

- N450—The sequence number.
- G00—Puts the machine in rapid traverse mode.
- Z coordinate—Raises the spindle to clear clamps.

N460

- N460—The sequence number.
- X coordinate—To move from #21 to #18. The move is in rapid traverse mode.

N470

- N470—The sequence number.
- Z0—Rapids the spindle to the start of the buffer zone.

N480

- N480—The sequence number.
- G01—Puts the machine in feedrate mode.
- Z coordinate—Feeds the spindle to depth.

N490

- N490—The sequence number.
- X/Y coordinates—Feed the tool from #18 to #22. This feedrate move positions the tool for the finish milling cut.

N500

- N500—The sequence number.
- Y coordinate—To feed from #22 to #23.

N510

- N510—The sequence number.
- G00—Puts the machine in rapid traverse mode.
- Z coordinate—Raises the spindle to clear clamps.

N520

- N520—The sequence number.
- X coordinate—To move from #23 to #24.

N530

- N530—The sequence number.
- Z0—Rapids the spindle to the start of the buffer height.

N540

- N540—The sequence number.
- G01—Puts the machine in feedrate mode.
- Z coordinate—To feed the tool to depth.

N550

- N550—The sequence number.
- Y coordinate—To feed from #24 to #25.

N560

- N560—The sequence number.
- G00—Puts the machine in rapid traverse mode.
- G49—Cancels the tool length offset.
- Z0—Retracts the spindle.
- M09—Turns off the coolant.

N570

- N570—The sequence number.
- X/Y coordinates—The location of the park position. This is a location, safely out of the way, at which to position the machine at the end of the program. In this case it is assumed that the park position is roughly the same place where tool changes occur. Upon a rerunning of the program, block N040 would initiate a tool change to select tool #1. The park position is then used to protect the operator and part during loading and unloading from the fixture.
- M05—Turns off the spindle.

N580

- N580—The sequence number.
- M30—Signals the end of the program. It also resets the computer memory to the start of the program.

OTHER G CODES USED IN CNC PROGRAMMING

In the examples presented in this chapter, the basic drill cycle G81 was used. There are a number of other G codes that can be used in CNC programs. A list of these is contained in Appendix 3. Some of the more common codes are explained below. These codes are diagrammed in Figures 7–8, 7–9, and 7–10.

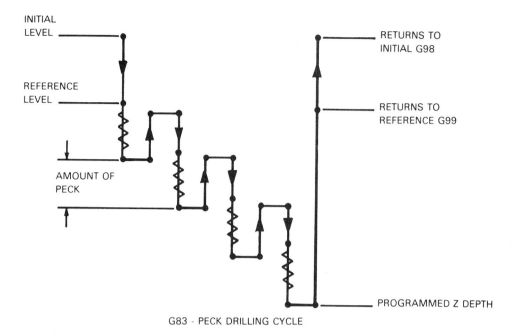

INITIAL LEVEL

REFERENCE LEVEL

RETURNS TO INITIAL G98

RETURNS TO REFERENCE G99

AMOUNT OF PECK

PROGRAMMED Z DEPTH

G83 - PECK DRILLING CYCLE

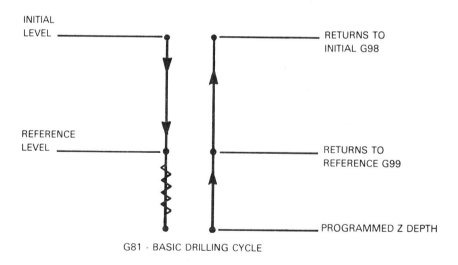

INITIAL LEVEL

REFERENCE LEVEL

RETURNS TO INITIAL G98

RETURNS TO REFERENCE G99

PROGRAMMED Z DEPTH

G81 - BASIC DRILLING CYCLE

━━━━━━ = RAPID MOVEMENT   ⋀⋀⋀⋀ = FEEDRATE MOVEMENT

FIGURE 7–8
G codes for peck drilling and basic drilling cycles

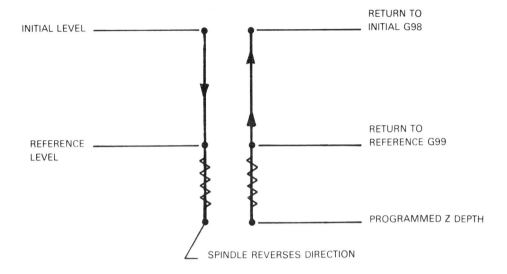

G84 - TAPPING CYCLE

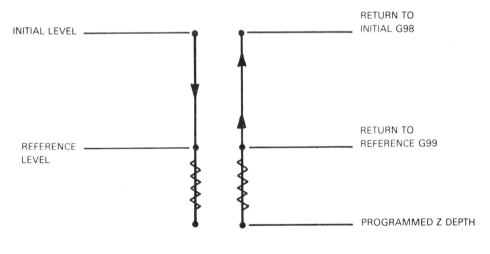

G85 - BORING CYCLE, TYPE A

FIGURE 7-9
G codes for tapping and boring cycles

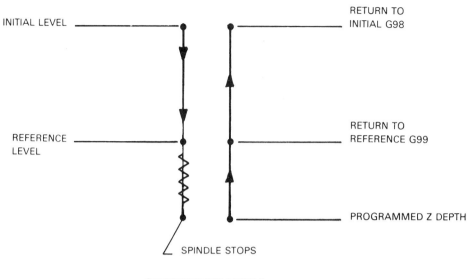

G86 BORING CYCLE TYPE B

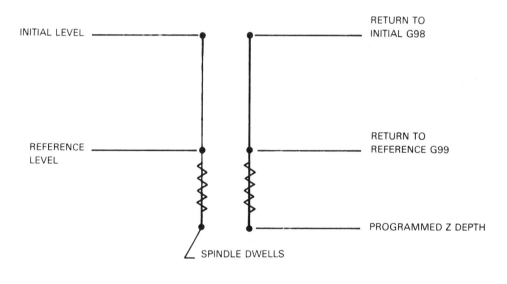

G89 - BORING CYCLE, TYPE C

 = RAPID MOVEMENT   = FEEDRATE MOVEMENT

FIGURE 7–10
G codes for boring cycles

G83 — Peck drilling cycle. The spindle rapids to reference level, feeds in .050, rapids out, feeds in .050 additional, and rapids out. This cycle repeats until the programmed Z-axis depth is reached. The spindle then rapids out to either the reference or initial level, depending on the program instructions. On some controllers, the amount of the peck is programmable using G codes.

G81 — Basic drilling cycle. The spindle rapids to reference level, feeds to Z-axis depth, and rapids out, returning to either reference or initial level.

G84 — Tapping cycle. The spindle rapids to the reference level and feeds to Z-axis depth; it then reverses direction, feeds to the reference level, and reverses direction again. If G98 is programmed, the spindle rapids to the initial level; if G99 is programmed, the spindle returns to reference level.

G85 — Boring cycle, type A. The spindle rapids to the reference level, feeds to Z-axis depth, and feeds to the reference level. If G98 is programmed, the spindle rapids to the initial level.

G86 — Boring cycle, type B. The spindle rapids to the reference level and feeds to Z-axis depth; it then stops and rapids to the reference level. If G98 is programmed, the spindle rapids to the initial level.

G89 — Boring cycle, type C. The spindle rapids to the reference level, feeds to Z-axis depth, dwells, and feeds to the reference level. If G98 is programmed, the spindle rapids to the initial level.

SPECIAL FANUC SECTION

Because of the growing popularity of Fanuc controls, this section has been included to aid those with access to CNC machinery with Fanuc controllers. Figure 7 – 11 is a program to drill and mill the part in Figure 7 – 1. It is written for a Bayer Industries ACROLOC vertical mill with a Fanuc 6M control. There are a few commands that should be noted.

- O0701 is the program "O" number that identifies this program to the MCU.
- N.. sequence numbers start at N1 and increment by one. There is no hardfast rule for sequence numbers. The control does not need them. Each programmer chooses their own numbering convention.
- An absolute zero shift is used in lines N2 and N3. G28X0.Y0.Z0. returns the spindle to home zero. G92X10.625Y7.5Z6. transfers the coordinate system to the part.
- The machine utilizes spindle direct tool change. An M06 is not necessary to change tools. It is used where it is the tool number (T01 for tool 1, T02 for tool 2, etc.).
- G44 is the command used for calling up the offsets.

```
%
O0701
(-----------------------------------------)
(COORDINATE SYSTEM ORIGIN)
(XO - CENTERLINE OF PART)
(YO - CENTERLINE OF PART)
(ZO - .100 ABOVE TOP OF PART)
(-----------------------------------------)
(TOOL LIST)
(TOOL 1 - 3/8 STUB DRILL)
(TOOL 2 - 1.0 END MILL 4-FLT. CENTER CUTTING)
(TOOL 3 - 1/2 END MILL 4-FLT.)
()
(-----------------------------------------)
(ABSOLUTE ZERO SHIFT TO PART SYSTEM)
N1G80G90G70
N2G28X0Y0Z0
N3G92X10.625Y7.5Z6.
()
(-----------------------------------------)
(3/8 STUBB DRILL - DRILL HOLES)
N4G00X-2.5Y1.5T01
N5S1066M03
N6G44Z0.H01M08
N7G81G99X-2.5Y1.5Z-.62R0.F12.8
N8X2.5
N9Y-1.5
N10X0.Y0.
N11X-2.5Y-1.5
N12G80G00Z0.
N13G49
()
(-----------------------------------------)
(1.0 DIA.4-FLT. END MILL)
(ROUGH MILL INSIDE SLOT)
N14G00G90X0.Y0.T02
N15S425M03
N16G44Z0.H02M08
(FEED TO DEPTH)
N17G01Z-.62F6.8
(ROUGH MILL INSIDE)
N18Y.48
N19X1.48
N20Y-.48
N21X-1.48
N22Y.48
N23X0.
N24G80G00Z0.M09
N25G49
()
(-----------------------------------------)
(1/2 DIA. 4-FLT. END MILL)
(FINISH INSIDE SLOT)
```

```
(ROUGH/FINISH OUTSIDE)
N26G00G90X0.Y0.T03
N27S800M03
N28G44Z0.H03M08
(FEED TO DEPTH)
N29G01Z-.62F12.8
(FINISH MILL INSIDE SLOT)
N30Y.75
N31X1.75
N32Y-.75
N33X-1.75
N34Y.75
N35X0.
(PULL AWAY FROM PART AND RETRACT SPINDLE)
N36Y.74
N37G00Z3.
(POSITION TO START OF OUTSIDE MILL CUT)
N38X-3.26Y2.26
N39Z0.
(FEED TO DEPTH AND ROUGH MILL 1ST SIDE)
N40G01Z-.62F12.8
N41Y-2.26
(RETRACT SPINDLE AND JUMP OVER CLAMP)
(POSITION FOR ROUGH CUT ON 2ND SIDE)
N42G00Z3.
N43X3.26
N44Z0.
(FEED TO DEPTH AND ROUGH MILL 2ND SIDE)
N45G01Z-.62F12.8
N46Y2.26
(RETRACT SPINDLE AND JUMP OVER CLAMP)
(POSITION FOR FINISH CUT ON 1ST SIDE)
N47G00Z3.
N48X-3.26Y2.26
N49Z0.
(FEED TO DEPTH - MOVE TO PART SURFACE)
(AND FINISH MILL 1ST SIDE)
N50G01Z-.62F12.8
N51X-3.25Y2.25
N52Y-2.25
(RETRACT SPINDLE AND JUMP OVER CLAMP)
(POSITION FOR FINISH CUT ON 2ND SIDE)
N53G00Z3.
N54X3.25
N55Z0.
(FEED TO DEPTH AND FINISH MILL 2ND SIDE)
N56G01Z-.62F12.8
N57Y2.25
(HOME SPINDLE - CANCEL TOOL OFFSETS AND END PROGRAM)
N58G80G00Z0
N59G49
N60X0Y6.T15M09
N61M30
%
```

FIGURE 7-11

- G80G00Z0. is the command sequence necessary to retract the spindle to the Z home position prior to issuing a G49 offset cancel.
- In block 60 the spindle is sent to X0 Y6 to allow easy access to the part. The tool carousel is also indexed to the last station. This is done so the tools only index one station at the start of the cycle, rather than all the way around.

SUMMARY

The important concepts presented in this chapter are:

- Three-axis hole operations in Machinist Shop Language are accomplished through the use of G codes.
- Tool lengths in three-axis machines must be preset by the operator or set in the program; in Machinist Shop Language, this is accomplished by using the TOOL command. The format for assigning offsets is TOOL 10##, where the first two digits (10) tell the MCU that tool information is to be assigned, and ## is the number of the tool being programmed.
- Tool lengths in word address format are programmed by using G10 H##, where ## is the tool register number.
- A buffer zone is the distance between the top of the part and the feed engagement point of the tool. The amount of buffer is determined by the programmer and is built into the tool lengths at setup.
- Feedrates and buffer zones for use with canned cycles are set by using V codes in Machinist Shop Language.
- On word address CNC machinery, the initial level is the Z-axis spindle position when an 80 series G code commences. A reference (or rapid) level is the Z-axis feedrate engagement point selected by the programmer. G98 selects a return to initial level, and G99 selects a return to reference level when using 80 series G codes.

REVIEW QUESTIONS

1. How are tool lengths called up in Machinist Shop Language? How are they canceled? How are they defined?
2. What is a buffer zone?
3. How are buffer zones set for canned cycles in Machinist Shop Language? For milling?

4. How are canned cycle feedrates set in word address format? In Machinist Shop Language?
5. How are tool lengths entered into a word address CNC? How are they called up? How are they canceled?
6. What is a modal command? What is a nonmodal command?
7. How are straight line feedrate moves initiated in word address?
8. What is meant by the terms initial level and reference level?
9. What command is used for initial level? For reference level?
10. How is absolute positioning specified on word address CNC machinery? How is incremental positioning specified?
11. Write a program to mill and drill the part in Figure 7 – 12:
    a.   Using Machinist Shop Language.
    b.   Using a word address CNC mill.

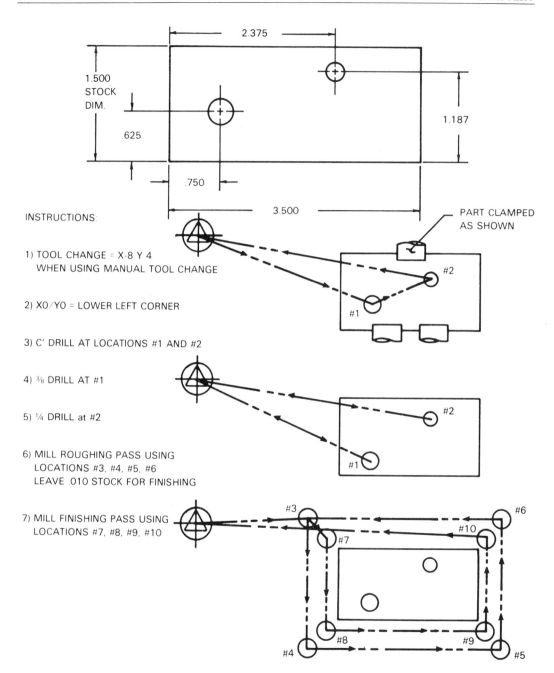

FIGURE 7–12
Part drawing for review question #11

Math for Numerical Control Programming

CHAPTER 8

OBJECTIVE Upon completion of this chapter, you will be able to:

- Use right-angle trigonometry to determine programming coordinates from part drawings.

In the following chapters, the machining of arcs and angles will be discussed. For students already possessing a good working knowledge of trigonometry, this chapter will serve as a review. It is included here for students who have either not taken a course in shop math or who feel a review is in order.

What the machinist is able to do by blending arcs and angles through skill and feel for the craft, the NC part programmer must put into numeric coordinates. It is necessary for the programmer to become proficient at trigonometry to accomplish this task. Trigonometry has applications not only in NC programming but also in other types of machining situations. It is easily mastered with a little practical experience.

BASIC TRIGONOMETRY

Trigonometry is the mathematical science dealing with the solution of triangles. For example, knowing one side plus one other angle or side of a right triangle, all other information concerning the triangle can be derived using trigonometry. For machine shop use, the types of triangles usually dealt with are right triangles (see Figure 8–1). Note that one of the angles in the triangle is 90 degrees. A 90-degree angle is called a right angle: hence the name right triangle. The following formulas are also given in Figure 8–1:

$$\text{SINE} = \frac{\text{OPPOSITE SIDE}}{\text{HYPOTENUSE}}$$

$$\text{COSINE} = \frac{\text{SIDE ADJACENT}}{\text{HYPOTENUSE}}$$

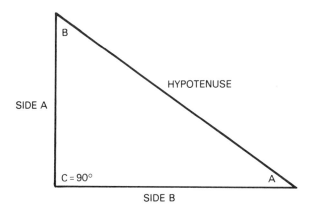

FIGURE 8-1
Right triangle

$$\text{TANGENT} = \frac{\text{SIDE OPPOSITE}}{\text{SIDE ADJACENT}}$$

The other formulas are the inverses of these three. In the machine shop these three will cover most situations.

Figure 8–2 will help to demonstrate the value of triangles in shop mathematics. If the part in Figure 8–2 is to be drilled without using a rotary table, it will be necessary to specify coordinates as dimensioned in Figure 8–2. These are known as *jig borer coordinates,* because they are a common way of locating hole patterns on jig borers. They are also commonly used with milling machines. They are especially important in CNC programming.

The immediate problem in looking at Figure 8–2 is that only the dimensions for holes #1 and #4 are known (they are located on the radius of the bolt circle). However, dimensions a and b can be determined by using trigonometry.

Note that a triangle has been constructed in the first quadrant of the part. If this triangle is solved for the length of its sides, it will supply the information

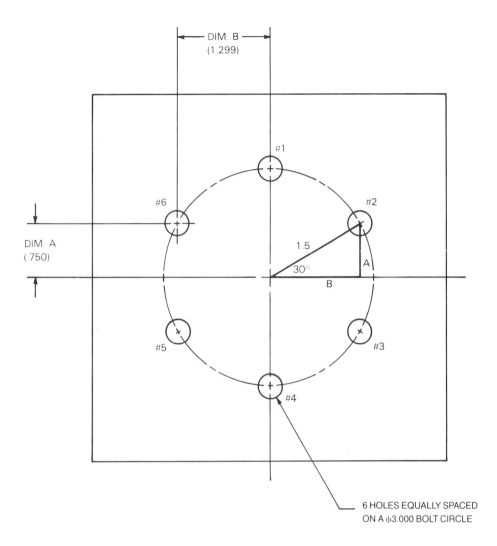

FIGURE 8–2

needed for the missing coordinates. The sides of this triangle are labeled a, b, and 1.5 (which is the hypotenuse of the triangle).

What is known about this triangle? The length of the hypotenuse is half the diameter of the bolt circle, or 1.500 inches. Angle A is also known; since the bolt circle (of 360 degrees) is divided into six equal spaces, the angle between each hole is 60 degrees. Half the distance to each hole lies on each side of the centerline. Therefore, the angle from the centerline to either hole #2 or hole #3 is 30 degrees, which is angle A. The formula for the sine is found to be most practical.

$$\text{SINE A} = \frac{\text{OPPOSITE SIDE}}{\text{HYPOTENUSE}}$$

$$\text{SINE 30} = \frac{a}{1.500}$$

Looking up the sine of 30 degrees in a trigonometric table or using a calculator:

$$.500 = \frac{a}{1.500}$$

$$.500 \times 1.500 = a$$

$$.750 = a$$

To solve for side b, the formula for the cosine is used:

$$\text{COSINE A} = \frac{\text{ADJACENT SIDE}}{\text{HYPOTENUSE}}$$

$$\text{COS A} = \frac{b}{1.500}$$

Using a calculator or table:

$$.866 = \frac{b}{1.500}$$

$$.866 \times 1.500 = b$$

$$1.299 = b$$

The part coordinates are now complete.

Another application of trigonometry is presented in Figure 8–3, where the value of the X dimension is needed. By constructing a triangle as shown in the figure, the length of side b can be determined. If this length is subtracted from the overall length of 3.000 inches, the X dimension is obtained. Two things are known about this triangle: the length of side a and angle A. By using the formula for the tangent, the triangle can be solved as follows:

$$\text{TAN 40} = \frac{\text{OPPOSITE SIDE}}{\text{ADJACENT SIDE}}$$

$$\text{TAN 40} = \frac{1.000}{b}$$

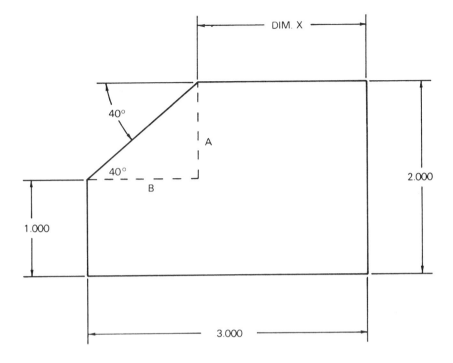

FIGURE 8-3

$$.839 = \frac{1.000}{b}$$

$$.839 \times b = 1.000$$

$$b = \frac{1.000}{.839}$$

$$b = 1.191$$

Dimension X equals 3.000 − 1.191, or 1.809.

USING TRIGONOMETRY FOR CUTTER OFFSETS

A common use for trigonometry in NC programming is calculating cutter offsets for use with linear or circular interpolation (discussed further in Chapter 9). Assume the angle in Figure 8–3 is to be milled. The coordinates of the cutter will have to be determined mathematically because, as Figure 8–4 illus-

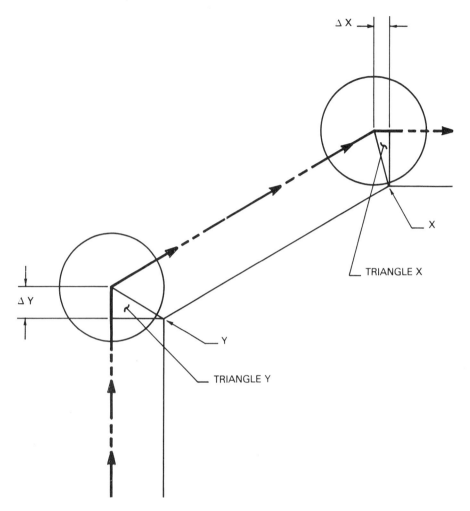

FIGURE 8–4

trates, the cutter cannot be positioned at point Y but must be positioned some unknown distance away. Similarly, the cutter cannot be moved to point X, but some unknown distance short of point X. By solving triangles Y and X, the proper coordinates can be determined.

The angles shown in Figure 8–5 can be found with little effort by looking at the angles formed around points Y and X. When determining angles by this method, three rules must be remembered:

1. The total number of degrees in a circle is 360.
2. The sum of the angles in a triangle is 180 degrees.
3. The complement of an angle is 90 minus the angle.

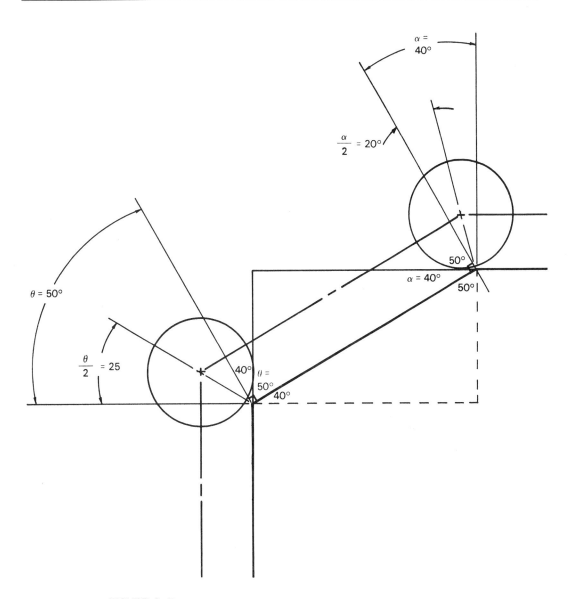

FIGURE 8–5

The angles that will be used for the calculation are 25 degrees for triangle Y and 20 degrees for triangle X, as shown in the figure.

Solving triangle Y for $\triangle$Y:

$$\frac{\triangle Y}{.250} = TAN\ 25$$

$$\triangle Y = TAN\ 25\ (.250)$$

$$\triangle Y = .46631\ (.250)$$

$$\triangle Y = .11658\ or\ .117$$

Solving triangle X for $\triangle$X:

$$\frac{\triangle X}{.250} = TAN\ 20$$

$$\triangle X = TAN\ 20\ (.250)$$

$$\triangle X = .36397(.250)$$

$$\triangle X = .09099\ or\ .091$$

The amount of offset can be added to or subtracted from points Y and X to arrive at the correct cutter coordinates. In the next chapter this sort of calculation will be performed for use with linear interpolation.

Refer to Figure 8 – 6 for right triangle solutions and to Figure 8 – 7 for oblique-angled triangle solutions.

A MILLING EXAMPLE

Figure 8 – 8 shows a goblet-shaped casting. A ledge .250 inch deep by 1.000 radius is to be milled in three places, blending to the .120 thick cast web with a .125 radius. A ¼ inch (.250 diameter) end mill can be used to mill the ledge. Six cutter locations must be calculated. Since three of these locations are mirrors of the other three, only the three locations P1, P2, and P3 need be calculated.

There are three triangles that must be solved to determine the proper co-ordinates. There are two pieces of information known that will allow the solution of P1, and determine angle θ: The radius from the center of the part to the center of the .250 diameter cutter (1.0 R − .125 R) and the .185 (.06 + .125) leg of triangle A. Angle θ can also be determined from this same information as shown in the figure. To determine angle θ, the SIN function is used:

$$SIN\ \theta = \frac{.185}{.875}\ or\ .21143$$

$$\theta = 12.206\ degrees$$

Once angle θ is known, angle α and angle β can be determined:

$$\alpha = 30° - \theta \quad and \quad \beta = 60° - \theta$$

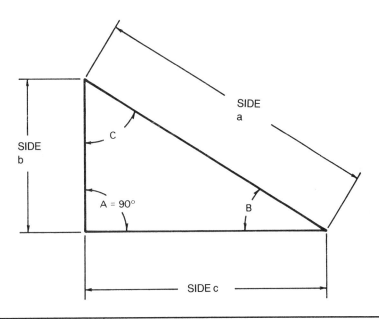

KNOWN VARIABLES	SOLUTION FORMULAS		
SIDE a, ANGLE B	$b = a \times SIN\ B$	$c = a \times COS\ B$	$C = 90° - B$
SIDE a, ANGLE C	$b = a \times COS\ C$	$c = a \times SIN\ C$	$B = 90° - C$
SIDE b, ANGLE B	$a = \dfrac{b}{SIN\ B}$	$c = b \times COT\ B$	$C = 90° - B$
SIDE b, ANGLE C	$a = \dfrac{b}{COS\ C}$	$c = b \times TAN\ C$	$B = 90° - C$
SIDE c, ANGLE B	$a = \dfrac{c}{COS\ B}$	$b = c \times TAN\ B$	$C = 90° - B$
SIDE c, ANGLE C	$a = \dfrac{c}{SIN\ C}$	$b = c \times COT\ C$	$B = 90° - C$
SIDES a AND b	$c = \sqrt{a^2 - b^2}$	$SIN\ B = \dfrac{b}{a}$	$C = 90° - B$
SIDES a AND c	$b = \sqrt{a^2 - c^2}$	$SIN\ C = \dfrac{c}{a}$	$B = 90° - C$
SIDES b AND c	$a = \sqrt{b^2 + c^2}$	$TAN\ B = \dfrac{b}{c}$	$C = 90° - B$

FIGURE 8–6
Solutions of right triangles

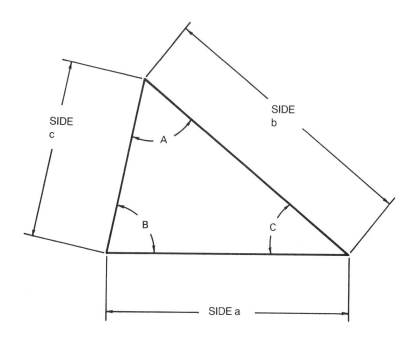

ONE SIDE AND TWO ANGLES KNOWN:
GIVEN: SIDE a, OPPOSITE ANGLE A, AND OTHER ANGLE B

$$C = 180° - (A + B) \qquad b = \frac{a \times \text{SIN B}}{\text{SIN A}} \qquad c = \frac{a \times \text{SIN C}}{\text{SIN A}}$$

TWO SIDES AND THE ANGLE BETWEEN THEM KNOWN:
GIVEN: SIDES a, b, AND ANGLE C

$$\text{TAN A} = \frac{a \times \text{SIN C}}{b - (a \times \text{COS C})} \qquad B = 180° - (A + C) \qquad c = \frac{a \times \text{SIN C}}{\text{SIN A}}$$

$$c = \sqrt{a^2 + b^2 - (2ab \times \text{COS C})}$$

TWO SIDES AND ANGLE OPPOSITE ONE SIDE KNOWN:
GIVEN: SIDE a, OPPOSITE ANGLE A, AND SIDE B

$$\text{SIN B} = \frac{b \times \text{SIN A}}{a} \qquad C = 180° - (A + B) \qquad c = \frac{a \times \text{SIN C}}{\text{SIN A}}$$

ALL THREE SIDES KNOWN:

$$\text{COS A} = \frac{b^2 + c^2 - a^2}{2\,b\,c} \qquad \text{SIN B} = \frac{b \times \text{SIN A}}{a} \qquad C = 180° - (A + B)$$

FIGURE 8–7
Solutions of oblique-angled triangles

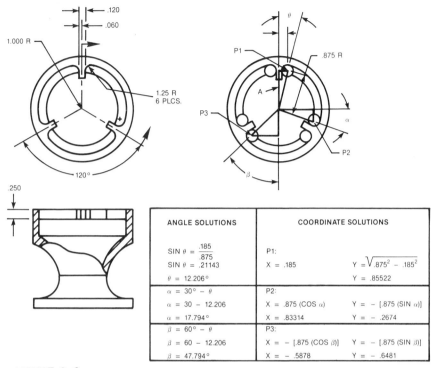

ANGLE SOLUTIONS	COORDINATE SOLUTIONS	
$\text{SIN } \theta = \dfrac{.185}{.875}$ $\text{SIN } \theta = .21143$	P1: X = .185	$Y = \sqrt{.875^2 - .185^2}$
$\theta = 12.206°$		Y = .85522
$\alpha = 30° - \theta$ $\alpha = 30 - 12.206$	P2: X = .875 (COS α)	Y = − [.875 (SIN α)]
$\alpha = 17.794°$	X = .83314	Y = − .2674
$\beta = 60° - \theta$ $\beta = 60 - 12.206$	P3: X = − [.875 (COS β)]	Y = − [.875 (SIN β)]
$\beta = 47.794°$	X = − .5878	Y = − .6481

FIGURE 8–8

The Pythagorean formula is used to determine the Y coordinate of P1:

$$X = .185 \text{ (known information)}$$
$$Y = \sqrt{.875^2 - .185^2} \text{ or } .85522$$

The coordinates of P2 are calculated by the SIN and COS of angle α.

P2:

$$X = .875(\text{COS } \alpha) \text{ OR } .83314$$
$$Y = - [.875(\text{SIN } \alpha)] \text{ or } - .2674$$

The coordinates of P3 are calculated by the SIN and COS of angle β.

P3:

$$X = - [.875(\text{COS } \beta)] \text{ or } - .5878$$
$$Y = - [.875(\text{SIN } \beta)] \text{ or } - .6481$$

A LATHE EXAMPLE

A typical lathe programming situation is depicted in Figures 8 – 9 and 8 – 10. A 37 degree angle is to be turned, intersecting with a 1.500 diameter. At the intersection points of the angle and part face and the intersection of the angle and the 1.500 diameter, a .005-inch radius is to be generated. This small radius serves to deburr the part. A .015 radius turning tool will be used to turn the part. Most programmers will deburr a part in this manner even if not specifically called for on the blueprint. Most companies have some type of engineering standard that applies in situations such as this and will determine the maximum edge break that is allowed.

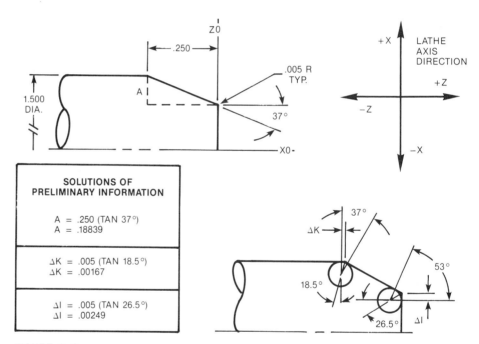

FIGURE 8–9

There are four cutter locations that must be calculated as shown in Figure 8 – 10: P1, P2, P3, and P4. To determine the cutter locations it is necessary to use a twofold approach. First, A, I, and K must be determined. Second, the co-ordinates of the four cutter locations are determined using A, $\triangle$I, and $\triangle$K.

To determine the preliminary information:

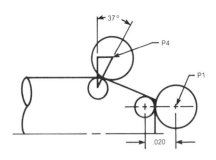

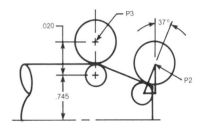

COORDINATE SOLUTIONS	
P1: X = .750 - A · Δl X = .55912	Z = -.005 + .020 Z = .015
P2: X × .55912 + .020 (COS 37°) X = .55912 + .01597 X = .57509	Z = -.005 + .02 (SIN 37°) Z = -.005 + .01204 Z = .00704
P3: X = .745 + .020 X = .765	Z = -(.250 + ΔK) Z = -.25167
P4: X = .745 + .020 (COS 37°) X = .745 + .01597 X = .76097	Z = -[.25167 - .020 (SIN 37°)] Z = (.25167 - .01204) Z = -.23963

FIGURE 8–10

$$A = .250(\text{TAN } 37) \text{ or } .18839$$

$$\triangle K = .005(\text{TAN } 18.5) \text{ or } .00167$$

$$\triangle I = .005(\text{TAN } 26.5) \text{ or } .00249$$

To determine the cutter locations:

P1:

$$X = .750 - A - \triangle I \quad Z = -.005 + .020$$

$$X = .55912 \qquad Z = .015$$

P2:
Note: .55912 in X formula from P1 X calculation.

$$X = .55912 + .020(\text{COS } 37°) \quad Z = -.005 + .02(\text{SIN } 37°)$$

$$X = .57509 \qquad Z = .00704$$

P3:

$$X = .745 + .020 \quad Z = -(.250 + \triangle K)$$

$$X = .765 \qquad Z = -.25167$$

P4:

Note: .25167 in Z formula from P3 Z calculation.

X = .745 + .020(COS 37°)   Z = −[.25167 − .020(SIN 37°)]

X = .76097   Z = −.23963

It is often necessary when determining coordinates to systematically solve a series of triangles in order to derive the necessary values for the final cutter solution, as this case demonstrates.

SUMMARY

The important concepts presented in this chapter are:

* Right-angle trigonometry is the mathematical science of solving right triangles.
* The sine of an angle equals the side opposite the angle divided by the hypotenuse of the triangle.
* The cosine of an angle equals the side adjacent to the angle divided by the hypotenuse of the triangle.
* The tangent of an angle equals the side opposite the angle divided by the side adjacent to the angle.
* The use of trigonometry is necessary for determining cutter offsets for linear and circular interpolation and for determining other part information from a blueprint.

REVIEW QUESTIONS

1. What is the sine of an angle? The cosine? The tangent?
2. What are the sine, cosine, and tangent of the triangle in Figure 8−11? What is angle B?
3. What are the coordinates of the holes in the part in Figure 8−12, assuming that the center is X0/Y0?
4. What are the coordinates of the four cutter locations indicated in Figure 8−13?

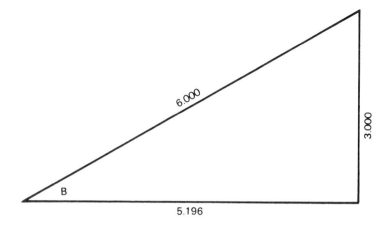

FIGURE 8–11
Triangle for review question #2

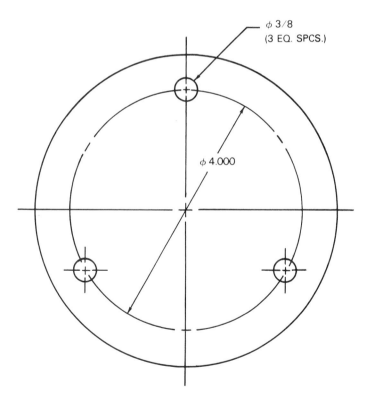

FIGURE 8–12
Part drawing for review question #3

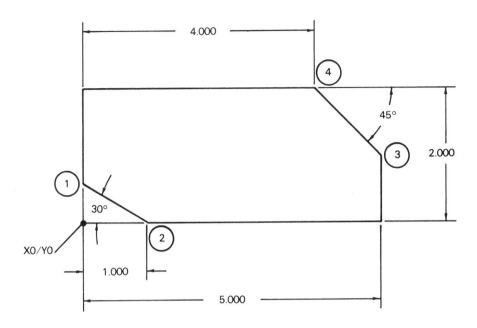

FIGURE 8–13
Part drawing for review question #4

CHAPTER 9

Linear and Circular Interpolation

OBJECTIVES Upon completion of this chapter, you will be able to:

* Write programs using linear interpolation to cut simple angles in both word address format and Machinist Shop Language.
* Write programs using circular interpolation to mill arcs in both word address format and Machinist Shop Language.

Simply put, *linear interpolation* is the ability to cut angles, and *circular interpolation* is the ability to cut arcs or arc segments. Without the ability to cut arcs and angles, a CNC machine is quite limited in its uses. In this chapter both concepts will be introduced.

LINEAR INTERPOLATION

Machines capable of linear interpolation have a continuous-path control system, meaning that the drive motors on the various axes can operate at varying rates of speed. When cutting an angle, the MCU calculates the angle based on the programmed coordinates. Since the MCU knows the current spindle location, it can calculate the difference in the X coordinate between the current position and the programmed location. The change in the Y coordinate divided by the change in the X coordinate yields the slope of the cutter center-line path. The computer then simply moves the drive motors in this ratio. Linear interpolation can be accomplished using the X/Y axes (X/Y plane), Z/X axes (Z/X plane), or Y/Z axes (Y/Z plane).

Calculating Cutter Offsets

Figure 9–1 shows a part on which an angle is to be milled. The cutter has already been positioned at location #1, Figure 9–2. A .500-inch-diameter end

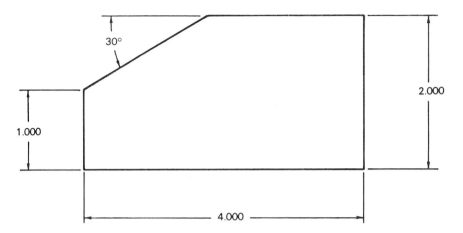

FIGURE 9–1
Part drawing

mill is being used. Before the angle can be cut, it is necessary first to position the spindle at location #2, Figure 9–2. Notice that the Y axis coordinate for location #2, as dimensioned on the part, is not the same point as the edge of the angle. In order to determine this Y axis cutter offset, it will be necessary to determine the amount that must be added to the dimension on the part print to

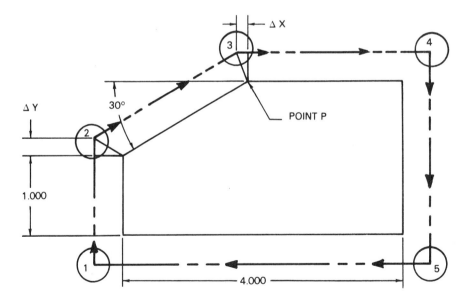

FIGURE 9–2
Cutter path for part in Figure 9–1

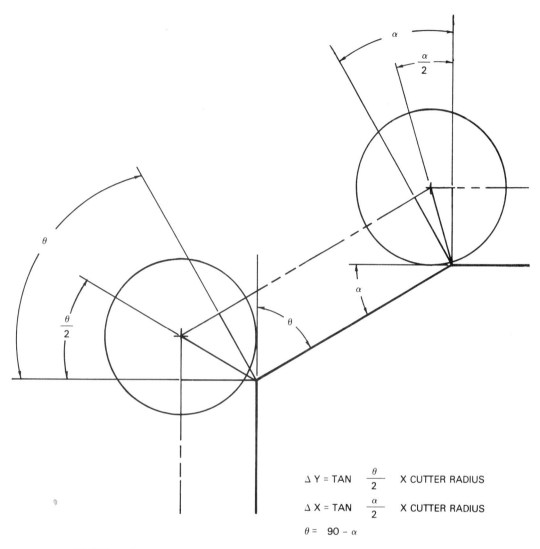

$\Delta Y = TAN \dfrac{\theta}{2} \; X \; CUTTER \; RADIUS$

$\Delta X = TAN \dfrac{\alpha}{2} \; X \; CUTTER \; RADIUS$

$\theta = 90 - \alpha$

FIGURE 9–3
Determining cutter offset

place the spindle at location #2. Similarly, it will be necessary to calculate an amount to be subtracted from the point on the part designated as "P" to arrive at the X axis coordinate for location #3. This is the same cutter offset situation presented in the last chapter. Figure 9–3 represents an enlarged view of locations #2 and #3, illustrating the triangles involved in determining the offsets. The formulas from Appendix 6, Figure 1, can be used to determine the offsets as follows:

$$\Delta Y = CR\left[TAN\left(\frac{\theta}{2}\right)\right] \quad (CR = \text{cutter radius})$$

$$\Delta Y = .25(TAN\ 30)$$

$$\Delta Y = .25(.5774)$$

$$\Delta Y = .144$$

The ΔY offset to be added to the part dimension to arrive at the Y coordinate for location #2 is .144.

The offset for location #3 can be determined as follows:

$$\Delta X = CR\left[TAN\left(\frac{\alpha}{2}\right)\right]$$

$$\Delta X = .25(TAN\ 15)$$

$$\Delta X = .25(.26794)$$

$$\Delta X = .067$$

Before using this information to determine the X axis coordinate, it will also be necessary to calculate the coordinate location of point "P" along the X axis. Again, using the trigonometry formulas, the coordinate can be calculated:

$$TAN\ 30 = \frac{1.000}{b}$$

$$.5774 = \frac{1.000}{b}$$

$$.5774 \times b = 1.000$$

$$b = \frac{1.000}{.5774}$$

$$b = 1.732$$

Subtracting .067 (the ΔX offset) from 1.732 produces the X-axis coordinate for the cutter, 1.665. The ΔY offset, which was found earlier to be .144, can now be added to the 1.000 Y-axis dimension on the part to arrive at a Y-axis coordinate of 1.144. This information can now be used to write a program to mill the angle specified on the part drawing.

Machinist Shop Language

In writing the Machinist Shop Language routine, absolute positioning will be used. It will be assumed that the spindle has been positioned at location #1. The events necessary to program the movement from location #1 to location #4 are:

Y 1.144   F A
X 1.665 Y 2.25 F A
X 4.25   F A

The spindle is first sent at feedrate to location #2, using the coordinate just calculated. The spindle is then sent to location #3. Two coordinates are used to specify the desired location, because a change in both the X and Y axes is required. Notice that the X coordinate is the coordinate calculated earlier, while the Y coordinate is the one required to position the cutter at the top of the part. After completing the cut, the cutter is sent to location #4. This move is a normal straight line milling cut.

Word Address Format

Milling an angle with word address is almost the same as with Machinist Shop Language. The necessary coordinates are simply programmed along with a G01, which causes a feedrate move. In both Machinist Shop Language and word address, any line is considered to be an angle. A move along the X axis would cut an angle of 0 degrees. A move along the Y axis would cut an angle of 90 degrees. The code G01 is technically defined as linear interpolation, with linear interpolation defined simply as a feedrate move between two programmed points. To illustrate linear interpolation with word address, the program is again picked up at location #1.

N . . . G01 Y1.144
N . . . X1.665 Y2.25
N . . . X4.25

A G01 is given to institute a straight milling cut to location #2. Next the X/Y coordinates calculated earlier are programmed, just as in Machinist Shop Language. The coordinates are then given to send the spindle to location #4.

Additional Example

Linear interpolation is not difficult. Aside from calculating the cutter offsets necessary to position the spindle, it is the same as straight line milling. The only real difference is that an X and a Y coordinate are specified for the ending point of the angle since there is a change in position in both axes.

Figure 9–4 shows another cutter offset situation. This part has two angles which intersect each other. In this case, the calculation of the cutter offsets becomes somewhat more complicated. In order to program locations #1 and #3, the formula in Appendix 6, Figure 1, can be used. For location #2, the formula from Appendix 6, Figure 2, can be used. A .500-inch cutter will be assumed.

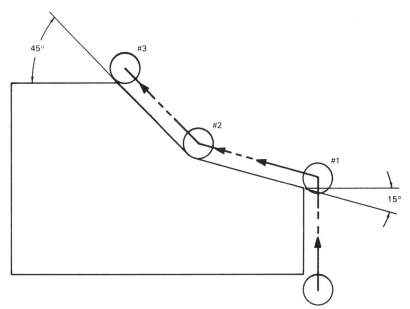

FIGURE 9–4
Part drawing with cutter path shown

Offset for location #1:

$$\Delta Y = CR\left[TAN\left(\frac{75}{2}\right)\right] \quad \Delta X = CR$$

$$\Delta Y = .25(TAN\ 37.5)$$

$$\Delta Y = .25(.7673)$$

$$\Delta Y = .1918$$

For location #2:

$$\Delta X = CR \times \frac{\left[SIN\left(\frac{(45+15)}{2}\right)\right]}{\left[COS\left(\frac{(45-15)}{2}\right)\right]}$$

$$\Delta X = .25 \times \left(\frac{SIN\ 30}{COS\ 15}\right)$$

$$\Delta X = .25 \times \left(\frac{.5}{.9659}\right)$$

$$\Delta X = .25 \times .5176$$

$$\Delta X = .1294$$

$$\Delta Y = CR \times \frac{\left[\cos\left(\frac{(45+15)}{2}\right)\right]}{\left[\cos\left(\frac{(45-15)}{2}\right)\right]}$$

$$\Delta Y = .25 \times \frac{(\cos 30)}{(\cos 15)}$$

$$\Delta Y = .25 \times \left(\frac{.866}{.9659}\right)$$

$$\Delta Y = .25 \times .8966$$

$$\Delta Y = .2241$$

For location #3:

$$\Delta X = CR\left[TAN\left(\frac{45}{2}\right)\right] \quad \Delta Y = CR$$

$$\Delta X = .25(TAN\ 22.5)$$

$$\Delta X = .25(.4142)$$

$$\Delta X = .1036$$

Other cutter situations will present themselves in NC part programming, such as arcs tangent to an angle, or arcs tangent to other arcs. A good working knowledge of the formulas listed in Figures 8 – 6 and 8 – 7 and the figures in Appendix 6 should be developed by the prospective NC part programmer.

CIRCULAR INTERPOLATION

In cutting arcs, the MCU uses its ability to generate angles to approximate an arc. Since the machine axes do not revolve around a centerpoint in a typical three-axis arrangement, the cutting of a true arc is not possible. The CNC machine calculates and then cuts a series of chord segments to generate an arc, as illustrated in Figure 9–5. These chord segments are very small and practically indistinguishable from a true arc.

Figure 9–6 shows a part with a radius to be machined. In order to generate the radius, circular interpolation will be used to send the cutter from location #3 to location #4, Figure 9–7. A .500-inch-diameter end mill will be used.

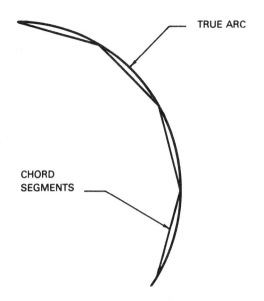

FIGURE 9–5
Circular interpolation

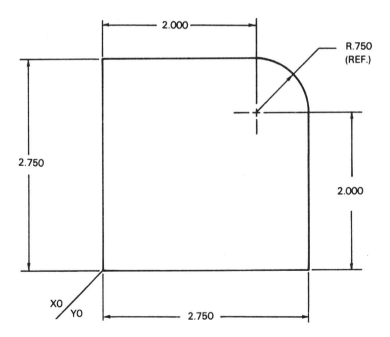

FIGURE 9–6
Part with radius to be machined

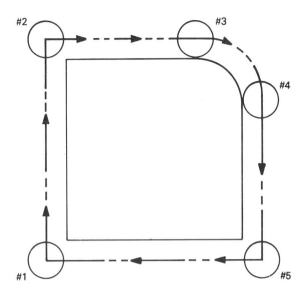

FIGURE 9–7
Cutter path for part shown in Figure 9–6

Machinist Shop Language

As with linear interpolation, it is necessary for the cutter to be positioned at the starting point of the cut (point of *arc tangency*) before the commands to generate the arc are given. To cut the arc, some new commands will be used:

ARC—This command tells the machine to cut an arc. It is also used with a direction command to define the arc direction to the machine's computer.

CW—This stands for clockwise direction. When used with ARC, it tells the machine's computer that a clockwise arc is to be cut.

CCW—This stands for counterclockwise. When used with ARC, it tells the machine's computer that a counterclockwise arc is to be cut.

A four-step procedure is used to cut an arc in Machinist Shop Language:

EVENT 1

- ARC CW/CCW—The ARC command combined with either CW or CCW tells the machine that an arc is to be cut and the direction in which it is to be cut.

EVENT 2

- X/Y coordinates of the center of the arc—Both an X and Y coordinate must be entered even though the cutter may be positioned at one of the coordinates when the cut starts.

EVENT 3

- X/Y coordinates of the endpoint of the arc cut—Both an X and Y coordinate must be entered. If no endpoint is entered, the computer will assume that the starting point is also the endpoint and generate a 360-degree arc.

EVENT 4

- ARC—The ARC command by itself constitutes the actual cutting of the arc.

Assuming the cutter has been positioned at location #1, Figure 9–7, the program routine to cut the arc will be as follows (absolute positioning is used):

```
Y3      FA
X2      FA
ARC/CW
X2    Y2 A
X3    Y2 A
ARC    FA
Y—.25 FA
```

First the cutter is sent from location #1 to #2 in a normal straight-line feed-rate move. The cutter is then sent to the start of the arc. This is also a straight feedrate move. The arc is then defined as being clockwise in direction. The X/Y coordinates of the arc centerpoint (X2, Y2) are then given. The X/Y coordinates of the endpoint of the arc cut (X3, Y2) follow. The ARC command is given last in the sequence. This initiates the cutting of the arc based upon the information the computer received in the previous three events, moving the cutter from location #3 to location #4.

Word Address

Circular interpolation can be accomplished in two ways using word address format, depending on the controller. Most controllers accept information defining an arc by the arc centerpoint and endpoint of the cut. In addition, some controllers allow an arc to be defined by the radius and the endpoint of the cut. To use circular interpolation, some new codes will be needed.

G02—This code tells the MCU to cut an arc in a clockwise direction.
G03—This code tells the MCU to cut an arc in a counterclockwise direction.
    These two G codes also institute the cutting of the arc.

I— This command defines the X-axis centerpoint of an arc. On some controllers I is the absolute X coordinate of the arc center. On others the I value is the incremental distance from the current cutter location to the center of the arc.

J— This command defines the Y-axis centerpoint of an arc. It can be used depending on the controller, as the absolute Y coordinate of an arc, or as the incremental distance from the current cutter location to the center of the arc.

K— This command, used like I and J, defines the Z-axis centerpoint of the arc if performing circular interpolation in either the X/Z or Y/Z planes.

R— This defines the arc radius when the radius is used instead of the centerpoint.

For consistency with Machinist Shop Language, this text will use I, J, and K as the absolute coordinates of an arc center. It should be noted that Fanuc and other similar controls use I, J, K as incremental distances.

As in Machinist Shop Language, the cutter must be positioned at the point of arc tangency before the commands are given to cut the arc. With some controllers, a 90-degree arc is the largest arc segment that can be cut. Cutting 360 degrees must be programmed as four arcs of 90 degrees each. Other controllers allow the cutting of a 360-degree arc in one block of information.

In word address format, a three-step process is followed to cut an arc. All three steps are usually contained in one program line.

For Centerpoint Programming

1. Give the G code for circular interpolation in the direction desired.
2. Give the X/Y coordinates of the endpoint of the arc, using X and Y to define the point.
3. Give the X/Y coordinates of the arc centerpoint, using I and J to define the point.

For Radius Programming

1. Give the G code for circular interpolation in the direction desired.
2. Give the X/Y coordinates of the arc endpoint, using X and Y to define the point.
3. Give the radius of the arc preceded by the R address.

The blocks to cut the arc moving from location #1 to #5 are as follows:

By the Centerpoint Method

```
N ... G01  Y3
N ... X2
N ... G02  X3  Y2  I2  J2
N ... G01  Y—.25
```

The first block is a straight milling cut to feed the cutter from location #1 to location #2. The second block is a straight milling cut to feed the cutter from location #2 to location #3 (the starting point of the arc cut). The third block initiates circular interpolation in a clockwise direction using G02. The X/Y coordinates of the arc endpoint and arc centerpoint are given, using I to define the X-axis centerpoint and J to define the Y-axis centerpoint. This block programs the entire arc, feeding the cutter from location #3 to location #4. The last block feeds the cutter from location #4 to location #5. Note that G01 was specified to put the machine back into the feedrate mode.

By the Radius Method

```
N ... G01  Y3
N ... X2
N ... G02  X3  Y2  R.75
N ... G01  Y—.25
```

The first block is a straight milling cut to feed the cutter from location #1 to location #2. The second block is a straight milling cut to feed the cutter from location #2 to location #3 (the starting point of the arc cut). The third block initiates circular interpolation in a clockwise direction using G02. X and Y define the endpoint of the arc cut. The arc's radius is given using the R address. As in the preceding example, this block programs the entire arc, feeding the cutter from location #3 to location #4. The last block feeds the cutter from location #4 to location #5.

Additional Example

The programs just given are for simple arcs which intersect a line parallel to a machine axis. In many cases, however, an arc will intersect an angle or another arc. Figures 9−8 and 9−9 are examples of such cases. The cutter offsets for these situations can be found by using the formulas from Appendix 6. The cutter radius (CR) in the following examples is .250 inch.

To calculate ΔX and ΔY in Figure 9−8, it is necessary to calculate Δi and Δj:

$$\Delta j = 1.25 - .75$$

$$\Delta j = .5$$

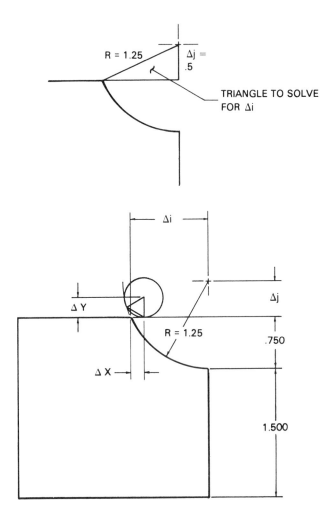

FIGURE 9-8

Using the Pythagorean theorem, i can be determined:

$$\Delta i = \sqrt{(1.25^2 - .5^2)}$$
$$\Delta i = \sqrt{(1.562 - .25)}$$
$$\Delta i = \sqrt{1.312}$$
$$\Delta i = 1.1454$$

This information can then be used to determine X and Y:

$$\Delta Y = CR$$

$$\Delta X = \sqrt{\Delta i - (R - CR)^2 - (\Delta j - CR)^2}$$

$$\Delta X = 1.1454 - \sqrt{(1.25 - .25)} - (.5 - .25)$$

$$\Delta X = 1.1454 - \sqrt{1 - .0625}$$

$$\Delta X = 1.1454 - \sqrt{.9375}$$

$$\Delta X = 1.1454 - .96825$$

$$\Delta X = .17715$$

To calculate ΔX and ΔY in Figure 9–9:

$$\Delta X = CR \times SIN\ 45$$

$$\Delta X = .25 \times .7071$$

$$\Delta X = .1769$$

$$\Delta Y = CR \times COS\ 45$$

$$\Delta Y = .25 \times .7071$$

$$\Delta Y = .1769$$

In this case, since the angle is 45 degrees, the offsets for ΔX and ΔY are the same. Had a different angle been used, the offsets would be different.

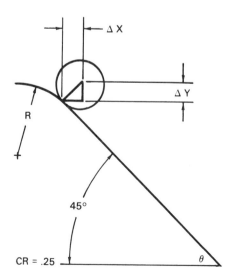

FIGURE 9–9   CR = .25

SUMMARY

The important concepts presented in this chapter are:

- Linear interpolation is the ability to cut angles. It is simply a feedrate move between two points.
- Circular interpolation is the ability to cut arcs or arc segments. Arcs are cut by means of a series of chordal segments generated by the MCU to approximate the arc curvature.
- It is necessary to calculate the cutter offset coordinates when using linear and circular interpolation.
- G01 is the code to institute linear interpolation in word address; in Machinist Shop Language it is treated as a feedrate move.
- G02 and G03 are used to institute circular interpolation in word address; in Machinist Shop Language the ARC command is used.
- The Machinist Shop Language format for circular interpolation is:

ARC CW/CCW
X/Y (centerpoint coordinates)
X/Y (endpoint coordinates)
ARC

- The word address format for circular interpolation is:

For centerpoint programming:

G02/G03 X . . . Y . . . I . . . J . . .

Where X and Y are the endpoint coordinates and I and J are the centerpoint coordinates.

For radius programming:

G02/G03 X . . . Y . . . R . . .

Where X and Y are the endpoint coordinates and R is the arc radius.

REVIEW QUESTIONS

1. What will be the result of cutting the angle in Figure 9–10 if the Y offset is not calculated, but the 4.000 dimension is used instead?
2. What two formulas can be used in calculating coordinates for simple angles where the angle intersects a line parallel to a machine axis?

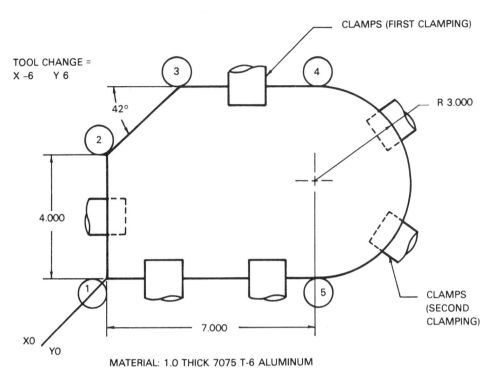

TOOL CHANGE =
X –6   Y 6

MATERIAL: 1.0 THICK 7075 T-6 ALUMINUM

INSTRUCTIONS: MILL FROM 1 TO 2
                MILL FROM 2 TO 3
                JUMP CLAMPS
                MILL FROM 4 TO 5
                MOVE TO TOOL CHANGE
                DWELL TO MOVE CLAMPS
                MILL FROM 3 TO 4
                MOVE TO 5
                MILL FROM 5 TO 1
                MOVE TO TOOL CHANGE

FIGURE 9–10
Part drawing for review questions #1 and #7

3. What code is used in word address format to initiate linear interpolation? In Machinist Shop Lanugage?
4. What is the format for circular interpolation in Machinist Shop Language? In word address?
5. What are I and J used for? What is R used for?
6. When is the ARC command used in Machinist Shop Language?
7. Write a program to mill the part in Figure 9–10:
   a. In Machinist Shop Language.
   b. In word address.

CHAPTER 10

Cutter Diameter Compensation

OBJECTIVES Upon completion of this chapter, you will be able to:

* Define cutter diameter compensation.
* Describe ramp on and ramp off moves and explain their importance.
* List the precautions necessary when using cutter diameter compensation.
* Write programs in word address and Machinist Shop Language that utilize cutter diameter compensation.

DEFINITIONS AND CODES

Programs presented in previous chapters required an allowance for the cutter radius in the programmed coordinates. Some types of CNC machinery have a built-in feature called *cutter diameter compensation (cutter comp)* that allows the part line to be programmed. (Confusion may be caused by use of the terms "offset" and "compensation." In this text, "compensation" refers to cutter diameter offset. The term "offset" refers to tool length offset and the change in axis coordinates when programming arcs and angles.) Cutter comp is also called cutter radius offset (CRO) by some controller manufacturers. In computer-aided programming languages (such as APT) it is also called cutcom. These terms all refer to the same thing: a built-in cycle in the MCU that, when activated, alters the tool path by an amount contained in the cutter comp register. The value in the register is entered in by the setup person when the job is being prepared.

In both word address and Machinist Shop Language formats, cutter comp is accomplished through the use of G codes: G40, G41, G42.

G40—Cutter diameter compensation cancel. Upon receiving a G40, cutter diameter compensation is turned off. The tool will change from a compensated position to an uncompensated position on the next X, Y, or Z axis move.

G41 — Cutter diameter compensation left. Upon receiving a G41, the tool will compensate to the left of the programmed surface. The tool will move to a compensated position on the next X, Y, or Z axis move after the G41 is received.

G42 — Cutter diameter compensation right. Compensates to the right of the programmed surface.

Machinist Shop Language allows for compensation on the X and Y axes. Word address format allows compensation on any axis and uses a G code to determine which axes combination is to be used. If the part is to be machined using the X and Y axes, compensation is desired in the X/Y plane. If using the X and Z axes, compensation in the X/Y plane is needed. If using the Y and Z axes, compensation is needed in the Y/Z plane. The X/Y plane is used most commonly, which may explain why Machinist Shop Language does not offer a choice of axes. The G codes used to select the desired work plane in word address are:

G17 — X/Y plane.
G18 — Z/X plane.
G19 — Y/Z plane.

Two terms are important for understanding cutter diameter compensation: *ramp on* and *ramp off*. Figure 10–1 will help to illustrate their meaning. If a tool is moved from point #1 to point #2 following a G41 command, the cutter will compensate in a plane perpendicular to the part surface and the spindle will move to point #3 rather than point #2. This initial compensation move is called the *ramp on* move. The machine is in the process of adjusting its path for the entire move from point #1 to point #2. By the time it reaches point #2, it has fully compensated its path.

In Figure 10–2 another type of situation is demonstrated. The cutter started at point #1 in this illustration, presenting the possibility that the corner of the part might be cut off in the process of moving to point #2 if the spindle were moving downward or already positioned there. If point #1 was the desired tool change location for this part, the spindle would need to be fully retracted and lowered after reaching the programmed location. When clamps or fixturing devices do not interfere, it is not uncommon to rapid the Z axis to depth on the same move that positions X and Y. Some controllers do not allow cutter comp to be instituted in two axes simultaneously. In these cases it is necessary to program a location away from the part surface, and ramp on the compensation 90 degrees from the desired part surface, as in Figure 10–3. Controllers are often particular about the manner in which cutter comp is ramped on. It is advisable when programming several different controllers or controllers of different ages to use the method in Figure 10–3. This is the most successful method of ramping on cutter comp.

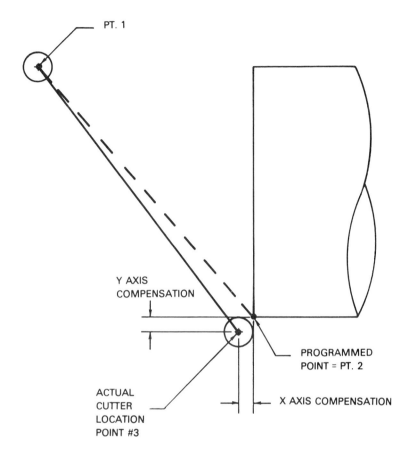

FIGURE 10–1
Ramp on move

The ramp off move is the opposite of the ramp on move and the same precautions are necessary. In Figure 10–2, assume that cutter comp is canceled and a move is made from point #2 to point #1. In this case, the corner of the part may also be cut off. Remember, *the compensation is not turned off completely until the ramp off move is completed.*

Two additional points should be noted. First, with many controllers, cutter comp must be turned on after the length offset is initiated. Similarly, cutter comp must be cancelled prior to cancelling the tool length offset. Failure to do this will result in the controller halting executing of the program, and an alarm signaling at the MCU console.

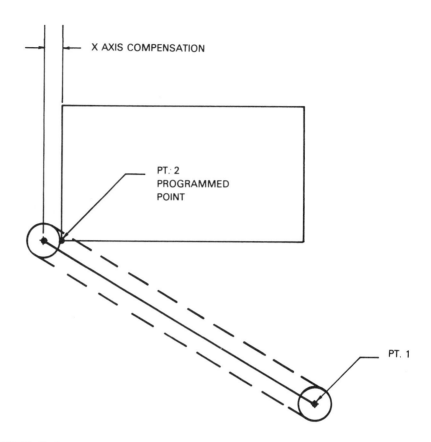

X AXIS COMPENSATION

PT. 2
PROGRAMMED
POINT

PT. 1

FIGURE 10-2

Second, in word address format controllers cutter comp must be commanded in rapid or feedrate modes, not in circular interpolation mode. If G40, G41, or G42 are commanded after G02 or G03, a machine controller alarm will result.

A short program to mill the part in Figure 10-4 is given in Figures 10-5 (Machine Shop Language) and 10-6 (word address format). The programs contain both a roughing and a finishing cut. One way to accomplish this without changing tools or programmed coordinates is to program the diameter of the cutter as two separate diameters. This program uses a .500-inch-diameter end mill. By defining it as both a .520-inch diameter and a .500-inch diameter and using the same coordinates for both passes, the result is that .010 inch of stock is left for the finish pass.

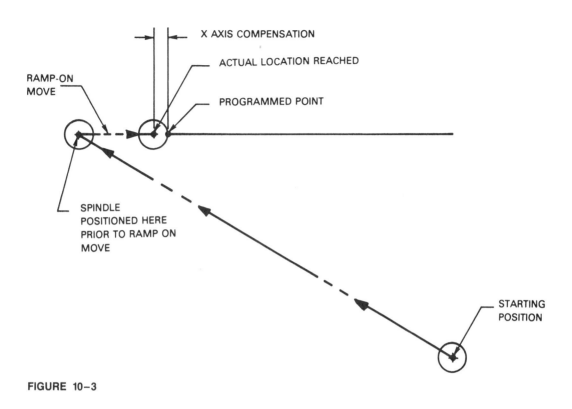

FIGURE 10-3

MACHINIST SHOP LANGUAGE

PROGRAM EXPLANATION

(Refer to Figure 10-5.)

EVENT 1
- TOOL 1 — Tells the MCU that tool information is being assigned.

EVENT 2
- X.52 — The diameter of tool #1. In reality a .500-inch-diameter end mill is being used. Defining the cutter as .520 inch will leave .010 of stock per side on the part.
- Z3 — The tool offset value for tool length offset.
- A — Specifies absolute positioning.

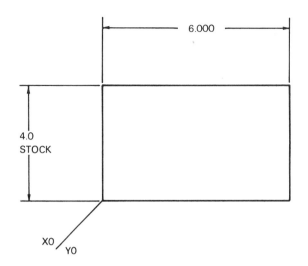

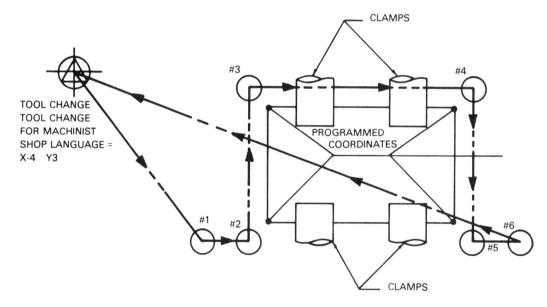

FIGURE 10-4
Part drawing and cutter path for cutter diameter compensation

```
X0/Y0 = LOWER LEFT CORNER
TOOL CHANGE = X-4 Y3
TOOLS:   .500 IN. END MILL
CLEARANCE OVER CLAMPS:  3.000 MIN.
BUFFER ZONE .100

 1 TOOL 1001
 2 X.52 Z3        A          REM:TOOL LENGTH/DIA FOR #1
 3 TOOL 1002
 4 X.5 Z3         A          REM:TOOL LENGTH/DIA FOR #2
 5 X-4 Y3         RA         REM:TCH
 6 TOOL 1
 7 FEED 12.8
 8 X-1 Y-.25      RA         REM:POSITION X/Y TO #1
 9 Z0             RA         REM:RAPID Z TO BUFFER START
10 Z-.62          FA         REM:FEED Z TO DEPTH
11 G41                       REM:CUTTER COMP LEFT
12 X0             FA         REM:RAMP ON MOVE TO #2
13 Y4             FA         REM:FEED FROM #2 TO #3
14 Z3             RA         REM:RAISE Z TO CLEAR CLAMP
15 X6             RA         REM:RAPID FROM #3 TO #4
16 Z0             RA         REM:RAPID Z TO BUFFER START
17 Z-.62          FA         REM:FEED Z TO DEPTH
18 Y0             FA         REM:FEED FROM #4 TO #5
19 G40                       REM:CUTTER COMP CANCEL
20 X7                        REM:RAMP OFF MOVE FROM #5 TO #6
21 TOOL 0                    REM:CANCEL TOOL OFFSET
22 Z0             RA         REM:RETRACT Z
23 X-1 Y-.25      RA         REM:MOVE TO #1
24 TOOL 2
25 Z0             RA         REM:RAPID Z TO BUFFER START
26 Z-.62          FA         REM:FEED Z TO DEPTH
27 G41                       REM:CUTTER COMP LEFT
28 X0             FA         REM:RAMP ON MOVE TO #2
29 Y4             FA         REM:FEED FROM #2 TO #3
30 Z3             RA         REM:RAISE Z TO CLEAR CLAMP
31 X6             RA         REM:RAPID FROM #3 TO #4
32 Z0             RA         REM:RAPID Z TO BUFFER START
33 Z-.62          FA         REM:FEED Z TO DEPTH
34 Y0             FA         REM:FEED FROM #4 TO #5
35 G40                       REM:CUTTER COMP CANCEL
36 X7                        REM:RAMP OFF MOVE FROM #5 TO #6
37 TOOL 0                    REM:CANCEL TOOL OFFSET
38 Z0             RA         REM:RETRACT Z
39 X-4 Y3         RA         REM:MOVE TO TCH
41 END
```

FIGURE 10-5
Cutter diameter compensation program to mill the part in Figure 10-4, Machinist Shop
Language

EVENT 3

- TOOL 1002—Tells the MCU that tool #2 is being defined.

EVENT 4

- X.5—The diameter of tool #2. This is the actual diameter of tool #1.
- Z3—The tool length offset value for tool #3.
- A—Specifies absolute dimensioning.

EVENT 5

- X/Y coordinates—To rapid to the tool change location.
- R—Specifies rapid traverse mode.
- A—Specifies absolute positioning.

EVENT 6

- TOOL 1 — Calls up tool #1's length and diameter. The length will be used for tool length offset; the diameter will be used for cutter diameter compensation.

EVENT 7

- FEED 12.8—Assigns a feedrate of 12.8 inches per minute to be used with all feedrate moves.

EVENT 8

- X − 1 Y − .25 — Positions the cutter at location #1, Figure 10 − 4. This location sets the cutter up for a ramp on move. Although the Anilam controller that uses Machinist Shop Language will permit compensation to occur in two axes simultaneously, it is not recommended by the manufacturer. Many controllers will not allow this; the type of ramp on move being used in this program is considered a safe move for all controllers.
- R—Specifies rapid traverse mode.
- A—Specifies absolute positioning.

EVENT 9

- Z0 — Rapids the cutter to the height of the buffer zone. A .100-inch buffer is being used with this program.
- R—Specifies rapid traverse mode.
- A—Specifies absolute positioning.

EVENT 10

- Z − .62—Feeds the cutter to the proper depth, which is .020 inch below the part.
- F—Specifies feedrate mode.
- A—Specifies absolute positioning.

EVENT 11

- G41 — Initiates cutter diameter compensation left. The machine will reach its compensated position at the end of the next programmed move.

EVENT 12

- X0 — The ramp on move, positions the cutter at location #2, Figure 10 −4.
- F — Specifies feedrate mode.
- A — Specifies absolute positioning.

EVENT 13

- Y4 — The coordinate necessary to position the machine at location #3. Note that the part dimension has been programmed, not the cutter centerline.
- F — Specifies feedrate mode.
- A — Specifies absolute positioning.

EVENT 14

- Z3 — Retracts the spindle to a height sufficient to clear the clamps. In this case, canceling the tool length offset to accomplish the Z-axis retraction is not possible. The same command that called up the tool length offset also defines the cutter diameter. The tool length offset should not be canceled until after a G40 cutter comp cancel has been completed. Not all controllers allow the Z axis to move once cutter comp has been initiated unless preceded by a particular G code. This will be demonstrated in the word address example.
- R — Specifies rapid traverse mode.
- A — Specifies absolute positioning.

EVENT 15

- X6 — The coordinate necessary to position the machine at location #4, again a part coordinate.
- R — Specifies rapid traverse mode.
- A — Specifies absolute positioning.

EVENT 16

- Z0 — Positions the cutter at the start of the buffer zone.
- R — Specifies rapid traverse mode.
- A — Specifies absolute positioning.

EVENT 17

- Z − .62 — Feeds the cutter to milling depth.
- F — Specifies feedrate mode.
- A — Specifies absolute positioning.

EVENT 18

- Y0—Feeds the cutter from location #4 to location #5.
- F—Specifies feedrate mode.
- A—Specifies absolute positioning.

EVENT 19

- G40—Cancels cutter diameter compensation. The cutter will be uncompensated by the end of the next programmed move.

EVENT 20

- X7—The ramp off move, moving the cutter from location #5 to location #6. During the course of this move, the cutter will become uncompensated. It is important to ramp off to a point at least one cutter radius away from the part. This move can be made in either rapid or feedrate mode, as the cutter is clear of the part. A feedrate move was selected here, since the distance is so small.
- F—Specifies feedrate mode.
- A—Specifies absolute positioning.

EVENT 21

- TOOL 0—Cancels the tool length offset. Notice that cutter comp was canceled first.

EVENT 22

- Z0—Retracts the spindle.
- R—Specifies rapid traverse mode.
- A—Specifies absolute positioning.

EVENT 23

- X − 1 Y − .25—The coordinates of location #1.
- R—Specifies rapid traverse mode.
- A—Specifies absolute positioning.

EVENT 24

- TOOL 2—Calls up tool #2's length and diameter. This time a .500-inch diameter will be used. The remaining .010 of stock will be removed from each side.

Events 25 to the end duplicate events 8 through 23. At the end of the program, the tool is sent back to the tool change location. The coordinates for the cutter locations could have been placed in a subroutine and so not repeated. Chapter 11 will deal with subroutines.

WORD ADDRESS FORMAT

PROGRAM EXPLANATION

(Refer to Figure 10–6.)

N010

- N010 — The sequence number.
- G00 — Puts the machine in rapid traverse mode.
- G40 — Cancels any active cutter comp. This code and the other codes in this line are used as a safety device. If any program was run on the machine prior to this one, there might be active codes detrimental to the execution of this program. This block eliminates any chance of accident.
- G49 — Cancels any active tool offset.
- G70 — Selects inch input.
- G80 — Cancels any active canned cycle.
- G90 — Selects absolute positioning.

N020

- N020 — The sequence number.
- G10 — Tells the MCU that tool information is being defined.
- H01 — Selects tool register #1 to hold the tool definition information.
- X.52 — Defines the tool diameter.
- Z3 — Defines the tool length offset.

N030

- N030 — The sequence number.
- G10 — Tells the MCU that tool information is being defined.
- H02 — Selects tool register #2.
- X.5 — Defines the tool diameter.
- Z3 — Defines the tool length offset.

N040

- N040 — The sequence number.
- M06 — Initiates an automatic tool change.
- T1 — Selects tool #1 for use.

N050

- N050 — The sequence number.
- G45 H01 — Calls up the offset in register #1 for use with the tool.

```
XO/YO = LOWER LEFT CORNER
TOOLS:   .500 IN. END MILL
CLEARANCE ABOVE CLAMPS:  3.000 MIN.
BUFFER ZONE:   .100

N010 G00 G40 G49 G70 G80 G90   REM:SAFETY LINE
N020 G10 H01 X.52 Z3
N030 G10 H02 X.5 Z3
N040 M06 T1
N050 G45 H01                   REM:TOOL #1 OFFSET
N060 X-1 Y-.25 S2500 M03       REM:RAPID TO LOCATION #1
N070 Z0                        REM:RAPID TO BUFFER START
N080 G01 Z-.62 F12.8 M08       REM:FEED TO DEPTH
N090 G17 G41 X0                REM:RAMP ON MOVE TO #2
N100 Y4                        REM:FEED FROM #2 TO #3
N110 G00 G18 Z3                REM:RETRACT SPNDL
N120 X6                        REM:RAPID FROM #3 TO #4
N130 Z0                        REM:RAPID TO BUFFER START
N140 G01 Z-.62                 REM:FEED TO DEPTH
N150 G17 Y0                    REM:FEED FROM #4 TO #5
N160 G40 X7                    REM:RAMP OFF TO #6
N170 G00 G49 Z0                REM:CANCEL OFFSET RETRACT SPNDL
N180 X-1 Y-.25                 REM:RAPID TO #1
N190 G45 H02                   REM:TOOL #2 OFFSETS
N200 Z0                        REM:RAPID TO BUFFER START
N210 G01 Z-.62                 REM:FEED TO DEPTH
N220 G17 G41 X0                REM:RAMP ON MOVE TO #2
N230 Y4                        REM:FEED FROM #2 TO #3
N240 G00 G18 Z3                REM:RETRACT SPNDL
N250 X6                        REM:RAPID FROM #3 TO #4
N260 Z0                        REM:RAPID TO BUFFER START
N270 G01 Z-.62                 REM:FEED TO DEPTH
N280 G17 Y0                    REM:FEED FROM #4 TO #5
N290 G40 X7                    REM:RAMP OFF TO #6
N300 G00 G49 Z0 M09            REM:CANCEL OFFSET RETRACT SPNDL
N310 X-12 Y8 M05               REM:RAPID TO PARK POSITION
N320 M30
```

FIGURE 10-6
Cutter diameter compensation program to mill the part in Figure 10-4, word address format

N060

- N060—The sequence number.
- X – 1 Y – .25—The coordinates of location #1.
- S2500—Sets the spindle speed to 2500 RPM.
- M03—Turns the spindle on clockwise.

N070

- N070—The sequence number.
- Z0—Rapids the spindle to the start of the buffer zone. A .100 buffer is being used with this program.

N080

- N080 — The sequence number.
- G01 — Puts the machine in feedrate mode.
- Z − .62 — The Z-axis coordinate to feed the cutter to depth.
- F12.8 — Assigns a feedrate to be used in feedrate moves.
- M08 — Turns the coolant on.

N090

- N090 — The sequence number.
- G17 — Selects the X/Y plane.
- G41 — Initiates cutter diameter compensation left.
- X0 — The coordinate of the part surface. Since a G41 was issued, the cutter will be positioned the distance of the tool radius away from the cutter at location #2. This tool was defined as having a .520-inch diameter. It is in reality .500 inch. This will leave stock for finishing on the part.

N100

- N100 — The sequence number.
- Y4 — Feeds the cutter from location #2 to location #3.

N110

- N110 — The sequence number.
- G00 — Puts the machine in rapid traverse mode.
- G18 — Selects the X/Z plane, since movement is required in Z to retract the spindle to clear the clamps on the next move.
- Z3 — Raises the spindle to clear the clamps. Since a 3-inch clearance was specified to clear the clamps in the setup sheet, this is a valid move. The disadvantage to this technique is that the setup man may not have established the correct clearance. An alternative move would be to cancel the cutter comp and tool offset, raise the spindle to Z0, and reinstitute the tool offset and cutter comp after repositioning the X and Y axes.

N120

- N120 — The sequence number.
- X6 — The coordinate to move from location #3 to location #4. The move is in rapid traverse mode.

N130

- N130 — The sequence number.
- Z0 — Rapids the cutter to the buffer height.

N140

- N140 — The sequence number.
- G01 — Puts the machine in feedrate mode.
- Z − .62 — Feeds the cutter to proper milling depth.

N150

- N150 — The sequence number.
- G17 — Selects the X/Y plane. It is used to reestablish movement along the X/Y axes canceled when the G18 was issued in block N110.
- Y0 — The coordinate for the feedrate move from location #4 to location #5.

N160

- N160 — The sequence number.
- G40 — Cancels the compensation.
- X7 — Moves the cutter from location #5 to location #6. This is a ramp off move.

N170

- N170 — The sequence number.
- G00 — Puts the machine in rapid traverse mode.
- G49 — Cancels the active tool offset. Note that the compensation was canceled first.
- Z0 — Retracts the spindle.

N180

- N180 — The sequence number.
- X − 1 y − .25 — The coordinates to rapid from location #6 to location #1.

N190

- N190 — The sequence number.
- G45 H02 — Calls up the tool information in register #2 for use with the tool.

Blocks N200 on repeat blocks N070 through N170. The coolant is turned off in block N300, and the spindle is turned off in block N310.

SPECIAL CONSIDERATIONS

Figure 10 – 7 illustrates the correct method for turning compensation on or off when machining an inside pocket. Point B must be a minimum of one cutter radius away from the corner of the pocket. If point C were programmed as the ramp on move, the cutter would cut into the corner as in Figure 10 – 8. The direction of the cut depends on whether the X or Y axis is programmed as the first move following the G41.

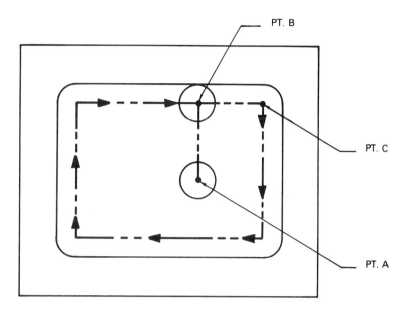

FIGURE 10-7
Turning compensation on or off when machining an inside pocket to prevent cutting into the corner

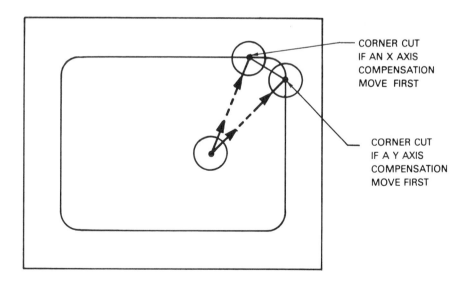

FIGURE 10-8

Figure 10−9 illustrates the precautions necessary when ramping on or off an angle. Point A should not be used for a ramp on or ramp off move since the corner of the angle will be cut off the part, and there may also be damage to the cutter. Point C, or some other point roughly perpendicular to the angle, should be used for the ramp on or ramp off move.

Two different methods of positioning are used for cutter comp with respect to angles, as demonstrated in Figure 10−10. On older CNC machinery, the machine positions the cutter tangent to point A. A G code is then used to initiate the rotation from Y1 to Y2. On newer machinery, the cutter is positioned directly to point P, tangent to both line A and line Y. No special G codes are necessary in this instance. The programming manual for a particular machine will tell the programmer whether a G code is required.

Approach Angles and Vectors

Another factor to consider when using cutter diameter compensation is the approach angle used when ramping on. As Figure 10−11 illustrates, there are three possible angles that can be used during a ramp−on move: 90 degrees to the next cut, less than 90 but greater than 45 degrees to the next cut and less than 45 degrees to the next cut. Some controllers will accept any of these approach angles, others will not. If an unacceptable approach angle is used, the cutter will move to the programmed coordinates, but the cutter compensation will not take place. When programming a number of controllers, or if the NC program will be run on more than one type of controller, it is best to use a 90 degree approach angle to eliminate problems when ramping on cutter comp.

Sometimes, a controller requires a vector to be commanded with the G41 or G42 to orient the cutter correctly prior to the ramp−on move. Technically, a vector is a geometric entity that has both magnitude (length) and direction. In NC programming, vectors are simply mathematical arrows that point the cutter in a given direction. To utilize a vector the I and J addresses are used. Figure 10−12 illustrates some cutter comp vectors. If cutter comp was to be initiated from point A (Figure 10−12) and ramped on to point B, the following program blocks would be used.

For Figure 10−12 (A):

    N010 G17 G42 I-.5 J-.866 D21
    N020 G00 X6.0 Y-.5

For Figure 10−12 (B):

    N010 G17 G42 I-.866 J-.5 D21
    N020 G00 X6.0 Y-.5

For Figure 10−12 (C):

    N010 G17 G42 I-1.0 J0 D21
    N020 G00 X6.0 Y-.5

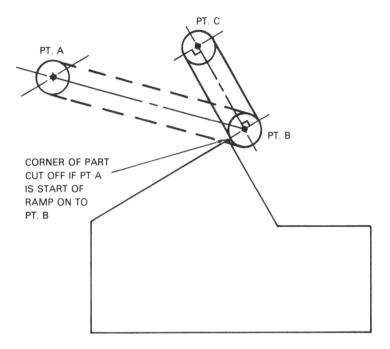

PT. C

PT. A

PT. B

CORNER OF PART
CUT OFF IF PT A
IS START OF
RAMP ON TO
PT. B

FIGURE 10−9
Ramping on or off an angle

Note that in each of these cases I is the X-axis component, and J is the Y-axis component of a vector 1.0 inch long. This is called a unit vector. In Figure 10−12 (A) the approach angle is 30 degrees, therefore, I equals the sine of 30 degrees, while J equals the cosine. In Figure 10−12 (B) the approach angle is 60 degrees. I equals the sine of 60 degrees, while J equals the cosine of 60 degrees. Since the approach angle in Figure 10−12 (C) is 90 degrees, I simply equals 1.0 and J equals zero.

Figure 10−13 shows a part to be milled using cutter diameter compensation. A program to mill the part is given in Figure 10−14. It is assumed that the part is clamped through two already existing holes. With the information given thus far, the student should be able to follow this program without further explanation.

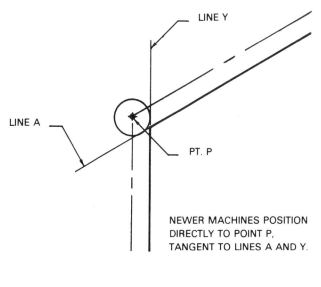

LINE Y

LINE A

PT. P

NEWER MACHINES POSITION
DIRECTLY TO POINT P,
TANGENT TO LINES A AND Y.

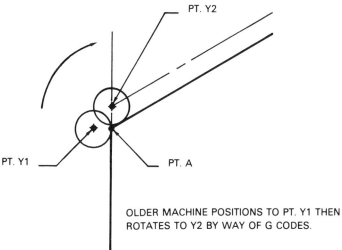

PT. Y2

PT. Y1

PT. A

OLDER MACHINE POSITIONS TO PT. Y1 THEN
ROTATES TO Y2 BY WAY OF G CODES.

FIGURE 10–10
Cutter diameter compensation of angles

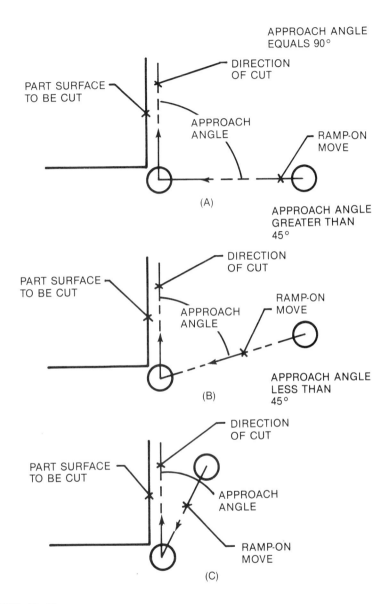

FIGURE 10–11
Cutter compensation approach angles

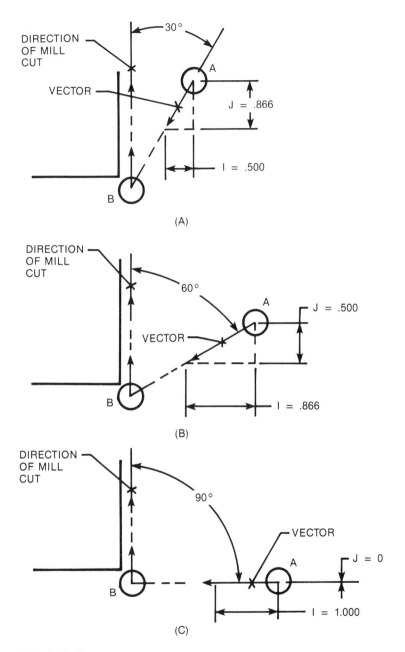

FIGURE 10–12
Cutter compensation vectors

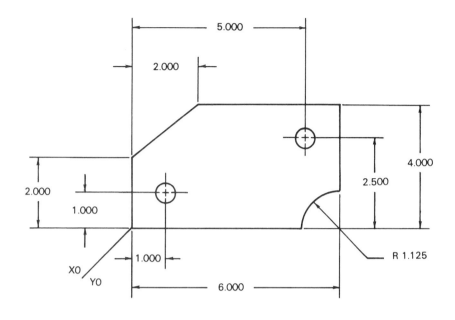

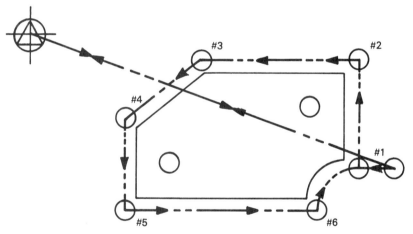

FIGURE 10–13
Part drawing and cutter path

```
XO/YO = LOWER LEFT CORNER
TOOLS:  1.000 IN. 4 FLUTE END MILL
CLEARANCE ABOVE CLAMPS:  3.000 MIN.
BUFFER ZONE:  .100 IN.

N010 G00 G40 G70 G90            REM:SAFETY LINE
N020 G10 H01 X1.02 Z3
N030 G10 H02 X1 Z3
N040 M06 T01                    REM:1 IN. E/M
N050 G45 H01                    REM:TOOL #1 OFFSET
N060 X7 Y.875 S400 M03          REM:POSITION FOR SAFE RAMP-ON
N070 Z0                         REM:RAPID TO Z0
N080 G01 Z-.89 F6.8 M08         REM:FEED Z AXIS TO DEPTH
N090 G17 G42 X6                 REM:RAMP ON TO #1
N100 Y4                         REM:FEED TO #2
N110 X2                         REM:FEED TO #3
N120 X0 Y2                      REM:FEED TO #4
N130 Y0                         REM:FEED TO #5
N140 X4.875                     REM:FEED TO #6
N150 G02 X6 Y1.125 I6 J0        REM:ARC FROM #6 TO #1
N160 G00 G40 X7                 REM:RAMP-OFF
N170 G00 G49 Z0                 REM:CANCEL OFFSET/RETRCT SPNDL
N180 G45 H02                    REM:TOOL OFFSETS #2
N190 Z0                         REM:RAPID TO Z0
N200 G01 Z-.89 F6.8             REM:FEED Z AXIS TO DEPTH
N210 G17 G42 X6                 REM:RAMP-ON TO #1
N220 Y4                         REM:FEED TO #2
N230 X2                         REM:FEED TO #3
N240 X0 Y2                      REM:FEED TO #4
N250 Y0                         REM:FEED TO #5
N260 X4.875                     REM:FEED TO #6
N270 G02 X6 Y1.125 I6 J0        REM:ARC FROM #6 TO #1
N280 G00 G40 X7.500 M09         REM:RAMP-OFF
N290 G00 G49 Z0                 REM:CANCEL OFFSET/RETRCT SPNDL
N300 X-8 Y6 M05                 REM:MOVE TO PARK/SPNDL OFF
N310 M30                        REM:END PROGRAM
```

FIGURE 10-14

FINE TUNING WITH CUTTER DIAMETER COMPENSATION

Up to this point, cutter diameter compensation has been used to program the part line; the program coordinates have matched the part dimensions. Another way cutter comp is employed is to fine tune the cutter path. In this type of programming, the part is programmed using the parallel path method used in Chapters 6, 7, and 9. Cutter comp is used to compensate for the difference between the programmed and actual cutter diameter. For example, if a program is written for a .500 diameter end mill, but a resharpened end mill measuring .490 diameter is used, the .020 diameter difference can be compensated for by using cutter comp.

In the fine tune method, cutter comp is usually used to compensate for a cutter which is smaller than the programmed diameter. When using the part line method exactly the opposite is the case. Cutter comp is used to compensate for a cutter which is larger than the zero diameter cutter programmed (the part line). For this reason it is necessary to use a minus (−) value in the cutter comp register when using the fine tune method.

Figure 10 − 15 is a word address program for the part in Figure 10 − 13, illustrating the fine tune method. Note that allowance is once again being made for the cutter radius. The cutter diameter compensation allows reground, undersize cutters to be used.

SPECIAL FANUC SECTION

Figure 10−16 contains a program to mill the part in Figure 10−8 using half the cutter diameter in the cutter comp register. Figure 10−17 contains a program to mill the part in Figure 10−13 using the fine tune method. The programs are

```
XO/YO = LOWER LEFT CORNER
TOOLS:   1.000 IN. 4 FLUTE END MILL
CLEARANCE ABOVE CLAMPS:   3.000 MIN.
BUFFER ZONE:   .100 IN.

NO10 G00 G40 G70 G90            REM:SAFETY LINE
NO20 G10 H01 X.990 Z3
NO30 M06 T01                    REM:1 IN. E/M
NO40 G45 H01                    REM:TOOL #1 OFFSET
NO50 X7.500 Y.875 S400 M03      REM:POSITION FOR SAFE RAMP-ON
NO60 ZO                         REM:RAPID TO ZO
NO70 G01 Z-.89 F6.8 M08         REM:FEED Z AXIS TO DEPTH
NO80 G17 G42 X6.510             REM:RAMP ON TO #1
NO90 Y4.510                     REM:FEED TO #2
N100 X1.7829                    REM:FEED TO #3
N110 X-.510 Y2.2171             REM:FEED TO #4
N120 Y-.510                     REM:FEED TO #5
N130 X5.3850                    REM:FEED TO #6
N140 G02 X6.500 Y.615 I6 JO     REM:ARC FROM #6 TO #1
N150 G01 Y4.500                 REM:FEED TO #2
N160 X1.7929                    REM:FEED TO #3
N170 X-.500 Y2.2071             REM:FEED TO #4
N180 Y-.500                     REM:FEED TO #5
N190 X5.375                     REM:FEED TO #6
N200 G02 X6.500 Y.626 I6 JO     REM:ARC FROM #6 TO #1
N210 G00 G40 X7.500 M09         REM:RAMP-OFF
N220 G00 G49 ZO                 REM:CANCEL OFFSET/RETRCT SPNDL
N230 X-8 Y6 M05                 REM:MOVE TO PART/SPNDL OFF
N240 M30                        REM:END PROGRAM
```

FIGURE 10−15

```
%
O1006
(---------------------------------------------------------)
(X0/Y0 - LOWER LEFT CORNER OF PART)
(Z0 - .100 ABOVE TOP OF PART)
(MIN. CLEARANCE ABOVE CLAMPS - 3.0 INCHEs)
(---------------------------------------------------------)
()
(MOVE TO TOOL CHANGE POSITION)
(ORIENT SPINDLE FOR TOOL CHANGE)
N001G90G17G40G80
N101G91G30X0Y0Z0M19
N102T01M06
()
(CALL UP WORK COORDINATE - PUT TOOL 2 IN STANDBY)
N103G90G00G54S2500T02B0
(MOVE TO START POSITION - PICK UP OFFSET ON Z MOVE)
N104X7.5Y.755M03
N105G45Z0H01M08
N106G01Z-.62F12.8
(INITIATE AND RAMP ON CRO - USES REGISTER D11)
N107G17G41X0.D11
N108Y4.
N109G00Z3.
N110X6.
N111Z0.
N112G01Z-.62
N113Y0.
(CANCEL AND RAMP OFF CRO)
N114G40X7.
N115G00Z3.
(MOVE TO START OF 2ND CUT )
N116X-1.Y-.25
N117Z0.
N118G01Z-.62F12.8
(INITIATE AND RAMP ON CRO - USES REGISTER D12)
N119G17G41X0D12
N120Y4.
N121G00Z3.
N122X6.
N123Z0.
N124G01Z-.62
N125Y0.
(CANCEL AND RAMP OFF CRO)
N126G40X7.
(CANCEL OFFSET AND RETRACT Z BY RETURNING TO TOOL CHANGE)
N127G91G30Z0M09
N128G30X0Y0Z0M19
N129M30
%
```

FIGURE 10-16

```
%
O1013
(------------------------------------------)
(XO/YO - LOWER LEFT CORNER OF PART)
(ZO - .100 ABOVE TOP OF PART)
(------------------------------------------)
()
N001G90G17G40G80
N101G91G30Z0Y0Z0M19
N102T01M06
N103G90G00G54S400T02B0
N104X7.5Y.875M03
N105G45Z0H01M08
N106G01Z-.89F6.8
(RAMP ON CRO - USES REGISTER D11)
N107G17G42X6.51D11
(BEGIN ROUGH PASS - LEAVE .01 STK. TO FINISH)
N108Y4.51
N109X1.7829
N110X-.51Y2.2171
N111Y-.51
N112X5.385
N113G02X6.5Y.615I.615J0
N114G01Y4.5
(BEGIN FINISH PASSS)
N115X1.7929
N116X-.5Y2.2071
N117Y-.5
N118X5.375
N119G02X6.5Y.625I.625J0
(CANCEL AND RAMP OFF CRO)
N120G00G40X7.5
(CANCEL OFFSETS BY RETURNING TO TOOL CHANGE)
N121G91G30Z0M09
N122G30X0Y0Z0M19
N123M30
%
```

FIGURE 10-17

written for a Hitachi Seiki HC-500 four-axis machining center with a Fanuc 11M control. It will be assumed that B0 positions the rotary table to the proper position to machine the part. The B axis (rotary motion around the Y axis) is commanded to B0 at the start of the program and left there throughout the program cycle. There are several important differences between this machine and the text examples that need to be understood.

Fanuc controls utilize a cutter comp called cutter radius offset (CRO). The radius value of the cutter is entered in a register rather than the cutter diameter. If using the fine tune method, it is the difference between the programmed cutter radius and the actual cutter radius that is entered in the register. A "D" address is used to specify the register used.

Like many horizontal machining centers, the HC-500 uses a preset tool change position. On this machine, the spindle must be commanded to the tool change position. The spindle must also be oriented to allow the tool changer to properly change the tools. The command used to accomplish this is G91G30Z0Y0Z0M19. The M19 orients the spindle. The length offset is cancelled by returning the Z axis to the tool change position. Once this is done, X and Y are returned to tool change as follows:

G91G30Z0M09
G30X0Y0M19

After changing tools, a T02 command is included on the line following the T01M06 command to place the next tool in the standby position, ready to be changed at the next tool change command.

This machine also uses work coordinates. The G54 command follows the tool change command line to call up the work coordinate. The work coordinate values have been manually entered.

SUMMARY

The important concepts presented in this chapter are:

- Cutter diameter compensation is the automatic calculation of the cutter path by the machine control unit, based on the part line and cutter information contained in the program.
- Cutter diameter compensation is instituted and canceled through use of the codes G40, G41, and G42. G41 is cutter compensation left, G42 is cutter compensation right, and G40 is cutter compensation cancel.
- The "ramp on" move is the initial compensation of the cutter. The compensation occurs 90 degrees to the next axis movement following the G41 or G42. Care must be taken with the spindle position prior to the ramp on move to avoid cutting the part in the wrong area.
- The "ramp off" move is the opposite operation. Ramp off will occur 90 degrees to the next axis movement following a G40. The compensation will be completely eliminated by the end of this move.

REVIEW
QUESTIONS

1. What is cutter diameter compensation? How does it differ from tool length offset?
2. What is a ramp on move? When does it occur?
3. What is a ramp off move? When does it occur?
4. Draw a sketch illustrating the proper technique for ramping on, assuming the machine does not have the capability to compensate in two axes simultaneously. Draw a sketch illustrating an improper ramp on.
5. What cautions must be observed when instituting cutter compensation inside a pocket? When milling angles?
6. What do the codes G40, G41, and G42 do?
7. Do all CNC machines directly position the cutter with respect to an angle? If not, how is the rotation accomplished?
8. Write a program to mill the part in Figure 10–18, using a roughing and a finishing pass with a 1.000-inch-diameter end mill:
   a.   In Machinist Shop Language.
   b.   In word address.

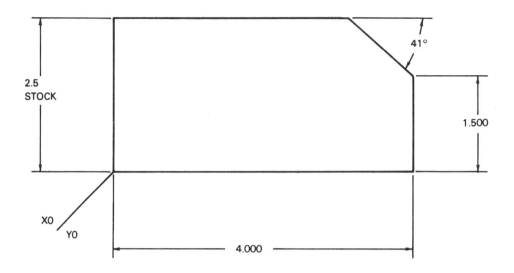

MATERIAL: 3/8 THICK 302 STAINLESS STEEL

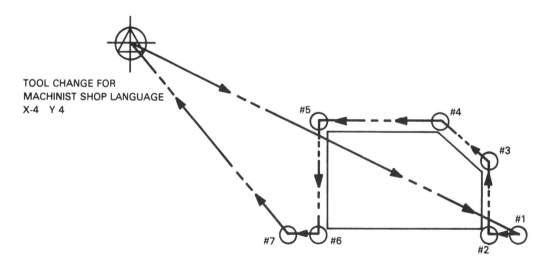

TOOL CHANGE FOR
MACHINIST SHOP LANGUAGE
X-4   Y 4

FIGURE 10–18
Part drawing for review question #8

CHAPTER 11

Do Loops and Subroutines

OBJECTIVES Upon completion of this chapter, you will be able to:

- Describe a do loop.
- Describe a subroutine.
- Describe nested loops.
- Write simple programs in word address and Machinist Shop Language using do loops, subroutines, and nested loops.

DO LOOPS

Figure 11−1 shows a part with a series of holes to be drilled, equally spaced. If an operation is to be repeated over a number of equal steps, it may be programmed in what is referred to as a do loop. In a *do loop,* the MCU is instructed to repeat an operation (in this case, drill a hole five times) rather than programmed for five separate hole locations.

Machinist Shop Language

The format for a do loop in Machinist Shop Language is:

1. DO n
2. X/Y/Z I
3. END

Where DO is the command to repeat the operation that follows, n is the number of times the operation is to be repeated, X/Y/Z is the coordinate information for the loop, I specifies incremental positioning, and END signals the end of the do loop.

Figure 11−2 is a program written in Machinist Shop Language to drill the five holes in Figure 11−1 using a do loop. The tool change location is at X − 2.000, Y2.000.

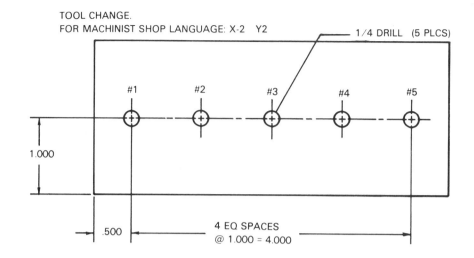

TOOL CHANGE.
FOR MACHINIST SHOP LANGUAGE: X-2   Y2

1/4 DRILL   (5 PLCS)

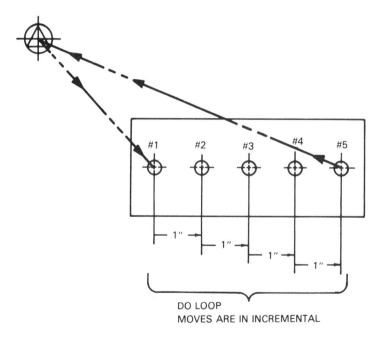

FIGURE 11-1

```
XO/YO = LOWER LEFT CORNER
TOOL CHANGE = X-2 Y2
TOOLS:   #3 C'DRILL, 1/4 DRILL
CLEARANCE OVER CLAMPS = 2.500 IN. MIN.

 1 TOOL 1001
 2 Z4              A
 3 TOOL 1001
 4 Z2.5            A
 5 X-2 Y2 ZO       RA
 6 TOOL 1                      REM:C'DRILL, 3500 RPM
 7 V20 10.5                    REM:FEEDRATE FOR DRILL CYCLE
 8 V21 .1          A           REM:DEFINE BUFFER ZONE
 9 G81                         REM:INITIATE DRILL CYCLE
10 X.5 Y1 Z-.1620 RA           REM:C'DRILL HOLE #1
11 DO 4                        REM:INITIATE LOOP
12 X1              RI          REM:INCREMENTAL COOR. IN LOOP
13 END                         REM:END LOOP
14 G80                         REM:CANCEL DRILL CYCLE
15 TOOL 0                      REM:CANCEL TOOL OFFSET
16 ZO              RA          REM:RETRACT SPNDL
17 X-2 Y2          RA          REM:RAPID TO TCH
18 TOOL 2                      REM:1/4 DRILL, 3500 RPM
19 V20 21                      REM:DRILL CYCLE FEEDRATE
20 V21 .1                      REM:DEFINE BUFFER
21 G81                         REM:INITIATE DRILL CYCLE
22 X.5 Y1 Z-.375  RA           REM:DRILL HOLE #1
23 DO 4                        REM:INITIATE LOOP
24 X1              RI          REM:INCREMENTAL COOR. IN LOOP
25 END                         REM:END LOOP
26 G80                         REM:CANCEL DRILL CYCLE
27 TOOL 0                      REM:CANCEL TOOL OFFSET
28 ZO              RA          REM:RETRACT QUILL
29 X-2 Y2          RA          REM:RAPID TO TCH
30 END                         REM:END OF PRGM
```

FIGURE 11-2
Do loop program for part in Figure 11-1, Machinist Shop Language

---

PROGRAM EXPLANATION

(Refer to Figure 11-2.)

EVENTS 1-4

- Assign tool information to the computer.

EVENTS 5-6

- Send the spindle to tool change and assign the tool length offset for tool #1 (as presented in Chapter 7).

EVENT 7

- V20 10.5—Assigns 10.5 as the feedrate for canned cycle operation to variable register 20. The MCU uses the value programmed in the V20 register for the Z-axis feedrate with a G81.

EVENT 8

- V21 .1—Assigns a buffer zone of .100 inch to be used for Z-axis moves with the G81 cycle.
- A—Specifies absolute positioning.

EVENT 9

- G81—Initiates the canned drilling cycle.

EVENT 10

- X.5 Y1—Coordinates of hole #1.
- Z − .1620—Z-axis depth for center drilling. The depth of the center drill into the part is .0620 inch. Since V21 defined a .100-inch rapid level, .100 inch must be added to .062 to arrive at the required Z axis coordinate.

EVENT 11

- DO 4—Tells the computer to repeat four times the operation defined in the events that follow.

EVENT 12

- X1—Incremental distance from one hole to the next on the X axis. The holes are in line, so no movement along the Y axis is necessary.
- R—Specifies that the moves from hole to hole be made at rapid traverse. Since this is a drilling operation, there is no need to move between holes at feedrate.
- I—Specifies incremental positioning. This is an example of the use of both absolute and incremental positioning in a program. The X1 must be an incremental dimension because each hole to be drilled is referenced to the previous one, not to the part X0/Y0 point.

EVENT 13

- END—Specifies the end of the do loop. END has three uses: end of program, end of do loop, and end of subroutine. The machine will act four times on the information contained between the DO statement and END. Since the G81 will be active from the first hole, it is not necessary to include it in the do loop.

EVENT 14

- G80—Cancels the active G81.

EVENT 15

- TOOL 0—Cancels the active tool offset.

EVENT 16

- Z0—Retracts the spindle.

EVENT 17

- X – 2 Y2—Coordinates of tool change location.

EVENT 18

- TOOL 2—Calls up the offset for the 1/4-inch drill.

EVENTS 19–21

- Initiate the drilling cycle and assign the feedrate and buffer zone.

EVENT 22

- X/Y coordinates—Position the machine to hole #1.
- Z coordinate — For drilling, determined by adding .3 times the drill diameter to the part thickness of .250 and the .100-inch buffer zone. (Note: An additional .050 was added to this result to allow for tooling and part tolerance.)

Events 23 through 29 duplicate events 11 through 17. Event 30 signals the end of the program.

Word Address Format

The same principle is now demonstrated using word address. Naturally there is a G code to institute a do loop. The format for a do loop in word address is:

1. G51 Nn
2. X/Y/Z
3. G50

Where G51 signals the start of a do loop, N is the address and n is the number of times an operation is to be repeated, X/Y/Z is the program information contained in the loop, and G50 signals the end of the do loop.

It should be noted here that the codes for do loops vary from one controller to another. The programming manual will need to be consulted for the proper codes. The coding used here is a type of General Numerics format used on Fanuc controllers also.

PROGRAM EXPLANATION

(Refer to Figure 11–3.)

N010

- Safety line to cancel any active G codes and bring the spindle to home position.

N020–N030

- Assigns tool information to tool length registers.

N040

- M06 T1—Initiates an automatic tool change, selecting tool #1 from the storage magazine.

```
X0/Y0 = LOWER LEFT CORNER
TOOLS:   TOOL #1 = #3 C'DRILL.
         TOOL #2 = 1/4 DRILL.
CLEARANCE OVER CLAMPS = 2.500 IN. MIN.

N010 G00 G40 G49 G80 G70 G90 X0 Y0 Z0      REM:SAFTEY LINE
N020 G10 H01 Z4                            REM:TOOL 1 OFFSETS
N030 G10 H02 Z2.5                          REM:TOOL 2 OFFSETS
N040 M06 T1                                REM:ATC #3 C'DRILL
N050 G45 H01                               REM:OFFSET 1
N060 S3500 F10.5 M03                       REM:SPNDL ON
N070 G81 G99 X.5 Y1 Z-.162 R0 M08          REM:C'DRILL HOLE #1,
COOL.ON
N080 G51 N4                                REM:INITIATE LOOP
N090 G91 X1                                REM:INC. COOR. IN LOOP
N100 G50                                   REM:END OF LOOP
N110 G80 G90 G49 Z0 M09                    REM:RETRACT Z, COOL. OFF
N120 M06 T2                                REM:ATC 1/4 DRILL
N130 G45 H02                               REM:#2 OFFSETS
N140 S3500 F21 M03
N150 G81 G99 X.5 Y1 Z-.375 R0 M08          REM:DRILL HOLE #1
N160 G51 N4                                REM:INITIATE LOOP
N170 G91 X1                                REM:INC. COOR. IN LOOP
N180 G50                                   REM:END LOOP
N190 G80 G90 G49 Z0 M09                    REM:RETRACT Z, COOL. OFF
N200 X-12 Y8 M05                           REM:TO SPNDL PARK/SPNDL
OFF.
N210 M30                                   REM:END PRGM
```

FIGURE 11–3
Do loop program for part in Figure 11–1, word address format

N050

- G45 H01—Calls up the tool offsets in register #1. Note that the codes for tool offsets also differ from machine to machine. This is just one example of tool offset coding. The important point is that tool offsets must be coded.

N060

- S3500—Assigns the spindle speed.
- F10.5—Assigns the feedrate. In this example, the feedrate moves are Z-axis movement during the canned cycle operation.
- M03—Turns the spindle on clockwise.

N070

- G81—Initiates the canned drilling cycle.
- G99—Selects a Z-axis return to the reference (rapid) level, which is the start of the buffer zone.
- X.5 Y1—Coordinates of the first hole.
- Z − .162—Z-axis center-drilling depth.
- R0—Sets reference level for the drilling cycle.
- M08—Turns the coolant on.

N080

- G51—Initiates the do loop.
- N4—Instructs the MCU to repeat the operation contained in the loop four times.

N090

- G91—Selects incremental positioning.
- X1—Incremental distance between the holes to be drilled inside the loop.

N100

- G50—Signals the end of the loop information.

N110

- G80—Cancels the G81.
- G90—Selects absolute positioning.
- G49—Cancels the active tool offset.
- Z0—Retracts the spindle.
- M09—Turns the coolant off.

N120

- M06 T2—Initiates an automatic tool change, selecting the #2 position in the storage magazine from which to take the new tool.

N130

- G45 H02—Calls up the offsets contained in register #2.

N140

- S3500—Sets the spindle speed.
- F21—Sets the feedrate.
- M03—Turns the spindle on clockwise.

N150

- G81—Initiates the canned drilling cycle.
- G99—Selects a return to reference (rapid level)A.
- X.5 Y1—Coordinates of hole #1.
- Z – .375—Z-axis depth to drill through the part.
- R0—Sets the reference level for the drilling cycle.
- M08—Turns the coolant on.

N160

- G51—Initiates the do loop.
- N4—Instructs the MCU to perform the loop four times.

N170

- G91—Selects incremental positioning.
- X1—Incremental distance between the holes.

N180

- G50—Signals the end of the loop information.

N190

- G80—Cancels the canned drilling cycle.
- G90—Selects absolute positioning.
- G49—Cancels the tool offsets.
- Z0—Retracts the spindle.
- M09—Turns off the coolant.

N200

- X – 12 Y8—Coordinates of the spindle park position. Any coordinates that safely position the cutter out of the way are adequate.
- M05—Turns off the spindle.

N210

- M30—Signals the end of the program, resetting the computer memory to the start of the sequence.

SUBROUTINES

A *subroutine* is a program within a program, placed at the end of the main program. For example, on the part in Figure 11−4, note that the holes occur in the same geometric and dimensional pattern in four different locations. A do loop could be programmed to drill the holes, but programming steps can be minimized by placing the pattern in a subroutine. The drill can be sent to hole #1 and the subroutine called to drill the four holes A, B, C, and D. Hole #2 can then be positioned and the subroutine called again, and so on.

One way to use a subroutine is to place one or more do loops in the subroutine. This is known as *nesting*. Subroutines may also be nested in other subroutines, or nested within do loops. This gives the programmer a great deal of flexibility and a powerful programming tool.

Machinist Shop Language

The format for a subroutine in Machinist Shop Language is:

1. SUBR n
2. Programming information
3. END

Where SUBR signals the start of the subroutine, n is the subroutine number, the programming information describes the operations required, and END signals the end of the subroutine. A program written in Machinist Shop Language for the part shown in Figure 11−4 is presented in Figure 11−5.

The subroutine at the end of the program is as follows:

SUBR 1
X − .5 Y.5 RI
Y − 1 RI
X1 RI
Y1 RI
END

SUBR 1—Signals that the following is subroutine #1.

X − .5 Y.5 RI—Causes an incremental movement − .500 inch in X and .500 inch in Y at rapid traverse.

Y − 1 RI—Causes an incremental rapid movement −1.000 inch in Y.

X1 RI—Causes an incremental rapid movement of 1.000 inch in X.

Y1 RI—Causes an incremental rapid movement of 1.000 inch in Y.

END signals the end of the subroutine.

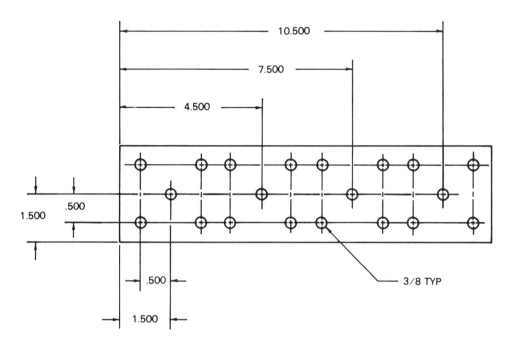

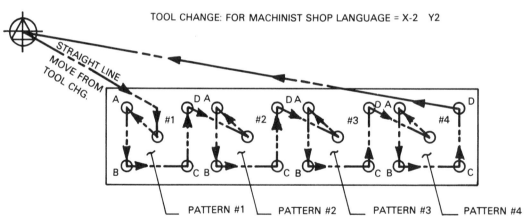

TOOL CHANGE: FOR MACHINIST SHOP LANGUAGE = X-2   Y2

MOVES TO HOLES #1, #2, #3, #4 IN ABSOLUTE

MOVES TO HOLES A, B, C, D IN SUBROUTINE ARE INCREMENTAL

FIGURE 11–4
Part drawing and tool path

```
XO/YO = LOWER LEFT CORNER
TOOLS:   TOOL #1 = #3 C'DRILL.
         TOOL #2 = 3/8 DRILL.
CLEARANCE OVER CLAMPS = 3.000 IN. MIN.

 1 TOOL 1001
 2 Z4                     A
 3 TOOL 1002
 4 Z3
 5 X-2 Y2 Z0              RA        REM:RAPID TO TCH
 6 TOOL 1                           REM:#3 C'DRILL
 7 V20 10.5                         REM:SET DRILLING FEED
 8 V21 .1                 A         REM:DEFINE BUFFER ZONE
 9 G81                              REM:INITIATE DRILLING CYCLE
10 X1.5 Y1.5 Z-.162 RA              REM:POSITION TO #1 AND C'DRILL
11 CALL 1                           REM:CALL SUBR 1
12 X4.5 Y1.5             RA         REM:POSITION TO AND C'DRILL#2
13 CALL 1                           REM:CALL SUBR #1
14 X7.5 Y1.5             RA         REM:POSITION TO AND C'DRILL #3
15 CALL 1                           REM:CALL SUBR 1
16 X10.5 Y1.5           RA          REM:POSITION TO AND C'DRILL #4
17 CALL 1                           REM:CALL SUBR 1
18 G80                              REM:CANCEL DRILL CYCLE
19 TOOL 0                           REM:CANCEL TOOL OFFSET
20 Z0                    RA         REM:RETRACT SPNDL
21 X-2 Y2                RA         REM:RAPID TO TCH
22 TOOL 2                           REM:3/8 DRILL
23 V20 21                           REM:SET DRILLING FEED
24 V21 .1                A          REM:DEFINE BUFFER
25 G81                              REM:INITIATE DRILL CYCLE
26 X1.5 Y1.5 Z-.375 RA              REM:POSITION TO AND DRILL #1
27 CALL 1                           REM:CALL SUBR 1
28 X4.5 Y1.5             RA         REM:POSITION AND DRILL #2
29 CALL 1                           REM:CALL SUBR 1
30 X7.5 Y1.5             RA         REM:POSITION AND DRILL #3
31 CALL 1                           REM:CALL SUBR 1
32 X10.5 Y1.5           RA          REM:POSITION AND DRILL #4
33 CALL 1                           REM:CALL SUBR 1
34 G80                              REM:CANCEL DRILL CYCLE
35 TOOL 0                           REM:CANCEL TOOL OFFSET
36 Z0                    RA         REM:RETRACT SPNDL
37 X-2 Y2                RA         REM:RAPID TO TCH
38 END                              REM:END OF MAIN PRGM
39 SUBR 1                           REM:DEFINES START OF SUBROUTINE 1
40 X-.5 Y.5              RI         REM:POSITIONS TO HOLE A
41 Y-1                   RI         REM:POSITIONS TO HOLE B
42 X1                    RI         REM:POSITIONS TO HOLE C
43 Y1                    RI         REM:POSITIONS TO HOLE D
44 END                              REM:END OF SUBROUTINE
```

FIGURE 11-5
Subroutine program for part in Figure 11-4, Machinist Shop Language

PROGRAM EXPLANATION

(Refer to Figure 11–5.)

EVENTS 1–5

- Assign the tool information and position the machine at tool change.

EVENT 6

- TOOL 1—Calls up tool #1's offsets. It also issues a dwell to allow the operator to put the center drill in the spindle. A combination drill can also be used to drill this part. By using two tools, the use of the subroutine can be better demonstrated.

EVENT 7

- V20 10.5—Assigns a feedrate of 10.5 inches per minute to be used with the G81 drill cycle.

EVENT 8

- V21 .1—Defines a .100-inch buffer zone between the top of the part and the end of the tool.
- A—Specifies absolute positioning.

EVENT 9

- G81—Initiates the drilling cycle.

EVENT 10

- X1.5 Y1.5—Coordinates for hole #1.
- Z – .162—Depth for drilling, with the buffer zone added.
- R—Specifies rapid traverse mode, which will be used throughout the drilling program.
- A—Specifies absolute positioning.

EVENT 11

- CALL 1—Instructs the MCU to go to subroutine #1 and carry out the instructions there. At the end of the subroutine, the MCU automatically resets to the line following the CALL statement.

EVENT 12

- X4.5 Y1.5—Coordinates for hole #2. Note that both an X and Y coordinate are necessary. At the end of the subroutine, the tool is positioned over hole D.
- A—Specifies absolute positioning, which must be reselected following the subroutine. The subroutine coordinates are incremental.

EVENT 13

- CALL 1—Instructs the MCU to carry out the instructions contained in subroutine #1.

EVENT 14
- X7.5 Y1.5—Positions the machine to hole #3. Since the G81 has not been canceled, a hole is drilled here.

EVENT 15
- CALL 1—Once again calls up the subroutine instructions, which drill holes A, B, C, and D.

EVENT 16
- X10.5 Y1.5—Positions the machine to hole #4.

EVENT 17
- CALL 1—Calls the subroutine to drill the other four holes in the pattern.

EVENT 18
- G80—Cancels the active drilling cycle.

EVENT 19
- TOOL 0—Cancels the tool offset.

EVENT 20
- Z0—Retracts the spindle.

EVENT 21
- X − 2 Y2—Coordinates to send the spindle to tool change, where the drill will be installed in the spindle.

EVENT 22
- TOOL 2—Calls up tool #2's offsets.

EVENT 23
- V20 21—Assigns a feedrate of 21 inches per minute to be used for drilling.

EVENT 24
- V21 .1—Establishes a buffer zone of .100 inch between the tool and part surface.

EVENT 25
- G81—Initiates the canned drilling cycle.

EVENTS 26–37
- Duplicate events 10 through 21.

EVENT 38
- END—Signals the end of the main program.

EVENT 39

■ SUBR 1—Defines the start of subroutine #1. This subroutine could have been given any number, but it makes sense to use 1 for the first subroutine, 2 for the second, and so on.

EVENT 40

■ X – .5 Y.5—Incremental coordinates to send the spindle from the hole in the center of the pattern to hole A.

■ R—Specifies rapid traverse mode, which has been used throughout.

■ I—Specifies incremental positioning. The coordinates within the subroutine must be incremental, in order to take advantage of the relationships common to all four hole patterns. Since the centerpoint of each pattern is in a different location, the absolute coordinates of the holes in each pattern are unique.

EVENT 41

■ Y – 1—Incremental coordinate to move from hole A to hole B.

EVENT 42

■ X1—Incremental coordinate to move from hole B to hole C.

EVENT 43

■ Y1—Incremental coordinate to move from hole C to hole D.

EVENT 44

■ END—Signals the MCU that this is the end of the subroutine.

This program could also have been written using the common incremental distance between hole pattern centers of 3.000 inches and a nested subroutine or a nested loop.

Word Address Format

In word address, the subroutine is not identified by a subroutine number as in Machinist Shop Language but is simply added with a sequence number following the main program. All blocks in a subroutine are then numbered from N010 consecutively as if the subroutine were an independent program. The codes associated with a word address subroutine are:

M98—Tells the machine to jump to a subroutine.
M99—Tells the machine to return to the main program.
P—The address P indicates a block sequence number when calling up or returning from a subroutine. P120 would mean N120, as will be demonstrated in a moment.
L—The address L tells the machine how many times to repeat the subroutine.

The format for a word address subroutine is:

1. SEQ # Pn1 Lnn M98
2. :PROG # N010
3. Programming information
4. SEQ # M99 Pn2

Where n1 is the block number that starts the subroutine, nn is the number of times the subroutine is to be repeated, :PROG # is the sequence number of the subroutine, N010 sets the sequence number of the subroutine equal to sequence number N010, and n2 is the sequence number of the main program to be returned to. Some controllers use an O (letter "O") address in place of the colon.

PROGRAM EXPLANATION

(Refer to Figure 11–6.)

N010–N030

- Assign the tool information to the tool registers.

N040

- M06 T1—Initiate an automatic tool change, with tool position #1 being used for the new tool.

N050

- S3500—Sets the spindle speed to 3500 RPM.
- F10.5—Sets the feedrate to 10.5 inches per minute.
- M03—Turns the spindle on clockwise.

N060

- G45 H01—Call up the offsets in register #1, which will be used for the center drill.

N070

- G81—Initiates the canned drilling cycle.
- G99—Selects a Z-axis return to the reference (rapid) level when the G81 is cycled.
- X1.5 Y1.5—Positions the center drill over hole #1.
- Z−.162—Z-axis depth for the center drilling.
- R0—Sets the reference level to Z0, which was set to be .100 off the top of the part at setup.
- M08—Turns on the coolant.

```
X0/Y0 = LOWER LEFT CORNER
TOOLS:    TOOL #1 = #3 C'DRILL.
          TOOL #2 = 3/8 DRILL.
CLEARANCE OVER CLAMPS = 3.000 IN. MIN.

N010 G00 G40 G49 G80 G70 G90 X0 Y0 Z0 REM:SAFTEY LINE
N020 G10 H01 Z4
N030 G10 H02 Z3
N040 M06 T1
N050 S3500 F10.5 M03                  REM:SET SPEED/FEED, SPNDL
ON
N060 G45 H01                          REM:CALL UP OFFSET #1
N070 G81 G99 X1.5 Y1.5 Z-.162 R0 M08  REM:POSITION AND C'DRILL #1
N080 P300 M98                         REM:JUMP TO SUBR 1
N090 G90 X4.5 Y1.5                    REM:POSITION AND C'DRILL #2
N100 P300 M98                         REM:JUMP TO  SUBR 1
N110 G90 X7.5 Y1.5                    REM:POSITION AND DRILL #3
N120 P300 M98                         REM:JUMP TO SUBR 1
N130 G90 X10.5 Y1.5                   REM:POSITION AND DRILL #4
N140 P300 M98                         REM:JUMP TO SUBR 1
N150 G49 G90 G80 Z0 M09               REM:RETRACT SPNDL
N160 M06 T2                           REM ATC, TOOL #2
N170 S3500 F21 M03                    REM:SET SPEED/FEED, SPNDL
ON
N180 G45 H02                          REM:OFFSET #2
N190 G81 G99 X1.5 Y1.5 Z-.375 R0 M08  REM:POSITION AND DRILL #1
N200 P300 M98                         REM:JUMP TO SUBR. 1
N210 G90 X4.5 Y1.5                    REM:POSITION AND DRILL #2
N220 P300 M98                         REM:JUMP TO SUBR 1
N230 G90 X7.5 Y1.5                    REM:POSITION AND DRILL #3
N240 P300 M98                         REM:JUMP TO SUBR 1
N250 G90 X10.5 Y1.5                   REM:POSITION AND DRILL #4
N260 P300 M98                         REM:JUMP TO SUBR 1
N270 G49 G80 G90 Z0 M09               REM:RETRACT SPNDL, COOL.
OFF
N280 X-12 Y8 M05                      REM:RAPID TO PARK POSTION
N290 M30                              REM:END OF PGRM
:300 N010 G91 X-.5 Y.5                REM:START OF SUBR, DRILL A
N020 Y-1                              REM:DRILL HOLE B
N030 X1                              REM:DRILL HOLE C
N040 Y1                              REM:DRILL HOLE D
N050 M99                              REM:JUMP TO MAIN PGRM
```

FIGURE 11-6
Subroutine program for part in Figure 11-4, word address format

N080

- P300—Sequence number of the subroutine starting location. P300 is used here instead of N300.
- M98—Tells the machine to jump to the subroutine. The MCU will go to the program block specified as P300, labeled :300, and execute the instructions listed there. The last command in the subroutine instructs the MCU to return to the main program.

N090

- G90—Selects absolute positioning. The subroutine coordinates are incremental; therefore absolute must be specified here.
- X4.5 Y1.5—Coordinates for hole #2. The G81 is still active so that a hole is drilled at every programmed location.

N100

- P300 M98—Again tell the MCU to carry out the instructions in the subroutine in block :300.

N110

- G90—Selects absolute positioning.
- X7.5 Y1.5—Coordinates for hole #3.

N120

- P300 M98—Initiate a jump to the subroutine.

N130

- G90—Selects absolute positioning.

- X10.5 Y1.5—Coordinates for hole #4.

N140

- P300 M98—Again causes a jump to the subroutine.

N150

- G49—Cancels the tool offset.
- G80—Cancels the drilling cycle.
- G90—Selects absolute positioning.
- Z0—Retracts the spindle.
- M09—Turns off the coolant.

N160

- M06 T2—Initiate an automatic tool change, taking tool #2 from the magazine and placing it in the spindle.

N170–N270

- Duplicate blocks N050–N150.

N290

- M30—Signals the end of the program, resetting the computer memory to the start.

:300

- :300—Identifies this block as N300, the beginning of a subroutine.
- N010—Identifies this block as block N010 of the subroutine.
- G91—Selects incremental positioning. Incremental coordinates are used throughout this subroutine.

- X − .5 Y.5—Incremental coordinates to move from the center of the hole pattern to hole A.

N020

- N020—Identifies this as block N020 of the subroutine.
- Y − 1—Incremental coordinate to move from hole A to hole B.

N030

- N030—Identifies this as block N030 of the subroutine.
- X1—Incremental coordinate to move from hole B to hole C.

N040

- N040—Identifies this as block N040 of the subroutine.
- Y1—Incremental coordinate to move from hole C to hole D.

N050

- N050—Identifies this as block N050 of the subroutine.
- M99—Instructs the MCU to return to the block in the main program following the M98 that sent it to the subroutine. If a different return spot is desired, a modifier can be added to this command using the P address to send the MCU to another program line. M99 P090, for example, would return the program to block N090 rather than the block following the M98 jump to the subroutine.

Subroutines for Cutter Diameter Compensation

In Chapter 10 it was pointed out that subroutines are often used with cutter diameter compensation. Figure 11 − 7 presents a program written in word address format for the part shown in Chapter 10, Figure 10 − 13. The program utilizes a subroutine to mill the part periphery. The first time the subroutine is called, the .52 tool compensation diameter is used with a .500-inch-diameter cutter. This procedure mills the roughing pass. The second time the subroutine is called, the .5 offset is active, resulting in the finish milling pass. The remark statements should make the program self-explanatory. The machining sequence is identical to that used in Chapter 10. The difference is that the duplication of coordinate locations is eliminated.

NESTED LOOPS

Do loops may nest inside other do loops or subroutines. Similarly, subroutines may nest inside other subroutines. This concept will be demonstrated using the part illustrated in Figure 11 − 8, with corresponding programs in Fig-

```
X0/Y0 = LOWER LEFT CORNER
TOOLS:   1.000 IN. 4 FLUTE END MILL
CLEARANCE ABOVE CLAMPS:  3.000 MIN.
BUFFER ZONE:   .100

N010 G00 G40 G49 G70 G90   REM:SAFETY LINE
N020 G10 H01 X.502 Z3
N030 G10 H02 X.5 Z3
N040 M06 T1                REM:1 IN. E/M
N050 G45 H01               REM:TOOL #1 OFFSETS
N060 S400 M03              REM:SET SPEED/FEED, SPNDL ON
N070 P110 M98              REM:JUMP TO SUBROUTINE.
N080 P110 M98              REM:JUMP TO SUBROUTINE
N090 X-8 Y6 M05            REM:MOVE TO PARK/SPNDL OFF
N100 M30                   REM:END PRGM
:110 N010 G00 X7 Y.875     REM:POSITION FOR RAMP ON
N020 Z0                    REM:RAPID TO BUFFER
N040 G01 Z-.89 M08         REM:FEED TO MILLING DEPTH
N050 G17 G42 X6            REM:RAMP ON TO #1
N060 Y4                    REM:FEED TO #2
N070 X2                    REM:FEED TO #3
N080 X0 Y2                 REM:FEED TO #4
N090 Y0                    REM:FEED TO #5
N100 X4.875                REM:FEED TO #6
N110 G02 X6 Y1.125 I6 J0   REM:CUT ARC FROM #6 TO #1
N120 G01 G40 X7 M09        REM:RAMP OFF/COOLANT OFF
N130 G00 G49 Z0            REM:CANCEL OFFSET/RETRACT SPNDL
N140 M99
```

FIGURE 11-7
Subroutine program for part in Figure 10–13, word address format

ures 11–9 and 11–10. These programs feature two loops nested inside a sub-routine. In Figure 11–8, the rows of holes have been labeled for easy reference. In writing CNC programs, a reference sketch such as this is a valuable aid in developing a machining strategy and provides a way for the programmer to check his or her work.

In the part programs, one of the loops will drill the holes in row A, moving from hole A1 to hole A6. Hole A1 will be drilled prior to instituting the do loop. After positioning to drill hole B6, another loop will be used to drill holes B6 to B1, moving in the −X direction. One do loop could drill all the holes, but that would require sending the spindle back to the first hole of each row prior to using the do loop. By using two do loops, machine motion is more efficient, drilling in both the positive and negative directions along the X axis. Nesting the loops in a subroutine allows drilling rows C and D with the same do loops.

MTL: .50 THICK 1018 CRS

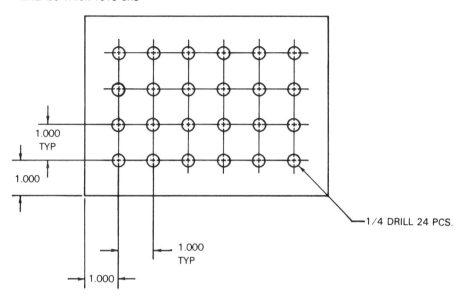

1/4 DRILL 24 PCS.

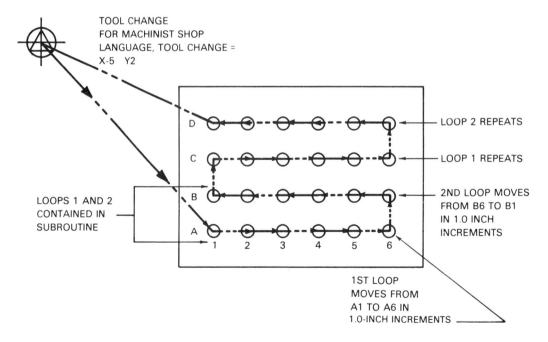

FIGURE 11–8
Part drawing and tool path

```
X0/Y0 = LOWER LEFT CORNER OF PART
TOOL CHANGE = X-5 Y2
TOOLS:  #3 C'DRILL, 1/4 DRILL
CLEARANCE OVER CLAMPS:  2.5 IN. MIN.

 1 TOOL 1001
 2 Z4                A      REM:C'DRILL
 3 TOOL 1002
 4 Z2.5              A      REM:1/4 DRILL
 5 X-2 Y5            RA     REM:TCH - C'DRILL
 6 TOOL 1                   REM:1700 RPM
 7 V20 5.1                  REM:C'DRILL FEEDRATE
 8 V21 .1            A      REM:C'DRILL BUFFER
 9 G81                      REM:INITIATE DRILL CYCLE
10 X1 Y1 Z-.162      RA     REM:C'DRILL HOLE 1A
11 CALL 1                   REM:CALL SUBR 1
12 Y1                RI     REM:C'DRILL HOLE C1
13 CALL 1                   REM:CALL SUBR 1
14 G80                      REM:CANCEL DRILLING CYCLE
15 TOOL 0                   REM:CANCEL TOOL OFFSET
16 Z0                RA     REM:RETRACT Z
17 X-5 Y2            RA     REM:TCH - 1/4 DRILL
18 TOOL 2                   REM:1200 RPM
19 V20 4.8                  REM:DRILL FEEDRATE
20 V21 .1            A      REM:DRILL BUFFER
21 G81
22 X1 Y1 Z-.7        RA     REM:DRILL HOLE A1
23 CALL 1
24 Y1                RI     REM:DRILL HOLE C1
25 CALL 1                   REM:CALL SUBR 1
26 G80                      REM:CANCEL DRILL CYCLE
27 TOOL 0                   REM:CANCEL TOOL OFFSET
28 Z0                RA     REM:RETRACT Z
29 X-2 Y5            RA     REM:TOOL CHANGE
30 END                      REM:END PROGRAM
31 SUBR 1
32 DO 5
33 X1                RI
34 END                      REM:END DO LOOP
35 Y1                RI     REM:START NEXT ROW
36 DO 5
37 X-1               RI
38 END                      REM:END DO LOOP
39 END                      REM:END OF SUBROUTINE
```

FIGURE 11-9

Nested loop program for part in Figure 11-8, Machinist Shop Language

Machinist Shop Language

In Machinist Shop Language, the do loop to drill the holes for row A (assuming the spindle is first positioned over hole A1) is:

DO 5
X1 R I (the incremental distance between holes)
END

```
XO/YO = LOWER LEFT CORNER OF PART
TOOLS:  #3 C'DRILL, 1/4 DRILL
CLEARANCE OVER CLAMPS:  2.5 IN.

%
N010 G00 G90 G80 G40 Z0                     REM:SAFETY LINE
N020 G10 H01 Z4.0                           REM:C'DRILL
N030 G10 H02 Z2.5                           REM:1/4 DRILL
N040 T01 M06                                REM:TOOL CHANGE
N050 S1700 M03
N060 G45 H01 M08                            REM:CALL OFFSET 01
N070 G81 G99 X1.0 Y1.0 Z-.162 R0 F5.1       REM:C'DRILL HOLE AT A1
N080 P240 M98                               REM:JUMP TO SUB
N090 Y1.0                                    REM:C'DRILL HOLE C1
N100 P240 M98                               REM:JUMP TO SUB
N110 G80 G90 M09                            REM:CANCEL G81
N120 G49                                     REM:CANCEL OFFSET

N130 T02 M06                                REM:TOOL CHANGE
N140 S1200 M03
N150 G45 H02 M08                            REM:CALL OFFSET 02
N160 G81 G99 X1.0 Y1.0 Z-.7 R0 F4.8         REM:DRILL HOLE AT A1
N170 P240 M98                               REM:JUMP TO SUB
N180 Y1.0                                    REM:DRILL HOLE C1
N190 P240 M98                               REM:JUMP TO SUB
N200 G80 G90 M09                            REM:CANCEL G81
N210 G49                                     REM:CANCEL OFFSET
N220 X-12.0 Y8.0 M05                        REM:PARK SPINDLE
N230 M30                                     REM:END PROGRAM

:240 N010 G51 N5                            REM:START SUB & LOOP
N020 G91 X1.0                                REM:CONTENTS OF LOOP
N030 G50                                     REM:END LOOP
N040 Y1.0                                    REM:START NEXT ROW
N050 G51 N5                                  REM:START LOOP
N060 X-1.0                                   REM:CONTENTS OF LOOP
N070 G50                                     REM:END LOOP
N080 M99                                     REM:PROGRAM RETURN
%
```

FIGURE 11-10

For row B (assuming the spindle is first positioned over hole B6), the do loop is:

DO 5
X – 1 R I (the incremental distance between holes)
END

These two loops are placed inside a subroutine along with some other programming information to drill the rows of holes. The subroutine is written as follows:

SUBR 1
DO 5   REM:DO LOOP FOR ROW A
X1 R I
END   REM:END OF LOOP

```
Y1 R I          REM:POSITION FOR START OF NEXT LOOP
DO 5            REM:DO LOOP FOR ROW B
X − 1 R I
END             REM:END OF LOOP
END             REM:END OF SUBROUTINE
```

By calling up this subroutine, rows A and B will be drilled. After positioning the machine to hole C1, calling up the subroutine a second time will drill rows C and D.

It cannot be overstressed that the examples presented here are only that. Codes vary from controller to controller, even in different controllers from the same manufacturer. The EIA standards are the guidelines manufacturers follow; however, many controllers break with standard coding on occasion for one reason or another. **When in doubt, check the programming manual.** Don't take chances with the machine's and, more importantly, the operator's safety.

SPECIAL FANUC SECTION

Figure 11−11 contains the program to drill the part in Figure 11−1 using a do loop. Figure 11−12 is the program to drill the part in Figure 11−4 using a subprogram. Fanuc controllers do not use subroutines as such. Each subroutine is actually a mini program, stored in the controller under its own "O" number. The subroutines are called subprograms for this reason. A colon (:) can be substituted for the letter "O" if desired, as was done in the text examples.

The programs in Figures 11−11 and 11−12 were written for a Tsugami "Lightning" horizontal machining center. A work coordinate is used to origin the part coordinate system using G54.

The length offsets are picked up using a G43 command. They are cancelled by returning the spindle to the home zero position, which is also the tool change position, using a G28 command.

This machine is a multipallet machine; up to 30 pallets can be used. The M91 command issued as part of the program startup statements is a pallet recognition command. This tells the controller the machine has been equipped with the multipallet option. In the examples given in this section, only one pallet is used.

```
%
01103
(--------------------------------)
(X0/Y0 - LOWER LEFT CORNER)
(Z0 - .100 ABOVE TOP OF PART)
(--------------------------------)
(TOOL NO. 1 - NO. 3 C-DRILL)
N001G17G90M91
N101T01M06
N102G54S3500M03T02
N103G00X.5Y1.M08
N104G43G90Z2.5H01
N105G81G99X.5Y1.Z-.162R0.F10.5
()
(BEGIN DO LOOP)
N106G51N4
N107G91X1.
N108G50
(END OF LOOP)
()
N109M09
N110G00G91G28Z0M05
N111G28X0Y0M01
()
(--------------------------------)
(TOOL NO. 2 - 1/4 DRILL)
N002G90
N201T02M06
N202G54S3500M03T01
N203G00X.5Y1.M08
N204G43G90Z2.5H02
N205G81G99X.5Y1.Z-.375R0.F21.
()
(BEGIN DO LOOP)
N206G51N4
N207G91X1.
N208G50
(END OF LOOP)
()
N209M09
N210G00G91G28Z0M05
N211G28X0Y0
N212M30
%
```

FIGURE 11-11

```
%
O1104
(------------------------------)
(XO/YO - LOWER LEFT CORNER)
(ZO - .100 ABOVE TOP OF PART)
(------------------------------)
(TOOL NO. 1 - NO. 3 C-DRILL)
NOO1G17G90M91
N101T01M06
N102G54S3500M03T02
N103G00X1.5Y1.5M08
N104G43G90Z3.H01
N105G81G99X1.5Y1.5Z-.162R0M08F10.5
N106P1000M98
N107G90X4.5Y1.5
N108P1000M98
N109G90X7.5Y1.5
N110P1000M98
N111G90X10.5Y1.5
N112P1000M98
N113G80M09
N114G91G28Z0M05
N115G28X0Y0Z0M01
()
(------------------------------)
(TOOL NO. 2 - 3/8 DRILL)
N002G90
N201T02M06
N202G54S3500M03T01
N203G00X1.5Y1.5M08
N204G43G90Z3.H02
N205G81G99X1.5Y1.5Z-.375R0.F12.
N206P1000M98
N207G90X4.5Y1.5
N208P1000M98
N209G90X7.5Y1.5
N210P1000M98
N211G90X105.Y1.5
N212P1000M98
N213G80M09
N214G00G91G28Z0M05
N215G28X0Y0
N216M30
()
()
O1000
(START OF SUB PROGRAM 1000)
NOO1G91X-.5Y.5
N002Y-1.
N003X1.
N004Y1.
N005M99
%
```

FIGURE 11-12

SUMMARY

The important concepts presented in this chapter are:

- A do loop instructs the MCU to repeat a series of instructions a specified number of times.
- The format for a do loop in Machinist Shop Language is:

DO #
Programming information
END

Where # is the number of times the loop is to be repeated.

- The format for a do loop in word address is:

G51 N#
Programming information
G50

Where # is the number of times the loop is to be repeated.

- A subroutine is a program within a program, placed at the end of the main program.
- The format for a Machinist Shop Language subroutine is:

SUBR #
Programming information
END

Where # is the number of the subroutine.

- The format for a word address subroutine as used in the text examples is:

SEQ # Pn1 Lnn M98
:PROG # N010
Programming information
SEQ # M99 Pn2

Where Pn1 = Sequence number at which subroutine starts, and Pn2 = the block number of the main program to be returned to. If no block number is specified, the MCU returns to the block following the M98 command last issued.

- Nested loops are loops placed inside other loops or inside subroutines.
- The codes for subroutines and do loops vary from controller to controller. To program a particular machine, it will be necessary to consult the programming manual for the machine in question.

REVIEW QUESTIONS

1. What is a do loop? A subroutine? A nested loop?
2. What is the format for a do loop in Machinist Shop Language? In word address?
3. What is the format for a subroutine in Machinist Shop Language? In word address?
4. Write a do loop to drill the hole patterns in Figure 11–13:
   a.   In Machinist Shop Language.
   b.   In word address.
5. Write a program using a subroutine to mill the slots in Figure 11–14:
   a.   In Machinist Shop Language.
   b.   In word address.

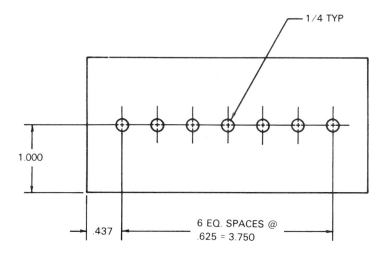

1/4 TYP

1.000

.437

6 EQ. SPACES @
.625 = 3.750

MATERIAL: 1/8 THICK 2024 T-3 ALUMINUM

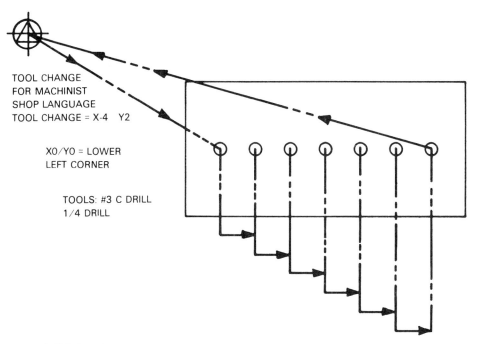

TOOL CHANGE
FOR MACHINIST
SHOP LANGUAGE
TOOL CHANGE = X-4   Y2

X0/Y0 = LOWER
LEFT CORNER

TOOLS: #3 C DRILL
1/4 DRILL

FIGURE 11–13
Part drawing for review question #4

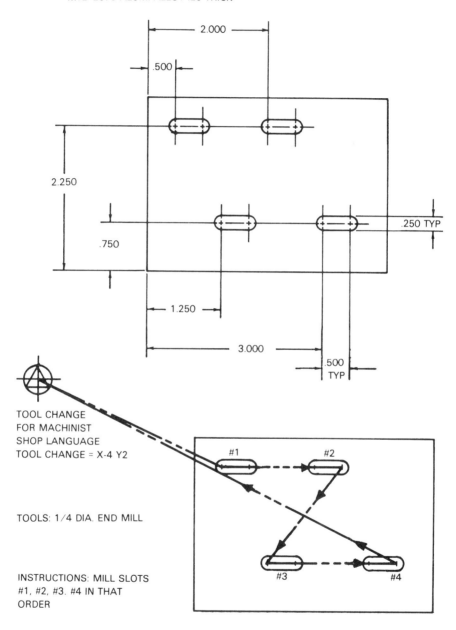

MTL: 2075 ALUM. ALLOY .25 THICK

TOOL CHANGE
FOR MACHINIST
SHOP LANGUAGE
TOOL CHANGE = X-4 Y2

TOOLS: 1/4 DIA. END MILL

INSTRUCTIONS: MILL SLOTS
#1, #2, #3. #4 IN THAT
ORDER

FIGURE 11–14
Part drawing for review question #5

CHAPTER 12

Advanced CNC Features

OBJECTIVES Upon completion of this chapter, you will be able to:

- Explain the concept of mirror imaging.
- Decide when the use of mirror imaging is appropriate.
- Write simple programs in Machinist Shop Language and word address that employ mirror imaging.
- Explain the concept of polar rotation.
- Decide when the use of polar rotation is appropriate.
- Write simple programs in Machinist Shop Language and word address that employ polar rotation.
- Write simple programs in Machinist Shop Language and word address that employ polar rotation used in a do loop.
- Explain the concept of helical interpolation.
- Decide when the use of helical interpolation is appropriate.
- Write simple programs in Machinist Shop Language and word address that employ helical interpolation.

MIRROR IMAGING

Mirror imaging is a simple concept that can be very useful in programming. In essence, *mirror imaging* reverses the sign (+ or −) of an axis direction. For example, mirror imaging can be employed to shorten the amount of programming required to make the part shown in Figures 12–1 and 12–2. Calling the centerline of this part X0/Y0, the pattern of holes to the right of the centerline can be programmed in a subroutine. After this pattern is drilled, mirror imaging along the X axis can be instituted and the subroutine called again. This will drill the same pattern of holes in the second quadrant, with no additional programming save the mirror imaging command. The process can be repeated, mirror imaging the Y axis to drill the pattern in the third quadrant. Canceling the mirror image on the X axis and leaving it active on Y will drill the pattern in the fourth quadrant.

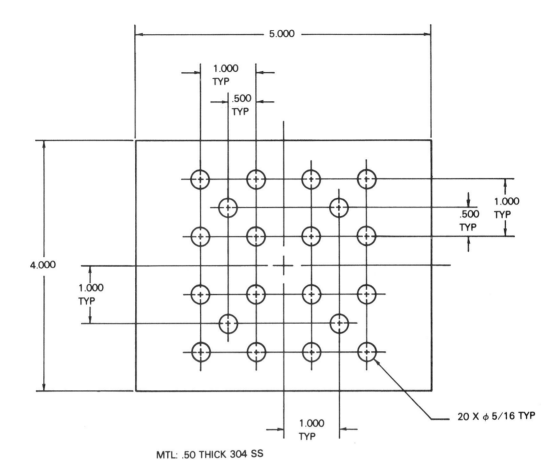

FIGURE 12–1
Part drawing

Machinist Shop Language

In Machinist Shop Language, mirror imaging is instituted by use of an auxiliary (AUX) code. Appendix 2 lists various auxiliary codes used with Machinist Shop Language. These codes are generally used for the same purposes for which miscellaneous functions are used in word address format. The Machinist Shop Language program used to drill this part (see Figures 12–1 and 12–2) is Figure 12 – 3. The program uses a combination drill/center drill. Notice how short the program is, considering the number of holes, when the right combination of machine features and tooling is used.

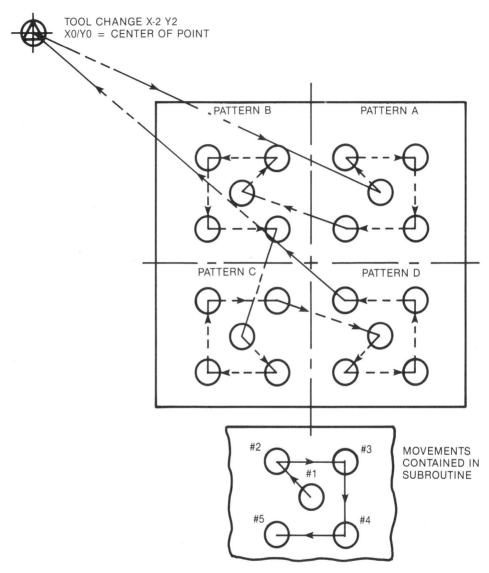

TOOL CHANGE X-2 Y2
X0/Y0 = CENTER OF POINT

PATTERN B   PATTERN A

PATTERN C   PATTERN D

#2   #3
#1
#5   #4

MOVEMENTS
CONTAINED IN
SUBROUTINE

FIGURE 12–2
Tool path for the part shown in Figure 12–1

```
X0/Y0 = CENTER OF PART
TOOL CHANGE = X-2 Y2
TOOLS:  5/16 DIA. COMB. DRILL
 1 TOOL 1001
 2 Z2              A
 3 X-2 Y2 Z0       RA              REM:TO TOOL CHANGE
 4 TOOL 1
 5 X0 Y0           RA              REM:MOVE TO X0/Y0
 6 V20 5
 7 V21 .1
 8 G81 Z-.7        RA
 9 CALL 1                          REM:DRILL PATTERN #A
10 AUX 100                         REM:MIRROR IMAGE IN X
11 CALL 1                          REM:DRILL PATTERN #B
12 AUX 800                         REM:MIRROR IMAGE OFF
13 AUX 300                         REM:MIRROR IMAGE X AND Y
14 CALL 1                          REM:DRILL PATTERN #C
15 AUX 800                         REM:MIRROR IMAGE OFF
16 AUX 200                         REM:MIRROR IMAGE Y
17 CALL 1                          REM:DRILL PATTERN D
18 AUX 800                         REM:MIRROR IMAGE OFF
19 G80
20 TOOL 0
21 Z0
22 X-2 Y2          RA              REM:TO TOOL CHANGE
23 END
24 SUBR 1                          REM:START SUBR 1
25 X1 Y1           RA              REM:DRILL HOLE #1
26 X-.5 Y.5        RI              REM:DRILL HOLE #2
27 X1              RI              REM:DRILL HOLE #3
28 Y-1             RI              REM:DRILL HOLE #4
29 X-1             RI              REM:DRILL HOLE #5
30 END                             REM:END OF SUBROUTINE
```

FIGURE 12-3

Mirror imaging program for the part in Figure 12-1, Machinist Shop Language

---

PROGRAM EXPLANATION

EVENTS 1-4

- These events assign and call up the tool information.

EVENT 5

- X0 Y0 — Position the part to the X0/Y0 point, which is the center of the part. The reason for this move is to allow the subroutine to control movement to the center hole in each of the hole patterns. The machine stops only momentarily at this location, with no spindle movement taking place.

- R — Specifies a rapid move. Rapid traverse will be used throughout since this is a drilling program.
- A — Specifies absolute positioning. This program will demonstrate the value of manipulating absolute and incremental positioning.

EVENT 6

- V20 — Sets the drilling feedrate to 5 inches per minute.

EVENT 7

- V21 .1 — Defines the buffer as .100 inches.

EVENT 8

- G81 — Initiates the drilling cycle.
- Z − .7 — Is the Z axis depth for drilling.

EVENT 9

- CALL 1 — Calls the subroutine; pattern A is drilled.

EVENT 10

- AUX 100 — Mirror images the X axis. When the subroutine is called again, the identical sequence of events will take place with the exception of the X axis moves, which will now be reversed.

EVENT 11

- CALL 1 – Calls the subroutine again. This time the AUX 100 is active, so that pattern B is drilled.

EVENT 12

- AUX 800 — Turns off the mirror image. Before issuing a mirror image command a second time, it is necessary to cancel the active command. Although some controllers may allow the commands to be issued one after another without a cancel command in between, it is always wise to play it safe.

EVENT 13

- AUX 300 — Mirror images the X and Y axes. Since the event preceding this one canceled any active mirror imaging, the movement with AUX 300 active will be reversed on both the X and Y axes.

EVENT 14

- CALL 1 — Calls the subroutine again. This time, with both X and Y reversed from pattern A, pattern C will be drilled.

EVENT 15

- AUX 800 — Cancels the mirror image.

EVENT 16

- AUX 200 — Mirror images the Y axis.

EVENT 17
- CALL 1 — Calls the subroutine again. This time through the Y axis movements are reversed from pattern A, and pattern D is drilled.

EVENT 18
- AUX 800 — Cancels the mirror imaging.

EVENT 19
- G80 — Cancels the active drilling cycle.

EVENT 20
- TOOL 0 — Cancels the active tool offset.

EVENT 21
- Z0 — Retracts the spindle.

EVENT 22
- X − 2 Y2 — Are the tool change coordinates.

EVENT 23
- END — Signals the end of the program.

EVENT 24
- SUBR 1 — Defines the start of subroutine #1.

EVENT 25
- X1 Y1 — Are the absolute coordinates from the X0/Y0 point to hole #1 in pattern A. When mirror imaging commands are active, the positioning will take place to the center holes of other patterns, depending on the active mirror imaging code.
- R — Specifies rapid traverse mode.
- A — Specifies that this is an absolute coordinate.

EVENT 26
- X − .5 Y.5 — Incremental coordinates to move from hole #1 to hole #2.
- R — Specifies rapid traverse mode.
- I — Specifies incremental positioning.

EVENT 27
- X1 — Incremental coordinate to move from hole #2 to hole #3.

EVENT 28
- Y − 1 — Incremental coordinate to move from hole #3 to hole #4.

EVENT 29
- X − 1 — Incremental coordinate to move from hole #4 to hole #5.

EVENT 30
- END — Signals the end of the subroutine.

Word Address Format

In word address, the same procedure is accomplished through either G codes or M functions, depending on the controller. In the following example, M functions are used as follows:

M21—Mirror image X axis.
M22—Mirror image Y axis.
M23—Mirror image off.

On some CNC machines, mirror imaging is selected at the MDI console by means of a switch. When programming such a machine, a dwell must be programmed at the place where mirror imaging is to be instituted, and instructions given for the operator to set the switches prior to restarting the program. The program to drill the part is shown in Figure 12–4.

```
XO/YO = CENTER OF PART
TOOL CHANGE = X-2 Y2
TOOLS:  5/16 DIA. COMB. DRILL
BUFFER:  2.5 IN. MIN.
N010 G00 G40 G49 G70 G90 ZO    REM:SAFTEY LINE
N020 G10 H01 Z2.5
N030 M06 T1
N040 S641 F5 M03               REM:SET SPEED/FEED
N050 G45 H01
N060 XO YO                     REM:POSITION TO XO/YO
N070 G81 G99 Z-.7 RO M08       REM:INITIATE DRILL CYCLE
N080 P190 M98                  REM:JUMP TO SUBROUTINE
N090 M21                       REM:MIRROR IMAGE X
N100 P190 M98                  REM:JUMP TO SUBROUTINE
N110 M22                       REM:MIRROR IMAGE Y
N120 P190 M98                  REM:JUMP TO SUBROUTINE
N130 M23                       REM:CANCEL MIRROR IMAGE
N140 M22                       REM:MIRROR IMAGE Y
N150 P190 M98                  REM:JUMP TO SUBROUTINE
N160 G80 G49 ZO M09            REM:CANCEL DRILLING, OFFSETS
N170 X--12 Y8 M05              REM:MOVE TO PARK, SPNDL OFF
N180 M30                       REM:END OF MAIN PGRM
:190 N010 X1 Y1                REM:START SUBROUTINE, DRILL #1
N020 G91 X-.5 Y.5             REM:DRILL #2
N030 X1                        REM:DRILL #3
N040 Y-1                       REM:DRILL #4
N050 X-1                       REM:DRILL #5
N060 M99
```

FIGURE 12–4
Mirror imaging program for the part in Figure 12–1, word address format

PROGRAM EXPLANATION

(Refer to Figure 12–4.)

N010
- This is the safety block, canceling any codes that may have been left active following a previous program.

N020–N030
- These blocks assign the tool information and select the tool.

N040
- S641—Sets the spindle speed to 641 RPM.
- F5—Sets the feedrate to 5 inches per minute.
- M03—Turns the spindle on clockwise.

N050
- G45 H01—Call up the tool offsets in register #1.

N060
- X0 Y0—Position the machine to the center of the part, where the sub-routine starts.

N070
- G81—Initiates the drilling cycle.
- G99—Selects a return to rapid level.
- Z–.7—Z-axis depth for drilling. Since a G81 code will not move the Z axis until after an X, Y, or X/Y move, no movement takes place along the Z axis yet.
- R0—Sets the start of the buffer (Z0 with a tool offset active) at the rapid level.
- M08—Turns the coolant on.

N080
- P190 M98—Instruct the MCU to jump to the subroutine that starts in block 190.

N090
- M21—Mirror images the X axis.

N100
- P190 M98—Causes a jump to the subroutine.

N110
- M22—Mirror images the Y axis.

N120
- P190 M98—Causes a jump to the subroutine.

N130

- M23—Cancels the active mirror image commands.

N140

- M22—Mirror images the Y axis. It was necessary to cancel the mirror image in block N130 because the X axis was mirror imaged along with the Y. Once canceled, an M22 is used to reestablish the mirror image on the Y axis.

N150

- P190 M98—Causes a jump to the subroutine.

N160

- G80—Cancels the drill cycle.
- G49—Cancels the tool offset.
- Z0—Retracts the spindle.
- M09—Turns off the coolant.

N170

- X – 12 Y8—Coordinates of the park position. As in other word address examples, any place that safely positions the tool out of the way can be used. It is assumed in these examples that the tool change location is at approximately X – 12, Y8 from the part X0/Y0.
- M05—Turns the spindle off.

N180

- M30 — Signals the end of the main program and resets the computer memory.

:190

- :190—Identifies this as block 190 of the main program.
- N010—Further identifies this as block N010 of the subroutine.
- X1 Y1 — Absolute coordinates to move from the center of the part to hole #1.

N020

- G91—Selects incremental positioning.
- X – .5 Y.5—Incremental coordinates to move from hole #1 to hole #2.

N030

- X1—Incremental coordinate to move from hole #2 to hole #3.

N040

- Y – 1—Incremental coordinate to move from hole #3 to hole #4.

N050

- X – 1—Incremental coordinate to move from hole #4 to hole #5.

N060

- G90 — Selects absolute positioning.
- M99 — Instructs the MCU to return to the main program.

POLAR ROTATION

Consider the part shown in Figure 12–5, in which four slots are to be milled. A machinist making this part on a conventional vertical milling machine would probably set up the workpiece on a rotary table, rotate 45 degrees from the nominal 0-degree location, and mill the first slot. The other three slots could then be milled, moving the various axes, or the machinist could simply index the part 90 degrees from the first slot to mill the second without excess movement along the X and Y axes. The same type of machining may be accomplished on a CNC machining center or CNC mill equipped with polar rotation.

A polar axis coordinate system is formed by constructing a line whose slope is not the same as either the X or Y axis. For example, in Figure 12–6, a line has been constructed between the origin (point #1) and point #2 on the graph. That line is a polar axis. Notice that point #2 is located 1.0 inch from the origin as measured along the polar axis. If point #2 is specified as (1,0) measured along the polar axis, then point #2 is called a polar coordinate. In mathematics more scientific definitions exist for a polar axis, but for the purposes of CNC programming, *polar rotation* can be thought of as rotating the Cartesian coordinate system.

When polar rotation is instituted in a CNC program, the MCU will triangulate the points necessary to position the tool to the desired coordinates from the program information that it is given. Polar rotation is supplied on most controllers as an optional feature. As with most options, the coding for polar rotation varies greatly from machine to machine. The examples given here can serve only to demonstrate the concept. The NC part programmer will have to consult the programming manual to program polar rotation successfully on a given machine.

Despite the differences in controllers, there is certain information that every MCU needs in order to carry out a polar rotation:

- The X axis coordinate of the center of rotation.
- The Y axis coordinate of the center of rotation.
- The *index angle,* or the angle as measured counterclockwise from the + X axis to the start of the rotation. In the case shown in Figure 12–5, the index angle is 45 degrees. This value is the angular rotation from the X axis to slot #1.

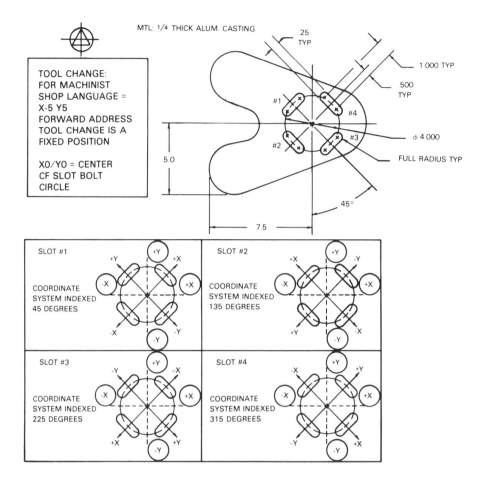

FIGURE 12–5
Part drawing

- The amount of the rotation. Following the initial rotation to the index angle, subsequent rotations may be specified as some angular value other than the index angle. The rotations will occur in a counterclockwise direction. In the case shown in Figure 12–5, this amount is 90 degrees. In other words, following the initial index of the coordinate system 45 degrees, subsequent rotations will be 90 degrees until the cancel command is given.
- A code to initiate polar rotation.
- A code to cancel polar rotation.

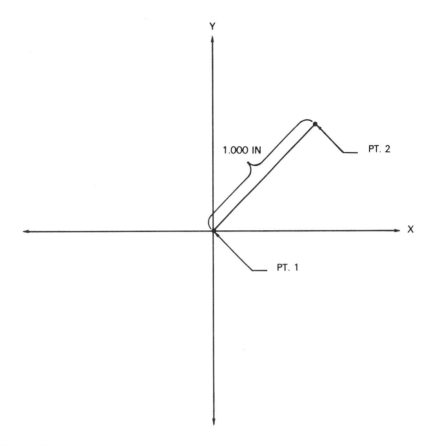

FIGURE 12-6

Machinist Shop Language

The format for instituting polar rotation in Machinist Shop Language is:

V11 — X-axis coordinate of the polar rotation.
V12 — Y-axis coordinate of the polar rotation.
V13 — Index angle.
V15 — Amount of rotation.
G51 — Code for instituting polar rotation. (Contains specific programming information)
G52 — Code for canceling the polar rotation.

The program to mill the aluminum casting shown in Figure 12-5 is presented in Figure 12-7. Only the slots need be milled.

```
X0/Y0 = CENTER OF SLOT ROTATION
TOOL CHANGE = X-5 Y5
TOOLS:   .250 END MILL
BUFFER:   .100 INCHES
CLEARANCE OVER CLAMPS:   3.000 IN. MIN.

 1 TOOL 1001
 2 Z3              A
 3 X-5 Y5 Z0       RA              REM:TOOL CHANGE
 4 TOOL 1
 5 FEED 28
 6 X0 Y0           RA
 7 V11 0                           REM:X AXIS POLAR CENTER
 8 V12 0                           REM:Y AXIS POLAR CENTER
 9 V13 45                          REM:INDEX ANGLE
10 V15 90                          REM:AMOUNT OF ROTATION
11 G51                             REM:INSTITUTES ROTATION
12 CALL 1                          REM:MILL SLOT #1
13 G51                             REM:INITIATE A POLAR ROTATION
14 CALL 1                          REM:MILL SLOT #2
15 G51                             REM:INITIATE A POLAR ROTATION
16 CALL 1                          REM:MILL SLOT #3
17 G51                             REM:INITIATE A POLAR ROTATION
18 CALL 1                          REM:MILL SLOT #4
19 G52                             REM:CANCEL POLAR ROTATION
20 TOOL 0                          REM:CANCEL TOOL OFFSET
21 Z0              RA              REM:HOME Z AXIS
22 X-5 Y5          RA              REM:TOOL CHANGE
23 END
24 SUBR 1
25 X-.5 Y2 Z0      RA              REM:POSITION SLOT 1
26 Z-.360          FA              REM:FEED Z TO DEPTH
27 X.5             FA              REM:MILL SLOT #1
28 Z0              RA              REM:RAISE SPINDL
29 END
```

FIGURE 12–7
Polar rotation program for part in Figure 12–5, Machinist Shop Language

PROGRAM EXPLANATION

(Refer to Figure 12–7.)

A program for this part could be written in several different ways. The example that follows is one fairly simple way to demonstrate not only the polar rotations involved but also the value of subroutine programming.

EVENTS 1–5

- These events assign the tool information and feed rate.

EVENT 6
- X0/Y0—Coordinates of the center of the slot bolt circle diameter.
- R—Specifies rapid movement.
- A—Specifies absolute positioning.

EVENT 7
- V11—Code for the X-axis coordinate of the center of the rotation.
- 0—X-axis coordinate for the rotation center. Had the X0/Y0 point been the lower left corner, the value 7.5 would have been entered with the V11. In this example, the center of rotation is conveniently the X0/Y0 point.

EVENT 8
- V12—Code for the Y-axis coordinate of the center of rotation.
- 0—Y-axis center coordinate.

EVENT 9
- V13—Code for the index angle.
- 45—Index angle in degrees. The angle is measured from the + X axis (3 o'clock position), counterclockwise to the first slot.

EVENT 10
- V15—Code for defining subsequent rotations.
- 90—All G51 commands issued after the initial one will rotate the co-ordinate system 90 degrees.

EVENT 11
- G51—Code to initiate a polar rotation. The coordinate system has now in effect rotated 45 degrees counterclockwise.

EVENT 12
- CALL 1—Calls up subroutine #1. The coordinates for milling the slot are contained in the subroutine. Slot #1 is milled in this event.

EVENT 13
- G51—Institutes the next polar rotation. The rotation will be 90 degrees, as specified in the V15 register, from the current coordinate system lo-cation.

EVENT 14
- CALL 1—Calls the subroutine; slot #2 is milled.

EVENT 15
- G51—Initiates the third polar rotation.

EVENT 16
- CALL 1—Calls the subroutine; slot #3 is milled.

EVENT 17
- G51—Initiates the fourth polar rotation.

EVENT 18
- CALL 1 — Calls the subroutine; slot #4 is milled.

EVENT 19
- G52 — Cancels polar rotation, returning the machine to its normal positioning.

EVENT 20
- TOOL 0 — Cancels the active tool offset.

EVENT 21
- Z0 — Retracts the spindle.

EVENT 22
- X − 5 Y5 — Tool change location coordinates.

EVENT 23
- END — Signals the end of the main program, resetting the computer memory.

EVENT 24
- SUBR 1 — Defines the beginning of subroutine #1.

EVENT 25
- X − .5 — X coordinate to position the tool at one end of the slot.
- Y 2 — Y coordinate to position the tool on the slot centerline.
- Z0 — Rapids the tool to the buffer zone.

EVENT 26
- Z − .36 — Z-axis coordinate to feed the tool to milling depth.
- F — Specifies a feedrate move.
- A — Specifies absolute positioning.

EVENT 27
- X.5 — Coordinate to mill from one end of the slot to the other.

EVENT 28
- Z0 — Z coordinate to retract the spindle to the start of the buffer zone (the tool offset is active).

EVENT 29
- END — Signals the end of the subroutine.

Word Address Format

To demonstrate polar rotation in word address, the same machining strategy just demonstrated in Machinist Shop Language will be used. The coding format here is designed to be generic for the purposes of instruction. Every controller uses a different coding method for polar rotations, and many controllers do not offer the capability. Polar rotation is used generally on three-axis machinery to compensate for the lack of a fourth rotary axis. The format for word address polar rotations used in this book is:

G61 X.... Y.... A... D... L..
Programming information
G60

Where G61 is the code to institute polar rotation, X.... is the X axis center of rotation, Y.... is the Y axis center of rotation, A.... is the index angle measured in degrees from the X axis, D.... is the subsequent amount of rotation measured in degrees, L.... is the number of rotations to be performed, and G60 is the code to cancel the rotation.

PROGRAM EXPLANATION

(Refer to Figure 12–8.)

N010–N040
- These blocks assign the tool information, speed, and feedrate and turn the spindle on clockwise.

N050
- X0 Y0 — Coordinates of the bolt circle diameter of the slots. The subroutine is designed to start from this location.

N060
- Z0 — Rapids the spindle to the rapid level.
- M08 — Turns on the coolant.

N070
- G61 — Initiates the first polar rotation. The first rotation will be to the index angle.
- X0 — Defines the X0 position as the X-axis center of the polar rotation.
- Y0 — Defines the Y0 position as the Y-axis center of the polar rotation.
- A45 — Defines the index angle as 45 degrees.
- D90 — Defines the rotations to occur after the initial rotation to the index angle as 90 degrees.
- L4 — Tells the MCU that four polar rotations will be performed.

N080
- P180 M98 — Instructs the MCU to jump to subroutine. Slot #1 is milled.

N090
- G61 — Initiates the second polar rotation.

```
X0/Y0 = CENTER OF SLOT BOLT CIRCLE
TOOLS:  .250 DIA. END MILL
BUFFER:  .100 IN.
CLEARANCE:  3.000 IN. MIN.


N010 G00 G40 G49 G90 G80              REM:SAFETY LINE
N020 G10 H01 Z3.0000
N030 G45 H01
N040 S3500 F28 M03
N050 X0 Y0
N060 Z0 M08
N070 G61 X0 Y0 A45 D90 L4             REM:INSTITUTE 1ST ROTATION
N080 P180 M98                         REM:JUMP TO SUBR MILL #1
N090 G61                              REM:INITIATE 2ND ROTATION
N100 P180 M98                         REM:JUMP TO SUBR MILL #2
N110 G61                              REM:INITIATE 3RD ROTATION
N120 P180 M98                         REM:JUMP TO SUBR :MILL #3
N130 G61                              REM:INITIATE 4TH ROTATION
N140 P180 M98                         REM:JUMP TO SUBR MILL #4
N150 G00 G60 G49 Z0 M09               REM:RETRACT Z, CANCEL ROTATION
N160 X-12 Y8 M05
N170 M30
:180 N010 G00 X-.5 Y2 Z0              REM POSITION TO SLOT
N020 G01 Z-.36                        REM:FEED Z TO DEPTH
N030 X.5                              REM:MILL SLOT
N040 G00 Z0                           REM:RAISE SPINDLE
N050 M99                              REM:RETURN TO MAIN PRGM
```

FIGURE 12-8
Polar rotation program for part in Figure 12-5, word address format

N100

- P180 M98—Second jump to subroutine. Slot #2 is milled.

N110

- G61—Initiates the third polar rotation.

N120

- P180 M98—Third jump to subroutine. Slot #3 is milled.

N130

- G61—Initiates the fourth rotation.

N140

- P180 M98—Jumps to subroutine to mill slot #4.

N150

- G00—Selects rapid traverse mode.
- G60—Cancels the polar rotation.
- G49—Cancels the tool offset.
- Z0—Retracts the spindle.
- M09—Turns off the coolant.

N160

- X – 12 Y8 — Coordinates of park position.
- M05 — Turns off the spindle.

N170

- M30 — Signals the end of the program.

:180

- :180 — Identifies this as main program block 180.
- N010 — Further identifies this as subroutine block 010.
- X – .5 Y2 — Polar coordinates of a slot, positioning the tool at one end.
- Z0 — Rapids the spindle to the rapid level.

N020

- G01 — Selects feedrate mode.
- Z – .36 — Z-axis milling depth.

N030

- X.5 — Polar coordinate to feed the tool from one end of the slot to the other.

N040

- G00 — Selects rapid traverse mode.
- Z0 — Retracts spindle to rapid level (tool offset is active).

N050

- M99 — Return to main program command.

HELICAL INTERPOLATION

Helical interpolation is another useful feature of CNC machinery. *Helical interpolation* allows circular interpolation to take place in two axes (usually X and Y), while subsequently feeding linearly with the third (usually Z). This makes possible the milling of helical pockets and threads.

Figure 12–9 shows a part on which a 1.000-20 thread is to be machined. An oddly shaped part like this can be cut on a CNC machine as easily as setting it up on a face plate on a lathe or a four-jaw chuck. In the case of a production run, machining this part on the mill eliminates the need for extra fixturing. It can be threaded in the same setup used to mill it to shape. The programs presented here will assume that the part has been cast separately, however, leaving only the thread to be milled. The thread will be cut by circular interpolation with the X and Y axes, while feeding with the Z axis.

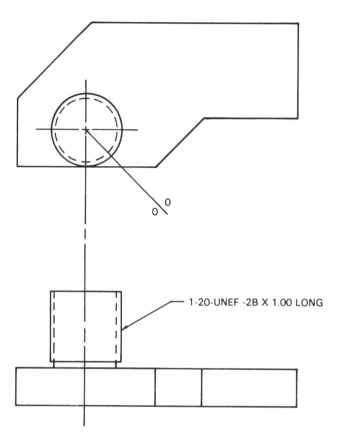

FIGURE 12–9
Part drawing

A special type of milling cutter, called a thread hob, will be used to mill the thread. The hob will be sent to a start position, fed into the workpiece, then helically interpolated for three turns. The cutter will then be withdrawn from the part. With each turn the hob makes around the part, the Z axis will advance downward an amount equal to the lead of the thread.

The lead of the thread can be determined by the formula:

$$L = P \times I$$

Where L is the lead of the thread, P is the pitch of the thread, and I is the number of leads on the thread. The pitch of a thread is 1 divided by N, where N is the number of threads per inch. For a 20 thread, the pitch is 1 divided by 20 or

.050 inch. The lead for a single lead 20 thread is .050 times 1, or .050. This means that the thread will advance .050 inch in one revolution. Note that the value of the lead and the pitch on a single lead thread are identical; however, the lead and pitch are not the same thing.

A 60-degree thread milling cutter is used to mill the thread on the part in Figure 12–9. Set up as shown in Figure 12–10.

Machinist Shop Language

The format for helical interpolation in Machinist Shop Language is:

ARC/DIRECTION—Clockwise or counterclockwise.
X.... Y....—Centerpoint of the arc.
V42—Number of 360-degree arcs to be cut (if less than 1, 0 is entered).
X.... Y.... Z....—Endpoint of the helix given in all three axes.
ARC—Code to initiate the arcs.

For the thread to be milled into this part, the coding for the helical interpolation is:

ARC/CW
X0 Y0
V42 2
X.9694
ARC

The program to mill the thread is given in Figure 12–11.

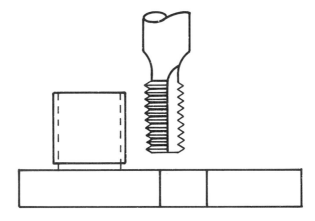

FIGURE 12–10

```
XO/YO = CENTER OF THREAD DIAMETER
TOOL CHANGE = X-2 Y2
TOOLS:  1.000 DIA. THREAD HOB
CLEARANCE:  3.000 MIN.
BUFFER:  ZERO BUFFER

 1 TOOL 1001
 2 X2 Z3              A          REM:TOOL CHANGE
 3 X-2 Y2             RA
 4 TOOL 1
 5 FEED 7
 6 X1.6 YO ZO         RA         REM:POSITION X/Y TO START
 7 Z-.818             RA         REM:POSTION Z TO START
 8 X.9694             FA         REM:FEED INTO PART
 9 ARC/CW
10 V42 3                         REM:SET NUMBER OF ARCS TO 3
11 XO YO                         REM:ARC CENTER
12 X.9694 YO Z.968 FA           REM:ARC ENDPOINTS
13 ARC                           REM:INITIATE ARC CUT
14 X1.6               FA         REM:WITHDRAW FROM PART
15 TOOL O
16 ZO                 RA         REM:RETRACT Z
17 X-2 Y2             RA         REM:TOOL CHANGE
18 END
```

FIGURE 12–11

Helical interpolation program for the thread in Figure 12–9, Machinist Shop Language

PROGRAM EXPLANATION

(Refer to Figure 12–11.)

This program mills the thread in one pass using two turns of the cutter.

To determine the coordinates for the thread depths, subtract the thread depth from half the major diameter of the thread. Thread depths can be found in a machinists' handbook. In this case, the depth of the thread is .03066. The final depth of .9694 is arrived at by subtracting the thread depth of .03066 from the radius of the thread (half the major diameter), which is .500. This leaves .4694 from the center of the arc to the root of the thread. Since the cutter is 1.000 inch in diameter, a radius of .500 must be added to .4694 to arrive at the proper cutter coordinate of .9694.

The number of turns of the thread is set to 20 using the V42 code.

EVENTS 1–5

- Assign the tool offset values and feedrate.

EVENT 6

- X1.6 Y0 Z0 RA — Position the cutter near the part.

EVENT 7

- Z − .818 RA — Positions the Z axis to the proper starting position.

EVENT 8

- X.9694 FA — Feeds the cutter into the workpiece to the thread minor diameter.

EVENT 9

- ARC/CW — Tells the MCU that a clockwise arc is to be cut.

EVENT 10

- V42 — V code signaling the number of arcs to be cut.
- 3 — The number arcs.

EVENT 11

- X0 Y0 — Arc centerpoint coordinates.

EVENT 12

- X.9694 Y0 Z.968 — Endpoint coordinates of the arc.

EVENT 13

- ARC — Initiates helical interpolation.

EVENT 14

- X1.6 FA — Withdraws the cutter from the part at feedrate.

EVENT 15

- TOOL 0 — Cancels the tool offset.

EVENT 16

- Z0 — Retracts the Z axis.

EVENT 17

- X − 2 Y2 RA — Rapids to the tool change position.

EVENT 18

- END — Signals end of program.

Word Address Format

Not every CNC machine has helical interpolation. It is usually an optional feature, purchased at additional cost. Helical interpolation in word address can be accomplished in any one of three plane combinations (X/Y, Z/X, or Y/Z). To select the planes, the following G codes are used:

G17 — X/Y plane.
G18 — X/Z plane.
G19 — Y/Z plane.

The format for helical interpolation in word address is as follows:

- *For the X/Y plane* — G17 G02/G03 X.... Y.... I.... J.... Z.... F...
- *For the X/Z plane* — G18 G02/G03 X.... Y.... I.... K.... Z.... F...
- *For the Y/Z plane* — G19 G02/G03 X.... Y.... J.... K.... Z.... F...

Where: G17, G18, and G19 select the plane, G02 and G03 select the direction of helical interpolation (G02 clockwise, G03 counterclockwise), X, Y, and Z are the arc endpoint coordinates, I, J, and K are the arc centerpoint coordinates, F sets the Z-axis feedrate.

To mill the part in Figure 12−9, the word address program in Figure 12−12 can be used. This program is identical in operation to the Machinist Shop Language example.

PROGRAM EXPLANATION

(Refer to Figure 12−12.)

N010−N040

- These blocks assign the tool offset and set spindle speed.

N050

- G00 X1 .6 Y0 — Position the cutter to the start position.

N060

- Z − .818 — Position Z to the starting depth.

```
XO/YO = CENTER OF THREAD DIA.
TOOLS:   1.000 IN. DIA. THREAD HOB
CLEARANCE:  3.000 MIN.
BUFFER:   ZERO BUFFER

N010 G80 G40 G49 G90 G98                    REM:SAFETY LINE
N020 G10 H01 Z2.0000
N030 G45 H01
N040 S800 M03
N050 G00 X1.6 YO ZO                         REM:POSITION TO START
N060 Z-.818                                 REM:Z TO START POINT
N070 G01 X.9694 F7.00                       REM:FEED INTO PART
N080 G17 G02 X.9694 YO Z-.8680 IO JO        REM:1ST TURN
N090 G02 X.9694 YO Z-.9180 IO JO            REM:2ND TURN
N100 G02 X.9694 YO Z-.9680 IO JO            REM:3RD TURN
N110 G01 X1.6                               REM:FEED OUT OF PART
N120 G00 G49 ZO M09                         REM:RETRACT Z
N130 X-12 Y8 M05                            REM:TOOL CHANGE
N140 M30
```

FIGURE 12−12
Helical interpolation program for the thread in Figure 12−9, word address format

N070

- G01 X.9694 F7.00 — Feeds the cutter into the workpiece to the minor diameter of the thread at a feedrate of 7 inches per minute.

N080

- G17 — Selects the XY plane for interpolation.
- G02 — Selects clockwise interpolation.
- X.9694 Y0 Z − .8680 — The endpoint coordinates of the first arc.
- I0 J0 — The centerpoint coordinates of the arc.

N090

- G02 — Selects clockwise interpolation.
- X.9694 Y0 Z − .9180 — The endpoint coordinates of the second arc.
- I0 J0 — The centerpoint coordinates of the arc.

N100

- G02 — Selects clockwise interpolation.
- X.9694 Y0 Z − .9680 — The endpoint coordinates of the third arc.
- I0 J0 — The centerpoint coordinates of the arc.

N110

- G01 X1 .6 — Withdraws the cutter from the part at feedrate.

N120

- G00 G49 Z0 — Cancels the tool offset, retracting the Z axis at rapid.
- M09 — Turns off the coolant.

N130

- X − 12 Y8 — Rapids to tool change position.
- M05 — Turns off the spindle.

N140

- M30 — Signals end of program.

SUMMARY

The important concepts presented in this chapter are:

- Mirror imaging means changing the sign (+ or −) of an axis movement.
- Mirror imaging is used in a program to save repetitive programming when the direction of movement is the only difference between part features.
- Mirror imaging is normally used in conjunction with subroutines or do loops.
- Polar rotation is an indexing of the NC machine's Cartesian coordinate system to some angle other than its normal state.

- Polar rotation may be used to perform operations that otherwise would require the use of a rotary axis or lengthy coordinate calculations.
- Polar rotations may be used in conjunction with do loops or subroutines.
- Helical interpolation is circular interpolation with two axes while simultaneously feeding at a linear rate with the third. The result of this type of operation is a helix.
- Care must be taken in calculating the number of turns and the lead of a helix, be it a thread or other type of part.
- Helical interpolation may be used inside of or in conjunction with do loops and subroutines.

REVIEW QUESTIONS

1. What is mirror imaging? Why is it used?
2. When would mirror imaging be used in a program?

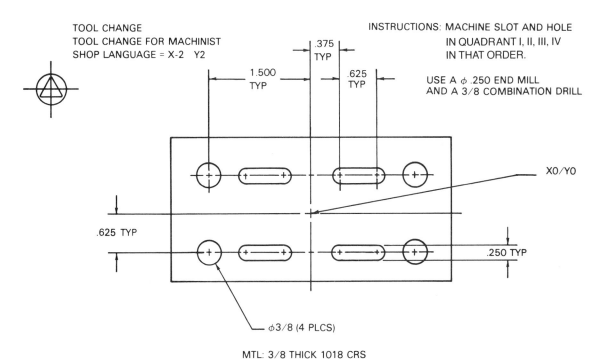

FIGURE 12-13
Part drawing for review question #3

3. Write a program to mill the slots and drill the holes in the part shown in Figure 12–13:
   a. In Machinist Shop Language.
   b. In word address.
4. What does polar rotation do?
5. What types of equipment can polar rotations substitute for?
6. Can polar rotations be used with subroutines and do loops?
7. What type of information must be given in the program for the MCU to perform polar rotation?
8. Write a program to mill the slots in the part shown in Figure 12–14:
   a. In Machinist Shop Language.
   b. In word address.
9. What is helical interpolation?
10. When would helical interpolation be useful in a program?

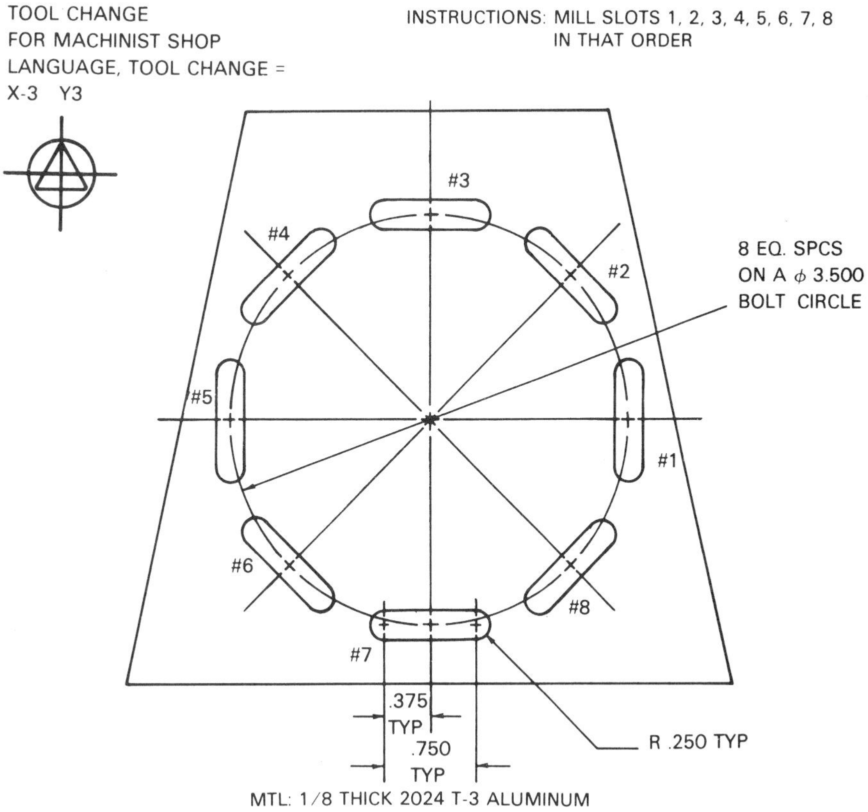

TOOL CHANGE
FOR MACHINIST SHOP
LANGUAGE, TOOL CHANGE =
X-3   Y3

INSTRUCTIONS: MILL SLOTS 1, 2, 3, 4, 5, 6, 7, 8
IN THAT ORDER

#3
#4
#2
8 EQ. SPCS
ON A φ 3.500
BOLT CIRCLE
#5
#1
#6
#8
#7
.375 TYP
.750 TYP
R .250 TYP
MTL: 1/8 THICK 2024 T-3 ALUMINUM

FIGURE 12–14
Part drawing for review question #8

CHAPTER 13

The Numerical Control Lathe

OBJECTIVES Upon completion of this chapter, you will be able to:

- Describe the difference between a conventional lathe bed arrangement and a slant bed arrangement, listing the advantages of the slant bed for NC.
- Explain axis movement on a CNC lathe.
- Describe the method of toolholding used on CNC turning machines.
- Explain what a tool offset number is.
- Describe two methods of tool selection used on CNC turning machines.
- Describe how spindle speed is designated on gear head and variable speed lathes.
- Explain how feedrates are specified on CNC turning equipment.
- Define TNR.

Up to this point, the programming features of CNC mills have been discussed, but numerical control is used for turning equipment as well. In the milling examples, both Machinist Shop Language and word address formats were given. For the turning programs discussed in Chapter 14, only word address format will be used. The coding will be a version used with Fanuc lathe controllers, designed to be generic and so to illustrate the basic programming steps involved. A numerical control lab in a school will have equipment that differs in one way or another from that presented here. Students are advised to familiarize themselves with the codes used for the machines they will be using.

LATHE BED DESIGN

Older NC lathes, and those that have been converted to numerical control with retrofit units, look like traditional engine lathes. The lathe carriage rests on the ways. The ways are in the same plane and are parallel to the floor, as illustrated in Figure 13–1. This arrangement allows the machinist to reach all the controls readily. Since the CNC lathe performs its operations automatically,

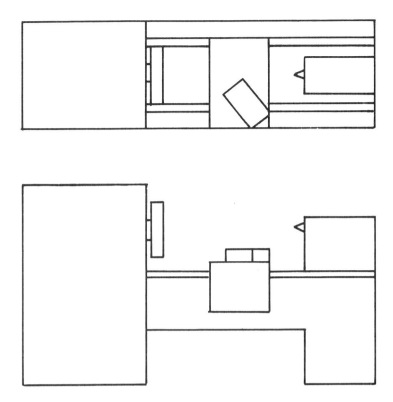

FIGURE 13-1
Bed arrangement on a conventional lathe

this type of arrangement is not necessary. In fact, it is quite awkward, as the operator will be busy with other responsibilities while the program is running and so will not necessarily be there to brush the chips off the ways. In a conventional lathe bed arrangement, the chips have nowhere to fall except on the ways. To overcome this problem, many CNC lathes make use of the slant bed design illustrated in Figure 13–2.

On many NC lathes, the turret tool post is mounted on the opposite side of the saddle, compared to a conventional lathe, to take advantage of the slant bed design. The slant bed allows the chips to fall into the chip pan, rather than on tools or bedways. Despite its odd appearance, the slant bed NC lathe functions just like a conventional lathe. Figures 13–3 and 13–4 show modern CNC turning machines. Notice the slant bed arrangement.

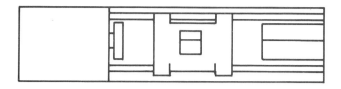

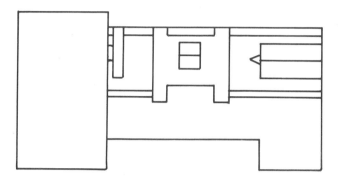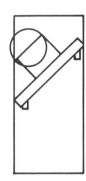

FIGURE 13-2
Slant bed for NC or CNC lathe

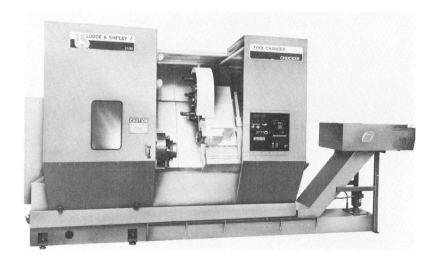

FIGURE 13-3
A modern CNC turning center employing automatic tool change *(Photo courtesy of Lodge and Shipley Co.)*

FIGURE 13−4
A four-axis CNC turning center *(Photo courtesy of Cincinnati Milacron)*

AXIS MOVEMENT

The axis movement of a basic CNC lathe is diagrammed in Figure 13−5. Some turning machines, such as that shown in Figure 13−4, are four-axis machines. In this book, only the basic two-axis machine is programmed. The programming concepts learned on a two-axis machine are the foundation necessary to program more complex machinery.

The basic lathe has only two axes, X and Z. Since the Z axis is always parallel to the spindle, longitudinal (carriage) travel is designated Z. The cross slide movement is designated X, since it is the primary axis perpendicular to Z. If it were possible to move the carriage up and down, that axis would be Y. There is, however, a potential problem with this arrangement. There appear to be two Z axes: the carriage movement and the tailstock movement. To eliminate this problem the tailstock is usually called the W axis on lathes with programmable tailstocks. Programmable tailstocks, which are rear turret assemblies on CNC equipment, are the third and sometimes fourth axis on more complex equipment. The turning center in Figure 13 − 4 has two programmable saddles. In such cases the axes of the second saddle are usually designated W and U, with

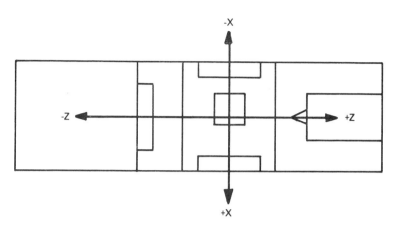

FIGURE 13–5
Lathe axis movement

W being saddle travel and U being cross slide travel. There are some imported lathes on which the X-axis direction is reversed. The programmer must determine if such a situation exists before writing the lathe program.

TOOLHOLDERS AND TOOL CHANGING

Either a rigid toolholder or a tool turret is used to hold the tools on an NC lathe. Figure 13–3 shows a CNC chucker employing a rigid toolholder. The turning center in Figure 13–4 employs a tool turret, in which the various tools needed for lathe operations are placed in toolholders. When a tool change is necessary, the appropriate turret is indexed to the next tool needed. Simple lathes use six-sided turrets; larger turning machines use eight-, ten-, and twelve-sided turrets.

With the development of robotics, new tool changing and work handling schemes are appearing. Figure 13–6 shows a robot arm used for handling workpieces, and Figure 13–7 illustrates the robot in operation. To teach the basics of CNC programming, this text will focus on nonrobotic tool change.

The toolholders used on NC turning machines are of very rigid design. The tools used for turning are of the carbide insert type, made to much more exacting tolerances than conventional lathe insert tooling.

A tool change command in a turning program either changes the turret position or causes an automatic tool change, depending on the type of machine used.

FIGURE 13-6
A robot arm used for part load and unload *(Photo courtesy of Cincinnati Milacron)*

FIGURE 13–7
A robot arm in action *(Photo courtesy of Cincinnati Milacron)*

Automatic Tool Change

In a CNC turning program for a machine with a rigid toolholder, M06 is used to initiate an automatic tool change. The T address is used (as it is in milling programs) to specify the desired tool. The T address also calls up the tool offsets. The format for automatic tool change is:

M06 T n1 n2

Where M06 initiates the tool change, T is the tool address, n1 is the tool number, and n2 is the tool offset number.

Turret Position

T is used in a similar manner with turret tool selection. The format is:

T n1 n2

Where the first number is the turret position and the second is the tool offset number.

Since one tool may be used in several positions, a turret position is used rather than a tool number. The turret position corresponds to the turret station number. T01 will index the tool in station one into position. Some NC lathes can utilize more than one tool on a single station. It is possible, therefore, for T0101 to refer to one tool and T0111 to refer to another. This is referred to as piggy-backing a tool station.

One other point should be kept in mind when changing tools: the carriage (or tailstock) does not necessarily move to a tool change location. It is often nec-essary, therefore, first to move the carriage or tailstock turret out of the way be-fore making a tool change. It may also be necessary to program a dwell (G04) to halt the program, giving the tool time to index to position safely.

Tool Offset Numbers.   Each turning tool used on a lathe has a radius. When programming the coordinates for a location, the centerline of the tool radius is programmed. Tool offsets allow the center of the tool nose to be programmed and thus compensate for minor differences in length that exist between tools, and the effects of tool wear.

When the tools are set up, the operator enters the offsets into the tool reg-isters. When the offsets are active, the MCU compensates for them, eliminating the need for premeasured tooling. In this manner, the programmer can treat all tools as being the same length, just as in milling.

The *offset number* is the number of the register in which a particular tool's offset is stored. Generally the register number will match the tool number.

The tool information entered manually prior to the start of the program is entered in this form:

X Offset
Z Offset
Tool Nose Radius
Standard Tool Nose Vector Number

Tool Nose Radius and Standard Tool Nose Vector Numbers

The tool nose radius and tool nose vector numbers are optional. They are entered if using cutter comp. Cutter diameter compensation is called *tool nose radius compensation* (TNR comp) on turning machines. The tool radius tells the MCU the amount of compensation that is to be used. With NC machining cen-ters this value was entered in a comp register.

TNR comp is utilized just as cutter comp was in Chapter 10. It can be used to program the part line, or fine tune the tool path to compensate for tool wear. The major difference is lathe tools are not completely circular as is a milling cut-ter. To aid in proper compensation of the tool path and correctly identify alarm conditions, a tool nose vector number is entered in the register. Tool nose vector numbers tell the MCU the orientation of the tool nose. Figure 13–8 shows the various directions in which a tool may be oriented. These directions are referred to as vectors. Each vector has a number associated with it that is used to de-scribe the tool orientation to the MCU.

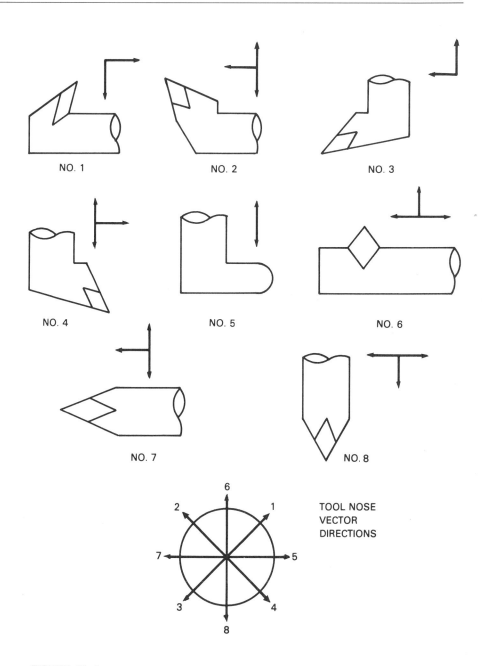

NO. 1   NO. 2   NO. 3

NO. 4   NO. 5   NO. 6

NO. 7   NO. 8

TOOL NOSE
VECTOR
DIRECTIONS

FIGURE 13–8
Tool nose vectors

Tool Edge Vs Centerline Programming

The tool nose may be programmed in one of two ways when TNR comp is not active: by the tool edge or by the tool nose radius centerline.

Tool edge programming is adequate for simple straight line cuts where the part surfaces intersect each other at right angles. Problems are encountered, however, when angles and especially arcs are programmed this way. Figure 13–9A illustrates this point. If the tool edge is programmed, the I and K center-points of the illustrated arc must be shifted. This results in a tool path that does not follow the desired arc exactly. The amount of error that is induced depends upon the size of the cutter and the radius of the arc. In any case, tool edge programming should not be used when encountering arcs and angles.

Tool centerline programming is identical to the centerline programming done when milling. Figure 13–9B demonstrates how the cutter centerlines and part surface centerlines coincide when the center of the tool nose radius is programmed. This type of programming is demonstrated in Chapter 14.

SPINDLE SPEEDS

Spindle speed is specified using an S address, just as in milling. On turning machines with a gear head design, the spindle speed is changed by shifting gears in the headstock. On gear head machinery, there are usually two or more gear ranges. An M function is used to select the gear range in which the desired speed is located. M40 through M46 generally serve this purpose. For gear head examples in this text, M40 will be used for low range, M41 for mid range, and M42 for high range.

The following chart shows a sample of speed ranges for gear head machines. This chart is not for a particular machine but is representative of the type of spindle speed spread found on a machine.

LOW RANGE

10	15	20	25
30	40	50	65
75	90	110	125

MEDIUM RANGE

55	70	95	120
140	155	175	200
235	260	290	300

HIGH RANGE

285	335	380	450
530	660	900	1200
1800	2100	2500	3000

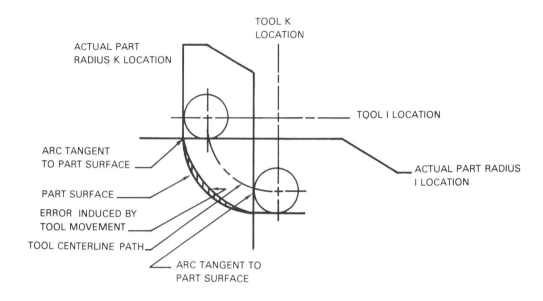

A. Error induced by programming tool edge

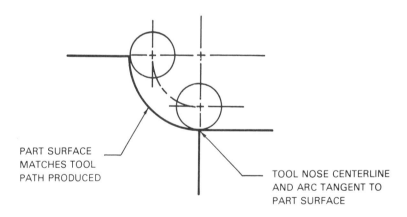

B. Tool nose centerline programming

FIGURE 13–9

Some CNC turning machines use a variable speed drive with which an infinite number of speeds are available between the highest and lowest speeds. In these cases the speed is selected using the S address as it is in milling.

FEEDRATES

With a CNC lathe, assigning feedrates is quite simple. A G98 or G94 code (depending on the controller) tells the MCU that the following feedrate is in inches per minute. For example, G98 F7 specifies a feedrate of 7 inches per minute. A G99 or G95 in a turning program specifies a feedrate in inches per revolution. For example, G99 F.015 specifies a feedrate of .015 inch per revolution. Appendix 3 contains a list of common G codes used in lathe programming.

MACHINE ORIGIN AND WORK COORDINATE SYSTEMS

An NC lathe generally has a fixed-zero position assumed by the executive program upon power-up. This position is known as the home zero or machine origin. The physical location of this position varies from controller to controller and machine model to machine model. It is usually one of two locations: X0 = centerline of the spindle, Z0 = the chuck mounting surface of the spindle, or X0 = extreme X+ location, Z0 = extreme Z+ location. It is usually necessary to establish a zero point on the part different from the machine origin location. This position is called the *work coordinate system* or *part zero.* There are two methods used to accomplish this.

The first method involves the use of an axis preset command—G50. The G50 transfers the zero point from the home zero to the coordinates specified with the command. The format for a G50 command is:

$$G50\ Xxx.xxxx\ Zzz.zzzz$$

Where: G50 = the axis preset command
          xx.xxxx = the X-axis distance to the part zero
          zz.zzzz = the Z-axis distance to the part zero

A G50 command is issued at the start of each tool. Since the programmer will not know the axis preset distances in advance, zeros should be used or some other prearranged value in the G50 line. The actual values will be determined by the setup person and edited in the control when the job is set up.

The second method uses registers called *work coordinates.* These are registers in the MCU that tell the MCU the distance from home zero to the part zero. If a machine has more than one available work coordinate, multiple zero points may be used for complex programming. Another advantage to multiple work coordinates is the ability to have more than one program loaded in the MCU, each with its own work coordinate. This is a decided advantage when running several repeating jobs through a turning center. Each work coordinate is called by a G code. If a program were to use four work coordinates, they would be selected by the codes G54, G55, G56, and G57. The first work coordinate (G54 in this case) is the default work coordinate. This work coordinate is automatically activated upon power-up. If using only the default work coordinate, the G code may be omitted. The work coordinate values are entered by the setup person when the job is prepared. The programmer must instruct the setup personnel the position on the part of the part zero location.

QUICKSETTERS

A fairly recent development has been the use of quicksetters — arms with tool sensors on them. During job setup the arm is lowered into position, the operator jogs a tool to the presetting position and touches it off on the sensor. The quicksetter automatically sets the values of the work coordinate and the tool offset registers.

SUMMARY

The important concepts presented in this chapter are:

- CNC turning machines often use a slant bed arrangement to protect the machine ways from chips. Although different in appearance, the functioning of a slant bed and conventional bed machine is identical.
- There are two basic axes, X and Z, on a CNC lathe. If the lathe has additional axes, they are generally designated U and W.
- TNR stands for tool nose radius compensation. TNR is the equivalent in CNC turning to cutter diameter compensation in milling.
- A tool turret or a rigid toolholder is used to hold the tools on an NC lathe.
- Tool offsets are entered into the MCU prior to running the program to compensate for minor setup adjustments.

- A standard tool nose vector number is used to identify the orientation of a particular tool when using TNR.
- A tool change command in turning programs will either change the turret position or cause an automatic tool change, depending on the type of machine used.
- The tool change format for turret changing is: T n1 n2
  Where T is the tool change command, n1 is the turret position and n2 is the tool offset number.
- The format for automatic tool change is: M06 T n1 n2
  Where M06 initiates the tool change, T is the tool address, n1 is the tool number, and n2 is the tool offset number.
- Spindle speeds are specified directly using the S address. On gear head machines, it is necessary to specify the gear range when selecting a range outside the active one.
- Feedrates on CNC lathes can be specified either in inches per minute (using G94 or G98), or in inches per revolution (using G95 or G99).
- To set a part at X0/Z0 point, it is necessary to transfer the machine origin to the workpiece using a G code.

REVIEW QUESTIONS

1. What is the difference between a slant and conventional lathe bed arrangement? What is the advantage of a CNC slant bed lathe?
2. Draw a sketch illustrating the axis movement on a lathe.
3. What type of toolholding is used on CNC turning machines?
4. What is the purpose of a tool offset register?
5. What is a standard tool nose vector number?
6. What types of turrets in addition to the four-sided turret are used on CNC lathes?
7. How are spindle speeds designated on CNC turning machines? What additional coding is required on gear head machines?
8. How are feedrates specified on CNC lathes? What codes are used?
9. What is the format for a turret tool change command? For an automatic tool change?
10. What does TNR stand for?
11. What is the machine origin?
12. How is an X0/Y0 point established on a workpiece?

CHAPTER 14

Programming CNC Turning Machines

OBJECTIVES Upon completion of this chapter, you will be able to:

- Write simple turning and facing routines.
- Write simple taper turning routines.
- Write simple routines to perform circular interpolation, using programmed arc centers and programmed radius value methods.
- Write simple thread-turning routines using single pass and multipass threading.

CNC lathe controllers vary in their coding to an even greater extent than mill controllers. It is, therefore, difficult to discuss programming practices. EIA standards specify axis movement, for example, but some lathes use a left-hand coordinate system, with the X and Z axes reversed from the standard configuration. Other lathes reverse the X axis direction and not the Z. On lathes using twin turrets, the X axis is often reversed. The uses of coding and the cycles available also differ to a large extent. The EIA codes pertaining to lathes are generally used, but many other codes may be added.

This chapter will discuss basic lathe programming routines for turning, facing, taper turning, circular interpolation, and thread cutting. Each routine is placed in a miniprogram. Each program can be thought of as a building block; to machine a complete part, these building blocks can be linked together in one program as will be demonstrated.

MACHINE REFERENCE POINT

A machine *reference point* is a fixed position on the machine. Upon receiving the proper G code, the machine automatically returns to the reference point location. This point is often the home zero location, used for tool changing and as a park position at the end of the program. Often it is necessary to send the tool back to the reference point by way of another point, called an *intermediate point.* The code used in this chapter to return the tool to reference is G28. An X and Z coordinate for the intermediate point are specified along with the G28. Upon receiving the G28, the tool moves to the intermediate point and then proceeds to the reference point.

DIAMETER VS RADIUS PROGRAMMING

The difference between radius programming and diameter programming is an important one. *Diameter programming* references the X-axis coordinate to the diameter of the workpiece. This means that every .001 inch programmed moves the tool .0005 inch as measured radially. If the X axis advances .500 inch into the part, .500 inch is removed from the diameter. To accomplish this, the X axis moves only .250 inch, or half the programmed amount.

In *radius programming,* the X axis moves the programmed amount. If .500 inch of movement along the X axis is programmed, the tool advances .500 inch. When the Z-axis move is made, 1.000 inch of material is removed from the part.

Canned cycles on a machine call for the information to be entered in either diameter or radius coordinates, depending on the cycle's function. The machine manual must be consulted to determine the type of coordinate expected. The coordinates may be either incremental or absolute, depending on whether G90 or G91 is active. As in milling, G90 selects absolute positioning and G91 selects incremental. Other controllers use a "W" address for incremental X and a "U" address for incremental Z.

TURNING AND FACING

Figure 14–1 shows a part to be turned and faced in a lathe. Note that the position of the tool turret relative to the X0/Y0 location and the machine origin is given. The machine coordinate system may be transferred to the workpiece either within the program by use of G codes or by the operator during machine setup. It is usually more efficient to define the work coordinate system during setup. For routines in this chapter, this will be assumed. Figure 14–2 shows a part similar to the one in Figure 14–1 but with metric dimensions. Figure 14–3 presents a short program to turn and face the part drawn in Figure 14–1. Figure 14–4 presents a metric version. To program this part, the following codes will be used:

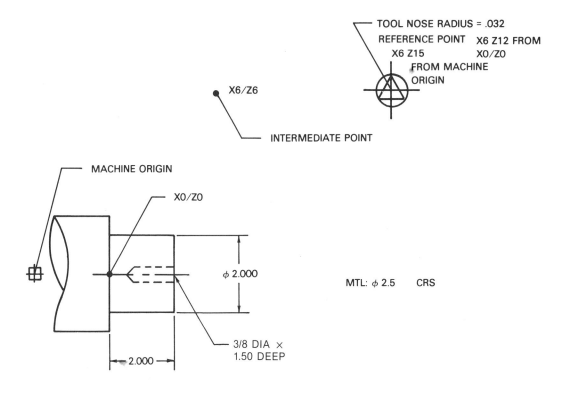

TOOL NOSE RADIUS = .032
REFERENCE POINT
X6 Z15
X6 Z12 FROM
X0/Z0
FROM MACHINE
ORIGIN

X6/Z6

INTERMEDIATE POINT

MACHINE ORIGIN

X0/Z0

φ 2.000

MTL: φ 2.5   CRS

3/8 DIA ×
1.50 DEEP

2.000

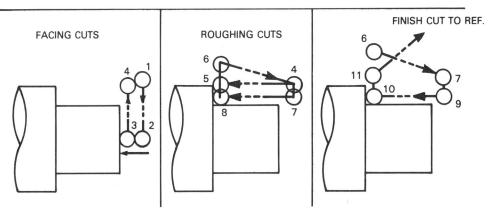

FINISH CUT TO REF.

FACING CUTS

ROUGHING CUTS

FIGURE 14–1
Part to be turned and faced in a lathe

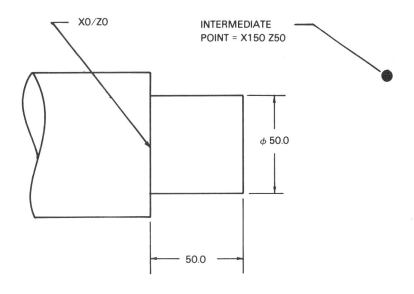

XO/ZO

INTERMEDIATE
POINT = X150 Z50

φ 50.0

50.0

MTL: φ 0.5 MM   CRS ≈ (2.5 INCHES)

FIGURE 14–2
Part from Figure 14–1 with metric dimensions

G00 — As in milling programs, G00 puts the machine in rapid traverse mode.
G01 — Linear interpolation. As with milling, the machine will position the tool to the programmed coordinates at feedrate, in a straight line.
G28 — Return to reference point. A G28 is programmed with an X and Z coordinate. Upon receiving the G28, the machine positions the tool at the fixed machine reference point, passing through the programmed X/Z location, called the intermediate point.
G70 — Selects inch input.
G71 — Selects metric input.
G90 — Selects absolute positioning.
G91 — Selects incremental positioning.
G94 — Selects inches per minute or millimeters per minute feedrate. Feedrates are treated just as milling feedrates are.

```
X0 = CENTERLINE OF SPNDL   Z0 = PART SHOULDER

N010 G00 G40 G90 G95 G28 X6 Z6 M08     REM:SAFETY LINE, COOLNT ON
N020 T0101 M42                         REM:TURRET POS, HIGH RANGE
N030 S1200 M03                         REM:SPNDL ON
N040 X2.6 Z2.042                       REM:POSITION TO #1
N050 G01 X0 F.007                      REM:FEED TO #2
N060 Z2.032                            REM:FEED TO #3
N070 X2.314 F.003                      REM:FEED TO #4
N080 Z.042 F.007                       REM:FEED TO #5
N090 X2.6                              REM:FEED TO #6
N100 G00 X2.320 Z2.132                 REM:RAPID TO #4
N110 G01 X2.0840                       REM:FEED TO #7
N120 Z.042 F.003                       REM:FEED TO #8
N130 X2.6                              REM:FEED TO #6
N140 G00 X2.084 Z2.132                 REM:RAPID TO #7
N150 G01 X2.062                        REM:FEED TO #9
N160 Z.032 F.003                       REM:FEED TO #10
N170 X2.55                             REM:FEED TO #11
N180 G00 G28 X6 Z6 M09                 REM:RETURN TO REF & COOLNT OFF
N190 M05                               REM:SPNDL OFF
N200 M30                               REM:END PRGM
```

FIGURE 14-3
Lathe facing and turning program for part in Figure 14-1, word address format, nonmetric

```
X0 = CENTERLINE OF SPNDL   Z0 = PART SHOULDER

N010 G00 G40 G90 G95 G28 X150 Z150 M08     REM:SAFETY LINE
N020 T0101 M42                             REM:TURRET POS, HIGH RANGE
N030 S1200 M03                             REM:SET SPEED
N040 X67 Z52                               REM:POSITION TO #1
N050 G01 X0 F.5                            REM:FEED TO #2
N060 Z51                                   REM:FEED TO #3
N070 X60 F.13                              REM:FEED TO #4
N080 Z2 F.5                                REM:FEED TO #5
N090 X67                                   REM:FEED TO # 6
N100 G00 X60 Z101                          REM:RAPID TO #4
N110 G01 X53                               REM FEED TO #7
N120 Z2 F.13                               REM:FEED TO #8
N130 X67                                   REM:FEED TO #6
N140 G00 X53 Z101                          REM:RAPID TO #7
N150 G01 X51                               REM:FEED TO #9
N160 Z1                                     REM:FEED TO #10
N170 X66                                   REM:FEED TO #11
N180 G00 G28 X150 Z150 M09                 REM:RETURN TO REF & COOLNT OFF
N190 M05                                   REM:SPNDL OFF
N200 M30                                   REM:END PGRM
```

FIGURE 14-4
Lathe facing and turning program for part in Figure 14-2, word address format, metric

G95—Selects inches per revolution or millimeters per revolution feedrates. The feedrates are the programmed value per revolution of the spindle. A G95 F.01 advances the tool .010 inch for every revolution of the spindle.

M40—Selects the low gear range.

M41—Selects the middle gear range.

M42—Selects the high gear range.

PROGRAM EXPLANATION

(Refer to Figures 14–3 and 14–4.)

A .032-inch tool nose radius is used on the tool in the nonmetric program. A 1-mm tool nose radius is used in the metric program. One roughing and one finish facing cut are used; two roughing and one finish turning cuts are used.

N010

- G00—Selects the rapid traverse mode.
- G40—Cancels any active tool nose radius compensation.
- G90—Selects absolute positioning.
- G95—Selects per revolution feedrate.
- G28—Causes a return to reference point.
- X/Z coordinates—Intermediate point location. The intermediate point should be chosen so that tool movement will be free of the lathe chuck and part.
- M08—Turns on the coolant.

N020

- T0101—Selects a tool number and calls the tool offset in register #1.
- M42—Selects high gear range.

N030

- S1200—Sets the spindle speed to 1200 RPM.
- M03—Turns on the spindle.

N040

- X/Z coordinates — Rapid the tool to location #1, Figure 14–1. The X axis coordinate is diameter programmed, as are all the X coordinates in this program.

N050

- G01—Selects feedrate movement.
- X0—Feeds the tool to location #2. This is the rough facing cut.
- F.007 — Sets the feedrate to .007 inch per spindle revolution (.5 mm metric.)

N060

- Z coordinate—Feeds the tool from location #2 to location #3. This sets the Z axis depth for the finish facing cut.

N070

- X coordinate—Feeds the tool from location #3 to location #4. The coordinate is diameter programmed.
- F.003 (F.13 metric)—Sets finish feedrate.

N080

- Z coordinate—Feeds the tool from location #4 to location #5. This is the first roughing pass.
- F.007 (F0.5 metric)—Sets the roughing pass feedrate.

N090

- X coordinate—To feed from location #5 to location #6. This cut rough faces the shoulder of the part and retracts the tool for the return move.

N100

- G00—Selects rapid traverse. This is a return to start of cut move. No feedrate is necessary.
- X/Z coordinates—Move the tool at rapid from location #6 to location #4.

N110

- G01—Selects linear interpolation (feedrate mode).
- X coordinate — Feeds the tool from location #4 to location #7. This move could also have been made in rapid traverse. Using a feedrate here eliminated the possibility of chipping the tool cutting edge on the corner of the stock.

N120

- Z coordinate—Feeds the tool from location #7 to location #8. This is the second rough turning pass.
- F.003 (F.13 metric)—Sets finish feedrate.

N130

- X coordinate—Rough faces the shoulder, retracting the tool.

N140

- G00—Selects rapid traverse.
- X/Z coordinate—Positions the tool to location #7.

N150

- G01—Selects feedrate movement.
- X coordinate—Feeds the tool from location #7 to location #9. This positions the X axis depth for the finish pass.

N160

■ Z coordinate — Feeds the tool from location #9 to location #10. This completes the turning.

N170

■ X coordinate — Feeds the tool from location #10 to location #11. This move finish faces the part shoulder.

N180

■ G00 — Selects rapid traverse.
■ G28 — Initiates a return to reference.
■ X/Z coordinates — Intermediate point.
■ M09 — Turns the coolant off.

N190

■ M05 — Turns the spindle off.

N200

■ M30 — Signals the end of program.

TAPER TURNING

Linear interpolation on a lathe is used to turn tapers. It is similar in use to linear interpolation to cut angles when milling. On the part pictured in Figure 14 – 5 is a taper to be bored. The part is a steel casting, requiring that the taper be rough and then finish machined. (The short program to perform these operations is shown in Figure 14 – 7.)

Cutter offset calculations necessary with taper turning are similar to those used when calculating angle cuts for milling. Figure 14 – 6 depicts the relationship of the lathe tool nose to the tapered part surfaces. Two coordinate locations require cutter offsets. Both locations present the identical situation, so that calculating one offset will automatically yield the other. This is the same simple cutter-to-angle relationship first discussed in Chapters 8 and 9, and the formula given in Appendix 6, Figure 1, can be used. In this case, the Y axis in the formula is the X axis on the lathe, and the X axis in the formula is the Z axis on the lathe. The offset is calculated as follows, where CR is the tool nose radius:

$$X = TAN \left(\frac{\theta}{2}\right) \times CR$$

$$X = TAN\, 40 \times .032$$

$$X = .8391 \times .032$$

$$X = .02685 \text{ or } .027$$

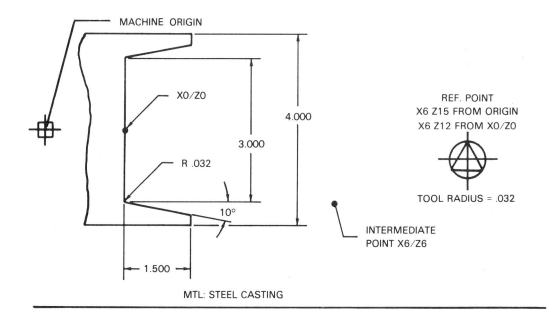

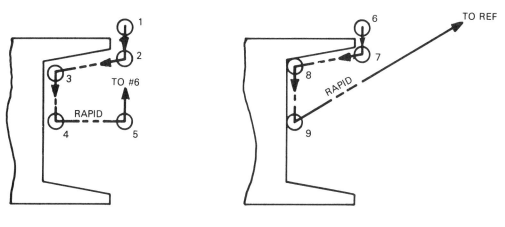

FIGURE 14–5
Taper turning

Before the cutter offset can be used, however, it is necessary to calculate the location of point B, Figure 14–6. By solving the indicated triangle for side b and adding that length to the known radius of the taper (1.5 inches), the radius dimension from the part center line to point B can be determined.

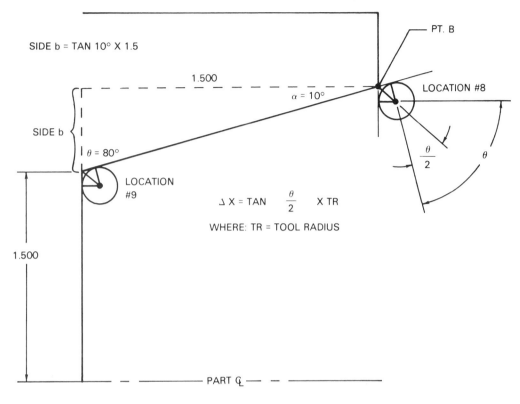

FIGURE 14–6
Determining cutter offsets

```
XO = CENTERLINE OF SPINDLE   ZO = PART SHOULDER

N010 G00 G40 G90 G95 G28 X6 Z6 M08    REM:SAFETY LINE, COOLNT ON
N020 T0101 M42                        REM:TURRET POS, HIGH RANGE
N030 S800 M03                         REM:SPNDL ON
N040 X4.1 Z1.51                       REM:POSITION TO #1
N050 G01 X3.454 F.007                 REM:FEED TO #2
N060 X2.974 Z.042                     REM:FEED TO #3
N070 X0                               REM:FEED TO #4
N080 G00 Z1.542                       REM:RAPID TO #5
N090 X4.1 Z1.532                      REM:RAPID TO #6
N100 G01 X3.474 F.003                 REM:FEED TO #7
N110 X2.946 Z.032                     REM:FEED TO #8
N120 X0                               REM:FEED TO #9
N130 G00 X6 Z6 M09                    REM:RETURN TO REF & COOLNT OFF
N140 M05                              REM:SPNDL OFF
N150 M30                              REM:END PRGM
```

FIGURE 14–7
Lathe taper turning program for part in Figure 14–5, word address format

$$\frac{b}{1.5} = TAN\ 10$$

$$b = TAN\ 10 \times 1.5$$

$$b = .1763 \times 1.5$$

$$b = .26445\ or\ .264$$

The value of .264 added to the 1.5 radius gives a distance of 1.764 from the part centerline to point B. The cutter offset can be subtracted from the 1.764 distance to find the dimension from the part centerline to cutter location #7. This distance is 1.737. The X coordinate for this location, however, will be diameter programmed. The 1.737 must now be doubled to arrive at the X coordinate to be programmed, or 3.474.

The calculated tool offset can also be subtracted from the 1.5 known radius to arrive at the 1.473 dimension from the part centerline to tool location #8. Doubling this distance gives 2.946, the X axis coordinate for location #8. The offset for the Z axis in both these cases is simply the radius of the tool nose.

PROGRAM EXPLANATION

(Refer to Figure 14 – 7.)

N010
- ■ G00—Selects rapid traverse.
- ■ G40—Cancels any active TNR comp.
- ■ G70—Specifies inch input.
- ■ G95—Specifies inches per revolution feedrate.
- ■ G90—Selects absolute positioning.
- ■ G28—Initiates a return to reference point.
- ■ X6 Z6—Intermediate point coordinates for the reference point return.
- ■ M08—Turns the coolant on.

N020
- ■ T0101—Select the tool and the offset.
- ■ M42—Selects high gear range.

N030
- ■ S800—Sets the spindle speed to 800 RPM.
- ■ M03—Turns the spindle on.

N040
- ■ X4.1 Z1.51—Position the tool to location #1, Figure 14–5.

N050

- G01 — Selects linear interpolation. The tool will feed in a straight line between the next coordinate programmed and the current tool location.
- X3.454 — Feeds the tool from location #1 to location #2. This coordinate was determined by adding approximately the desired amount of finished stock to the cutter coordinate of location #8, calculated previously.
- F.007 — Sets the feedrate.

N060

- X2.974 Z.042 — Coordinates to feed the tool from location #2 to location #3. The X coordinate was determined by subtracting .020 from the calculated finished location coordinate. Although this coordinate will not leave exactly .010 inch of stock per side to be removed during finishing, the amount left will be close to that.

N070

- X0 — Feeds the tool from location #3 to location #4.

N080

- G00 — Selects rapid traverse.
- Z1.542 — Sends the tool at rapid to location #5. This is an intermediate location used before sending the tool to location #6. If the tool were moved from location #4 to location #6 directly, the corner of the part would be cut off. Laying a straightedge between location #4 and location #6 will demonstrate the point.

N090

- X4.1 Z1.532 — Feeds the tool from location #5 to location #6 at rapid (G00 is active).

N100

- G01 — Selects linear interpolation.
- X3.474 — Feeds the tool from location #6 to location #7. This is the coordinate location calculated earlier.
- F.003 — Sets the finish pass feedrate to .003 inch per revolution.

N110

- X2.946 Z.032 — Coordinates of location #8.

N120

- X0 — Feeds the tool from location #8 to location #9.

N130

- G00 — Specifies rapid traverse.
- G28 — Initiates a return to reference.
- X6 Z6 — Intermediate point coordinates for the reference return.
- M09 — Turns the coolant off.

N140

- ■ M05—Turns the spindle off.

N150

- ■ M30—Ends the program.

CIRCULAR INTERPOLATION

Circular interpolation on a lathe does not differ significantly from circular interpolation when milling. There are two ways that an arc center can be programmed using CNC turning machines. The centerpoint can be programmed using I and K, or the center may be specified on some machinery as a radius value. Some machining centers may have an arc centerpoint specified by the radius method also.

When I and K are used, I is programmed as the X-axis coordinate of the arc centerpoint, and K is programmed as the Z-axis coordinate. The format is:

N... G02/G03 X.... Z.... I.... K....

Where G02 is clockwise circular interpolation, and G03 is counterclockwise circular interpolation; X is the X axis endpoint of the arc; Z is the Z axis endpoint of the arc; I is the X axis coordinate of the arc centerpoint; and K is the Z axis coordinate of the arc centerpoint.

When the center is specified using a radius, the R address is used. R is programmed as an incremental value from the current tool position. The format is:

N... G02/G03 X.... Z.... R....

Two programs are presented here for turning a spherical end on a 2.000-inch-diameter piece of 304 stainless steel (see Figure 14−8). Figure 14−9 is a program to turn the end using I and K; Figure 14−11 is identical except that R is used instead.

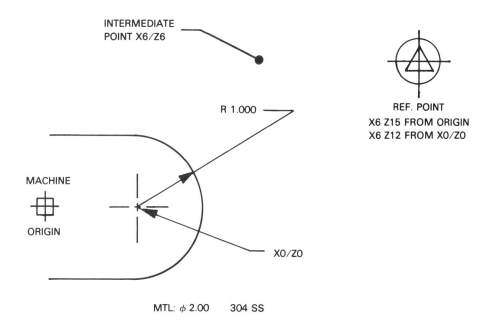

INTERMEDIATE
POINT X6/Z6

R 1.000

MACHINE
ORIGIN

X0/Z0

REF. POINT
X6 Z15 FROM ORIGIN
X6 Z12 FROM X0/Z0

MTL: ϕ 2.00   304 SS

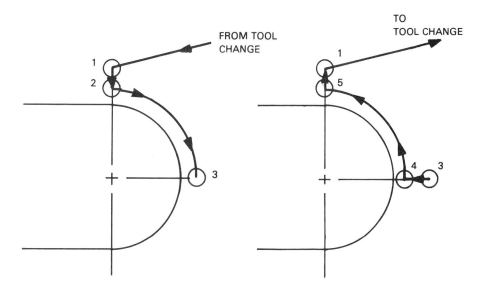

FROM TOOL
CHANGE

TO
TOOL CHANGE

FIGURE 14–8
Turning a spherical end

```
X0/Z0 = CENTERLINE OF PART RADIUS

N010 G00 G40 G70 G90 G28 X6 Z6 M08    REM:SAFETY LINE
N020 T0101 M42                        REM:TURRET POS, HIGH RANGE
N030 S150 M03                         REM:SPNDL ON
N040 X2.1 Z0                          REM:POSITION TO #1
N050 G01 X2.084 F.003                 REM:FEED TO #2
N060 G02 X0 Z1.042 I0 K0              REM:CW ARC TO #3
N070 G01 Z1.032                       REM:FEED TO #4
N080 G03 X2.062 Z0 I0 K0             REM:CCW ARC TO #1
N090 G00 X2.084 M09                   REM:RAPID TO #1, COOLNT OFF
N100 G28 X6 Z6 M05                    REM:RETURN TO REF
N110 M30
```

```
X0/Z0 = CENTERLINE OF PART RADIUS

N010 G00 G40 G70 G90 G28 X6 Z6 M08    REM:SAFETY LINE
N020 T0101 M42                        REM:TURRET POS, HIGH RANGE
N030 S150 M03                         REM:SPNDL ON
N040 X2.1 Z0 M03                      REM:POSITION TO #1
N050 G01 X2.084 F.003                 REM:FEED TO #2
N060 G02 X0 Z1.042 R1.042             REM:CW ARC TO #3
N070 G01 Z1.032                       REM:FEED TO #4
N080 G03 X2.062 Z0 R1.032             REM:CCW ARC TO #1
N090 G00 X2.084 M09                   REM:RAPID TO #1, COOLNT OFF
N100 G28 X6 Z6 M05                    REM:RETURN TO REF
N110 M30
```

FIGURE 14-9

PROGRAM EXPLANATION

(Refer to Figure 14-9.)

N010

- Safety line to cancel any active codes, returns the tool to the reference point. Turns coolant on.

N020

- T0101 — Selects tool #1, offset #1.
- M42 — Selects high gear range.

N030

- S150 — Sets the spindle speed to 150 RPM.
- M03 — Turns the spindle on.

N040

- X2.1 Z0 — Positions the tool to location #1, Figure 14−8.

N050

- G01 — Selects feedrate movement.
- X 2.084 — Feeds the tool from location #1 to location #2.
- F.003 — Assigns the feedrate.

N060

- G02 — Selects clockwise circular interpolation.
- X0 Z1.042 — Arc endpoint coordinates, location #3.
- I0 K0 — Centerpoints of the arc, Figure 14−9, top.
- R1.042 — Radius value, Figure 14−9, bottom. The 1.042 value is the incremental distance from the arc start point (location #2) to the arc center.

N070

- G01 — Selects feedrate movement.
- Z1.032 — Feeds the tool from location #3 to location #4.

N080

- G03 — Selects counterclockwise circular interpolation.
- X2.062 Z0 — Endpoint coordinates of the arc.
- I0 K0 — Centerpoints of the arc, Figure 14−9, top.
- R 1.032 — Radius of the arc, Figure 14−9, bottom.

N090

- G00 — Selects rapid traverse.
- X2.084 — Rapids the cutter from location #5 to location #1.
- M09 — Turns the coolant off.

N100

- G28 X6 Z6 — Returns the tool to the reference point.
- M05 — Turns the spindle off.

N110

- M30 — Signals end of program.

DRILLING

Drilling on NC lathes is accomplished in similar manner to turning and boring. The tool is sent to a desired start position, and the coordinates are given to move along the proper path. When drilling, the tool point is programmed since there is no tool radius involved. Canned cycles like those used for drilling on NC mills will be discussed in a later section.

To drill a ⅜ diameter hole 1.500 inches deep in part Figure 14−1, a center-drill and a ⅜ drill can be added to the program in Figure 14−3. This has been done in Figure 14−10. The program explanation follows.

```
XO = CENTERLINE OF SPINDLE  ZO = PART SHOULDER

N010 G00 G40 G90 G95 G28 X6 Z6 M08
N020 T0101 M42
N030 S1200 M03
N040 X2.6 Z2.042
N050 G01 X0 F.007
N060 Z2.032
N070 X2.314 F.003
N080 Z.042 F.007
N090 X2.6
N100 G00 X2.320 Z2.132
N110 G01 X2.0840
N120 Z.042 F.003
N130 X2.6
N140 G00 X2.084 Z2.132
N150 G01 X2.062
N160 Z.032 F.003
N170 X2.55
N180 G00 G28 X6 Z6 M09
N190 M01                              REM:OPSTOP

N200 M08                              REM:COOLNT ON
N210 T0202 M42                        REM:TURRET POS & HIGH RANGE
N220 S1800 M03                        REM:SPNDL ON, 1800 RPM
N230 G00 X0 Z2.100                    REM:POSITION TO START
N240 G01 Z-1.850 F.003                REM:FEED TO DEPTH
N250 G00 Z2.100                       REM:RAPID TO START POS.
N260 G28 X6 Z6 M09                    REM:RETURN TO REF, COOLNT OFF
N270 M01                              REM:OPSTOP

N280 M08                              REM:COOLNT ON
N290 T0303 M42                        REM:TURRET POS & HIGH RANGE
N300 S1600 M03                        REM:SPNDL ON, 1600 RPM
N310 G00 X0 Z2.100                    REM:RAPID TO START POS.
N320 G01 Z1.625 F.003                 REM:FEED TO 1ST PECKING DEPTH
N330 G00 Z2.500                       REM:RAPID OUT OF PART
N340 Z1.630                           REM:RAPID TO START OF PECK
N350 G01 Z1.375                       REM:FEED TO 2ND PECKING DEPTH
N360 G00 Z2.500                       REM:RAPID OUT OF PART
N370 Z1.380                           REM:RAPID TO START OF PECK
N380 G01 Z1.000                       REM:FEED TO 3RD PECKING DEPTH
N390 G00 Z2.500                       REM:RAPID OUT OF PART
N400 Z1.005                           REM:RAPID TO START OF PECK
N410 G01 Z.625                        REM:FEED TO 4TH PECKING DEPTH
N420 G00 Z2.500                       REM:RAPID OUT OF PART
N430 Z.630                            REM:RAPID TO START OF PECK
N440 G01 Z.387                        REM:FEED TO FINISH DEPTH
N450 G00 Z.100                        REM:RAPID TO START POSITION
N460 G28 Z6 Z6 M09                    REM:RETURN TO REF, COOLNT OFF
N470 M05                              REM:SPNDL OFF
N480 M30                              REM:END PRGM
```

FIGURE 14−10

PROGRAM EXPLANATION

N010 – N180 are identical to Figure 14 – 3.

N190

- Optional stop code. This code aids the operator during setup. If the optional stop switch is turned on at the console, the program will stop at this line. The operator can then inspect the workpiece during setup. It is common practice to include an M01 at the end of each tool.

N200 – N220

- Selects the tool, offset, gear range. Turns the spindle and coolant on.

N230

- G00 — Rapid traverse mode.
- X0 Z2 .100 — Rapids the centerdrill to the start position, .100 away from the workpiece face.

N240

- G01 — Feedrate mode.
- Z – 1.850 — Depth of centerdrilling (.150 deep).
- F.003 — Sets feedrate at .003 IPR.

N250

- G00 — Rapid traverse mode.
- Z2.100 — Returns tool to the start position.

N260

- Returns tool to the reference point and cancels the tool offset.

N270

- M01 — Optional stop code.

N280 – N300

- Selects tool, offset, gear range. Turns on spindle and coolant.

N310

- Rapids tool tip to the start point.

N320

- G01 — Feedrate mode.
- Z1.625 — Depth of first drill peck.
- F.003 — Sets the feedrate to .003 IPR.

N330

- G00—Rapid traverse mode.
- Z2.500—Sends the tool tip .500 away from the part face. The .500 distance gives the coolant sufficient area to enter the section of hole just drilled to lubricate the drill point on the next drill peck.

N340

- Z1.630—Sends the tool tip to the start of the next peck, .005 from the end point of the previous drill peck.

N350

- G01—Feedrate mode.
- Z 1.375—End point to the second drill peck.

N360

- Rapids tool .500 out of part.

N370

- Rapids tool tip to start of third peck.

N380–N440

- The pecking cycle is repeated until final hole depth is achieved.

N450

- Tool rapids out of part to original start position.

N460

- Returns to reference line.

N470

- Spindle off.

N480

- END of program.

THREADING

When threading on CNC lathes, one of three threading cycles is used: single pass threading (G33), multiple pass threading (G92), or multiple pass threading (G76). When a G33 is issued, the tool travels the length of the thread and stops. The tool then has to be retracted from the thread, returned to the starting point, and the whole procedure repeated. When a G92 command is issued, the tool moves to a programmed X coordinate, feeds across the length of the thread to the programmed Z coordinate, and returns to the start point. This process is automatically repeated with the X-axis moving to a new programmed

X coordinate until the final X coordinate has been executed. When a G76 is issued, the machine makes a threading pass, then automatically retracts the tool to the X axis reference position and returns it to the Z axis start position. It then automatically repeats the procedure until the final depth of the thread is achieved.

Three types of threads can be cut using a CNC lathe: constant lead, increasing lead, and decreasing lead. The lead of a thread is the distance that the thread advances in one revolution. Some CNC lathes are capable of cutting only constant lead threads, depending on the thread-cutting options selected when the machine is purchased. Threads of increasing and decreasing lead are specialized applications and will not be dealt with in this text.

When cutting threads, the relationship between spindle speed and tool feedrate is very important. When a G code is used for thread cutting, the feedrate override controls on the MCU console, which allow the operator to adjust the feedrate during machining, will not function. When beginning a threading pass, a certain distance (A in Figure 14–11) must be allowed ahead of the part face, to give the lathe carriage time to accelerate to the proper feedrate. Failure to allow this distance will result in improper leads on the first several threads.

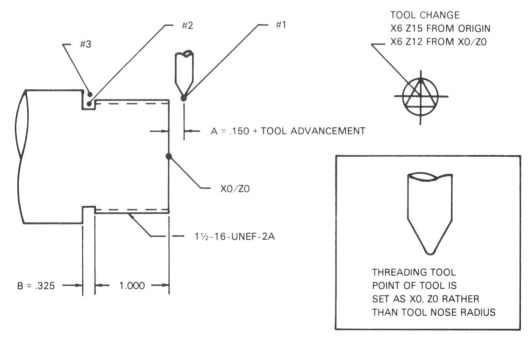

FIGURE 14–11
Part to be threaded

Starting distance A varies from machine to machine. Charts giving the distance for a particular thread on a particular machine will be found in the programming manual. If a chart is not available, the following formula can be used:

$$A = (RPM \times LEAD \times .006) + Z$$

Where Z is the amount of tool advancement in the Z axis. *Tool advancement* occurs, prior to the start of a threading cut, along two axes, as illustrated in Figure 14–12. Advancement along the Z axis is calculated by the formula:

$$Z = X (TAN\ 30)$$

Some programmers prefer to feed the tool in at a 29-degree angle instead of 30. In this case the formula would be:

$$Z = X (TAN\ 29)$$

The stopping distance is similar to the starting distance. This distance is shown in Figure 14–11 as dimension B. The minimum stopping distance can be calculated by the following formula if a chart is not available:

$$B = RPM \times LEAD \times .013$$

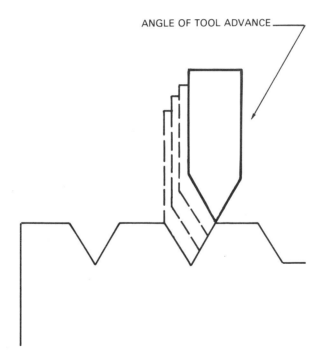

ANGLE OF TOOL ADVANCE

FIGURE 14–12
Tool advancement

Three threading programs have been written for the part shown in Figure 14−10. The program in Figure 14−13 cuts the thread using single pass threading. The program in Figures 14−14 and 14−15 cut the thread using multiple pass threading. The format for single pass threading is:

n... G33... Z.... F....

```
XO = SPINDLE CENTERLINE ZO = PART FACE

N010 G00 G40 G70 G90 G95 G28 X6 Z6   REM:SAFTEY LINE, REF. RETURN
N020 M06 T0101
N030 S400 M03
N040 X1.47 Z.015                     REM:POSITION TO #1
N050 G91 G33 Z-1.15 F.0625           REM:1ST THREAD PASS
N060 G00 X.015                       REM:RETRACT X
N070 Z1.168                          REM:RETURN TO START
N080 X-.032 Z-.018                   REM:ADVANCE TOOL
N090 G33 Z-1.168 F.0625              REM:2ND THREAD PASS
N100 G00 X.032                       REM:RETRACT X
N110 Z1.186                          REM:RETURN TO START
N120 X.032 Z-.018                    REM:ADVANCE TOOL
N130 G33 Z-1.186 F.0625              REM:FINISH PASS
N140 G00 X.032 M09                   REM:RETRACT X
N150 G90 G28 X6 Z6 M05               REM:RETURN TO REF.
N160 M30
```

FIGURE 14−13
Single pass threading program for part in Figure 14−11, word address format

```
XO = SPINDLE CENTERLINE  ZO = PART FACE

N010 G00 G40 G70 G90 G95 G28 X6 Z6 M08
N020 T0101 M42
N030 S700 M03
N040 G00 X1.600 Z.150                REM:THD. START POINT
N050 G92 X1.580 Z-1.150             REM:1ST PASS
N060 X1.570                          REM:2ND PASS
N070 X1.550                          REM:3RD PASS
N080 X1.530                          REM:4TH PASS
N090 X1.510                          REM:5TH PASS
N100 X1.490                          REM:6TH PASS
N110 X1.470                          REM:7TH PASS
N120 X1.460                          REM:8TH PASS
N130 X1.455                          REM:9TH PASS
N140 X1.450                          REM:10TH PASS
N150 X1.445                          REM:11TH PASS
N160 X1.443                          REM:12TH PASS
N170 X1.440                          REM:13TH PASS
N180 X1.438                          REM:14TH PASS
N190 X1.437                          REM:15TH PASS
N200 X1.436                          REM:16TH PASS
N210 G28 X6 Z6 M09
N220 M05
N230 M30
```

FIGURE 14−14

```
XO = SPINDLE CENTERLINE ZO = PART FACE

NO10 GOO G40 G70 G90 G95 G28 X6 Z6                    REM:SAFTEY LINE,
REF. RETURN
NO20 MO6 TO101
NO30 S400 MO3
NO40 Z.15
NO50 G76 X1.436 Z1 IO K.032 F.0625 D.015 A60 REM:THREADING PASS
NO60 GOO G28 X6 Z6 MO9                                 REM:RETURN TO REF.
NO70 MO5
NO80 M30
```

FIGURE 14–15
Multipass threading program for part in Figure 14–11, word address format

Where G33 is the thread-cutting G code, Z is the length of the threading cut, and F is the lead of the thread. (Some lathe controllers use K to specify the lead of the thread.)

The format for G92 multipass threading is:

N... G92 X.... Z.... F....
N... X...
N... X...
    .
    .
    .
N... X...

Where: G92 = multipass threading code
       X = X coordinate of the first threading pass
       Z = Z coordinate of the threading end point
       F = the feedrate (lead) of the thread

       X = depth of second pass
       X = depth of third pass

    etc. until

       X = depth of final pass

Usually the lead can be given to only four decimal places, so that some round-off error will occur. This is usually so slight that it will affect only threads several feet long. Some machines have the capacity to accept thread leads to five or six decimal places.

The format for G76 multiple pass threading is:

N... G76 X.... Z.... I.... K.... D.... F.... A..

Where: G76 = multipass threading G code

X = minor diameter of the thread

Z = length of thread

I = difference in thread radius from one end of the thread to the other. This value is used for cutting tapered threads. For straight threads, a value of zero is entered.

K = height of the thread (a radius value, given from the crest of the thread to the root)

D = depth of cut for the first pass

F = lead of the thread

A = angle of the tool tip. (For Unified, American National, and IFI metric threads, the angle is 60 degrees.)

The main differences between G92 and G76 are first, G76 requires only one line of program code, and second, G92 plunges the tool straight into the workpiece rather than feeding in at an angle. The infeed direction with G76 is 30 degrees. The infeed direction with G33 is controlled by the programmer.

PROGRAM EXPLANATION

(Refer to Figure 14–13.)

N010

■ Safety line, returns tool to reference.

N020

■ M06 T0101 — Selects tool #1, offset #1.

N030

■ S400 — Sets the spindle speed to 400 RPM.

■ M03 — Turns on the spindle.

N040

■ X1.47 Z.15 — Coordinates of location #1, Figure 14–11. The X coordinate is diameter programmed and positions the tool to the depth of the first pass. The Z coordinate is the starting distance. Subsequent passes will add to the starting distance the amount of Z-axis tool advancement.

N050

■ G91 — Selects incremental positioning.

■ G33 — Initiates single pass threading.

■ Z1.15 — Feeds the tool from location #1 to location #2, Figure 14–11.

■ F.0625 — Lead of the thread.

N060

- G00—Selects rapid traverse.
- X.015—Incremental coordinate to rapid the tool from location #2 to location #3.

N070

- Z – 1.168—Incremental distance to rapid the tool back to the starting point. This coordinate also compensates for the additional starting distance required by the tool advancement for the next pass.

N080

- X – .032—Incremental coordinate to advance the tool for the next cut. Two .015-inch roughing cuts are being made. This coordinate advances the X axis the .015 inch the tool was retracted at the end of the first pass, plus the .015 inch desired for the second.
- Z – .018—Calculated Z-axis tool advancement to cause the tool to advance on a 30-degree angle.

N090

- G33—Initiates the threading cycle.
- Z – 1.168—Feeds the tool from the start point (location #1) to the end of the thread point (location #2).
- F.0625—Lead of the thread.

N100

- G00—Selects rapid traverse.
- X.032—Retracts the X axis from the thread.

N110

- Z1.168—Returns the tool to the starting point of the thread.

N120

- X – .32 Z – .018—Advances the tool to final thread depth.

N130

- G33—Initiates thread cutting.
- Z – 1.168—Feeds the tool from #1 to #2.
- F.0625—Lead of the thread.

N140

- G00—Selects rapid traverse.
- X.032—Retracts the tool from the thread.
- M09—Turns off the coolant.

N150

- G90—Selects absolute positioning.
- G28—Returns the tool to the reference point.
- X6 Z6—Intermediate point coordinates.
- M05—Turns off the spindle.

N160

- M30—Signals end of program.

PROGRAM EXPLANATION

(Refer to Figure 14–14.)

N010

- Safety line, returns to reference.

N020

- T0101—Selects tool and offset.
- M42—Selects high gear range.

N030

- S700 M03—Turns the spindle on at 700 RPM.

N040

- X1.600 Z.150—Start position of the thread.

N050

- G92—Initiates threading cycle.
- X1.580—X coordinate of first threading pass.
- Z – 1.150—Z coordinate of the ending point.
- F.0625—The thread lead.

N060–N200

- X coordinates of the succeeding thread passes. N200 is the last pass. Note the passes gradually remove less and less stock per pass to eliminate tearing of the thread.

N210–N220

- Returns the tool to reference. Turns off coolant and spindle.

N230

- END of program.

PROGRAM EXPLANATION

(Refer to Figure 14–15.)

N010

- Safety line.

N020

- M06 T0101 — Selects tool #1, offset #1.

N030

- S400 — Sets the spindle speed.
- M03 — Turns the spindle on.

N040

- Z1.5 — Positions the Z axis at the start of the thread.

N050

- G76 — Initiates multipass threading.
- X1.436 — Minor diameter of the thread.
- Z1 — Length of the thread.
- I0 — Difference in radius of the thread from the starting point to the finish point.
- K.032 — Height of the thread measured from the crest to the root.
- D.015 — Specifies a .015-inch first pass.
- F.0625 — Lead of the thread.
- A60 — Specifies a 60-degree thread.

N060

- G00 — Selects rapid traverse.
- G28 — Initiates a return to reference.
- X6 Z6 — Intermediate point coordinates.
- M09 — Turns off the coolant.

N070

- M05 — Turns off the spindle.

N080

- M30 — Signals the end of program.

Note how the amount of programming is reduced when using the multipass cycle.

Do loops and subroutines may also be used in lathe programming; they are programmed in just as when milling. Tool nose radius compensation may also be used. TNR comp has not been discussed here in order to concentrate on the basics of tool nose centerline programming. It is used in similar fashion to cutter diameter compensation in CNC milling programs. Once tool nose centerline programming is understood, there should be no problem in using TNR comp. The same ramp on/ramp off precautions apply in turning as in milling.

A COMPLETE LATHE EXAMPLE

Up to this point, small lathe programming routines have been presented. These routines illustrate various lathe operations which usually are parts of a single lathe program. Figure 14-16 is a part for which a program has been written. The program is contained in Figure 14-17. A brief program explanation follows. There are several codes used in this program that should be noted.

G98—used to select inch per revolution feedrates.
G97—used to select direct RPM programming.
M24—used when threading to cause the tool to pull straight out of the part. The default condition for a thread cycle is for the tool to pull out at a 60 to 45 degree angle.

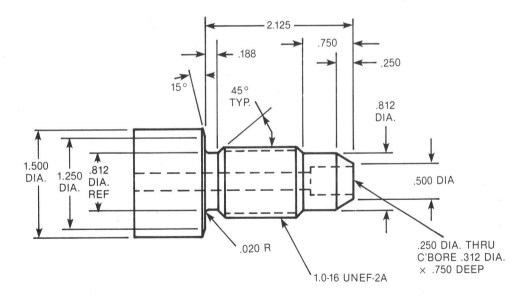

FIGURE 14-16
Part drawing

```
LATHE PROGRAMMING EXAMPLE
X0 = CENTERLINE OF PART
Z0 = FACE OF PART


( .031R X 80 DEG. TURNING TOOL)
N1 G97
N2 G99
N3 M08
N4 G00 T0101
N5 S2133 M03
( ROUGH FACE PART - LEAVE .005 STK.)
N6 X1.0000 Z.0310
N7 G01 X.0000 F.0070
N8 G00 Z.1000
( ROUGH TURN 1.0 DIA. IN 2 PASSES - LEAVE .005 STK./SIDE)
N9 X1.1720
N10 G01 Z-2.0890 F.0070
N11 X1.6720
N12 G00 Z.1000
N13 X1.0720
N14 G01 Z-2.0890 F.0070
N15 X1.2594
N16 X1.6720 Z-2.1443
N17 G00 Z.1360
N18 G28 X6.0000 Z6.0000
N19 M01


( .007R X 35 DEG. TURNING TOOL)
N20 G99
N21 M08
N22 G00 T0202
N23 S2133 M03
( ROUGH THREAD RELIEF AREA)
N24 X1.0740 Z-1.8230
N25 G01 X.8360 Z-1.9420 F.0030
N26 Z-2.1050
N27 G02 X.8520 Z-2.1130 I.8520 K-2.1050
N28 G01 X1.0840
( FINISH O.D.)
N29 G00 Z.0070
N30 G01 X.0000 F.0030
N31 X.4890
N32 G03 X.5178 Z-.0010 I.4889 K-.0100
N33 G01 X.8208 Z-.2439
N34 G03 X.8260 Z-.2529 I.7920 K-.2529
N35 G01 Z-.7471
N36 X1.0040 Z-.8361
N37 G03 X1.0140 Z-.8481 I.9800 K-.8481
N38 G01 Z-1.8389
N39 G03 X1.0040 Z-1.8509 I.9800 K-1.8389
N40 G01 X.8260 Z-1.9399
N41 Z-2.1050
```

```
N42 G02 X.8520 Z-2.1180 I.8520 K-2.1050
N43 G01 X1.2518
N44 X1.6140 Z-2.1665
N45 G00 Z.1070
N46 G28 X6.0000 Z6.0000
N47 M01

( THREADING TOOL)
( THREAD O.D. 1-16-2A)
N48 G99
N49 M08
N50 G00 T0303
N51 S900 M03
N52 X-.5000 Z.6000 M74
N53 G92 X.9900 Z-2.1000 F.0625
N54 X.9800
N55 X.9718
N56 X.9654
N57 X.9600
N58 X.9552
N59 X.9510
N60 X.9470
N61 X.9434
N62 X.9400
N63 X.9368
N64 X.9336
N65 X.9308
N66 X.9278
N67 X.9252
N68 X.9234
N69 X.9234 Z-2.1000
N70 G28 X6.0000 Z6.0000
N71 M01

( NO. 4 C'DRILL)
( C'DRILL TO .260 DIA.)
N72 G99
N73 M08
N74 G00 T0404
N75 S3000 M03
N76 X.0000 Z.1000
N77 G01 Z-.2780 F.0030
N78 G00 Z.1000
N79 G28 X6.0000 Z6.0000
N80 M01

( 1/4 DRILL)
( DRILL .250 DIA. THRU)
N81 G99
N82 M08
N83 G00 T0505
N84 S2000 M03
N85 X.0000 Z.1000
N86 G01 Z-.3000 F.0030
```

```
N87 G00 Z.5000
N88 Z-.2950
N89 G01 Z-.6000 F.0030
N90 G00 Z.5000
N91-.5950 $
N92 G01 Z-.9000 F.0030
N93 G00 Z.5000
N94 Z-.8950
N95 G01 Z-1.2000 F.0030
N96 G00 Z.5000
N97 Z-1.1950
N98 G01 Z-1.5000 F.0030
N99 G00 Z.5000
N100 Z-1.4950
N101 G01 Z-1.8000 F.0030
N102 G00 Z.5000
N103 Z-1.7950
N104 G01 Z-2.1000 F.0030
N105 G00 Z.5000
N106 Z-2.0950
N107 G01 Z-2.4000 F.0030
N108 G00 Z.5000
N109 Z-2.3950
N110 G01 Z-2.7000 F.0030
N111 G00 Z.5000
N112 Z-2.6950
N113 G01 Z-3.0000 F.0030
N114 G00 Z.5000
N115 Z-2.9950
N116 G01 Z-3.2500 F.0030
N117 G00 Z.1000
N118 G28 X6.0000 Z6.0000
N119 M01

( .005R BORING BAR)
N120 G99
N121 M08
N122 G00 T0606
N123 S3500 M03
( ROUGH C'BORE - LEAVE .005 STK/SIDE)
N124 X.2920 Z.0350
N125 G01 Z-.7400 F.0020
N126 X.1520
N127 G00 Z.0400
( FINISH C'BORE - DEBURR EDGE WITH .01R)
N128 X.3320
N129 G01 Z.0105 F.0020
N130 G02 X.2920 Z-.0095 I.3320 K-.0095
N131 G01 Z-.7400
N132 X.1320
N133 G00 Z.1100
N134 G28 X6.0000 Z6.0000 M09
N135 M05
N136 M30
```

FIGURE 14-17

Program for the part in Figure 14-16

PROGRAM EXPLANATION

First Tool:

N1–N5
- Selects first tool. Turns on spindle and coolant.

N6–N8
- Part is rough faced with .005 stock left for finishing.

N9–N11
- First roughing pass on o.d.

N12–N17
- Second roughing pass on o.d. The 15 degree angle is also rough turned at this time.

N18
- Tool is returned to reference point. Tool offset cancelled.

Second Tool:

N20–N23
- Selects second tool. Turns on spindle and coolant.

N24–N28
- Thread relief area is rough turned. .005 stock is left for finishing.

N29–N31
- Face of part is finished.

N32
- Deburring radius is turned at the intersection of the first angle and the face of the part.

N33
- First angle is finish turned.

N34
- Deburring radius is turned at the intersection of the first angle and the .812 diameter.

N35
- The .812 diameter is finish turned.

N36
- The front thread chamfer is finish turned.

N37

- A radius is turned at the intersection of the thread chamfer and major diameter.

N38

- The major diameter of the thread is turned.

N39

- A radius is turned at the intersection of the back thread chamfer and major diameter.

N40

- The back thread chamfer is turned.

N41

- The .812 diameter thread relief is turned.

N42

- The .020 radius is turned.

N43

- The 2.125 dimension is faced.

N44

- The 15 degree angle is finish turned.

N45 – N47

- The tool is returned to reference. The offset is cancelled.

Third Tool:

N48 – N51

- Tool and offset selected, spindle and coolant turned on.

N52

- The tool is sent to the start position for threading.
- The M74 turns off the thread chamfering at the end of a thread pass.

N53

- G92 multi-pass thread cycle initiated.

N54 – N68

- Succeeding X values for the G92 cycle. Each X value is used on a separate thread pass.

N69

- Last threading pass which is a repeat pass. The Z coordinate is optional.

N70 – N71

- Returns the tool to reference. Offset is cancelled.

Fourth Tool:

N72 – N75
- Tool, offset, spindle speed selected.

N76 – N78
- Drill sent to start point, fed to depth, and rapids back to start position.

N79 – N80
- Returns to reference.

Fifth Tool:

N81 – N84
- Tool, offset, spindle speed selected.

N85 – N116
- Peck drilling of ¼ inch through hole. Each peck is .300 deep. At end of peck the tool is sent at rapid z.500 to clear out chips and allow coolant into the hole. The tool sequence repeats until final depth is achieved in N116.

N117
- Tool is returned to the starting position.

N118 – N119
- Returns to reference.

Sixth Tool:

N120 – N123
- Tool, offset, spindle speed selected.

N124 – N127
- The c'bore is rough bored. .005 stock is left for finishing.

N128 – N130
- A deburring radius is turned at the intersection of the c'bore and the part face.

N131 – N133
- The c'bore is finish bored and tool retracted from part.

N134
- Return to reference line, coolant off.

N135
- Spindle off.

N136
- END of program.

CANNED CYCLES

Most modern CNC lathe controllers contain a number of built-in canned cycles. The threading cycles G33, G92, and G76 are standard from controller to controller. Other canned cycles are options offered by the controller manufacturer. These cycles are often unique to a given controller manufacturer (sometimes unique to a given model of controller) and therefore not transportable from controller to controller. With the current CNC lathe investment strategies by small and midsized companies, canned cycles will become as standardized as mill cycles at some future point. It is not possible to cover the number of cycle variations in a text of this size. The student should be aware, however, that these cycles exist. Documentation on the use of these cycles will be contained in the programming and operational manuals for a given machine.

How much a company relies on canned cycles for lathe programming is dependent on their use or non-use of computer-aided programming. Where computer-aided or graphics programming is utilized, there is little need for canned cycles aside from the standard lathe threading cycles. Where MDI programming is used, canned cycles can save many hours of programming time. The cycles used in these situations usually include: rough turning and boring cycle, rough facing cycle, finish turning and boring cycle, finish facing cycle, peck drilling cycle, step drilling cycle, chamfering cycle, and grooving cycle.

One caution should be noted by the programmer: Canned cycles valid for one controller can cause a crash situation if run on an incompatible controller if the controller does not stop and put out an alarm message when the canned cycle is encountered.

Appendix 7 contains a sample program utilizing canned cycles, along with an explanation of how the cycles are used.

SUMMARY

The important concepts presented in this chapter are:

- In diameter programming, the X-axis coordinates are one-half the actual tool movement.
- In radius programming, the X-axis coordinates and the tool movement are the same.
- G01, linear interpolation, is used for feedrate moves.
- Coordinates for taper turning must be calculated using trigonometry (or other math methods), just as when milling angles.
- G02 and G03 are used for circular interpolation.
- I and K are the addresses used to program the center points of an arc.

- The R address is used in place of I and K to program an arc using the arc radius instead of the arc centerpoints.
- Single pass threading cycles produce one threading cut. The cycle must be reinitiated for each threading pass.
- Multipass threading can produce an entire finished thread without additional programming.
- When threading, the Z axis tool advance must be calculated from the X-axis depth of cut by the formula $Z = X\,TAN(30)$.
- Minimum starting and stopping distances must be calculated for use in a threading program.

REVIEW QUESTIONS

1. What G codes are used for feedrate moves?
2. What is the difference between diameter and radius programming?
3. Write a program to turn and face the part in Figure 14−18.
4. Write a program to turn the taper on the part in Figure 14−19.
5. What codes are used to institute circular interpolation?
6. What addresses are used to define the X and Z axis center point of an arc?
7. What address is used to define the arc using the arc radius?
8. Write a program to machine the part in Figure 14−20.
9. What is the code for single pass threading? What is the format?
10. What is the code for multipass threading? What is the format?
11. Write a program to thread the part in Figure 14−21:
    a. Using single pass threading.
    b. Using multipass threading G92.
    c. Using multipass threading G76.

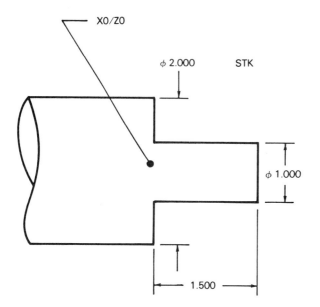

XO/ZO

φ 2.000   STK

φ 1.000

1.500

MATERIAL: CARPENTER STENOR TOOL STEEL

X6/Z12 FROM XO/ZO =
REFERENCE POINT
INTERMEDIATE POINT =
X6/Z6

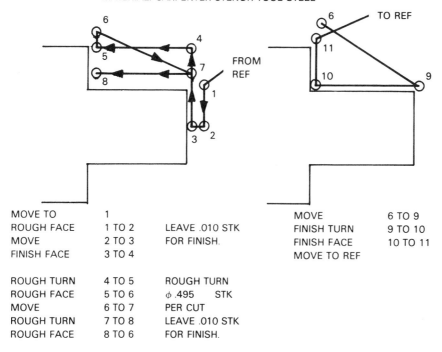

MOVE TO	1	
ROUGH FACE	1 TO 2	LEAVE .010 STK
MOVE	2 TO 3	FOR FINISH.
FINISH FACE	3 TO 4	
ROUGH TURN	4 TO 5	ROUGH TURN
ROUGH FACE	5 TO 6	φ .495 STK
MOVE	6 TO 7	PER CUT
ROUGH TURN	7 TO 8	LEAVE .010 STK
ROUGH FACE	8 TO 6	FOR FINISH.

MOVE	6 TO 9
FINISH TURN	9 TO 10
FINISH FACE	10 TO 11
MOVE TO REF	

FIGURE 14–18
Part drawing for review question #3

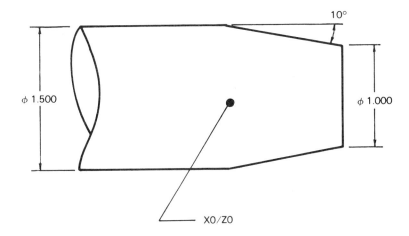

10°

ϕ 1.500

ϕ 1.000

REFERENCE
POINT = X6/Z12
FROM X0/Z0
INTERMEDIATE
POINT = X6/Z6

X0/Z0

MATERIAL: ϕ 1.500   4140 STEEL

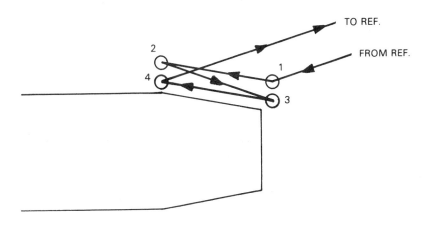

TO REF.

FROM REF.

MOVE TO 1
ROUGH TURN   1 TO 2
MOVE   2 TO 3
FINISH TURN   3 TO 4
MOVE   TO REF

FIGURE 14–19
Part drawing for review question #4

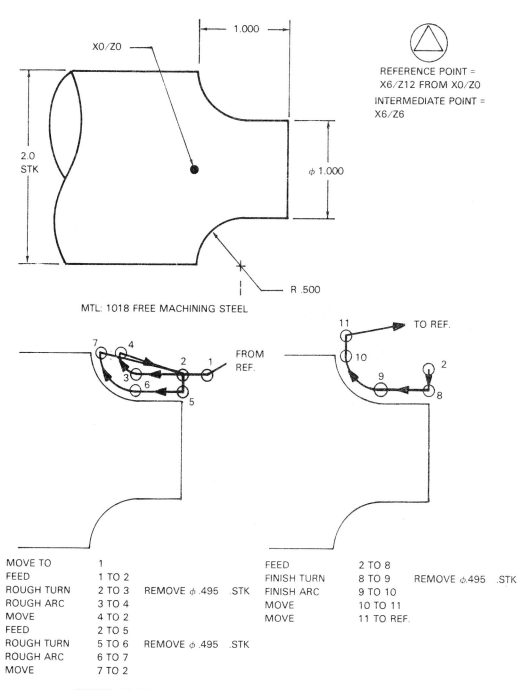

REFERENCE POINT =
X6/Z12 FROM X0/Z0

INTERMEDIATE POINT =
X6/Z6

MTL: 1018 FREE MACHINING STEEL

MOVE TO	1		
FEED	1 TO 2		
ROUGH TURN	2 TO 3	REMOVE φ.495	.STK
ROUGH ARC	3 TO 4		
MOVE	4 TO 2		
FEED	2 TO 5		
ROUGH TURN	5 TO 6	REMOVE φ.495	.STK
ROUGH ARC	6 TO 7		
MOVE	7 TO 2		

FEED	2 TO 8		
FINISH TURN	8 TO 9	REMOVE φ.495	.STK
FINISH ARC	9 TO 10		
MOVE	10 TO 11		
MOVE	11 TO REF.		

FIGURE 14–20
Part drawing for review question #8

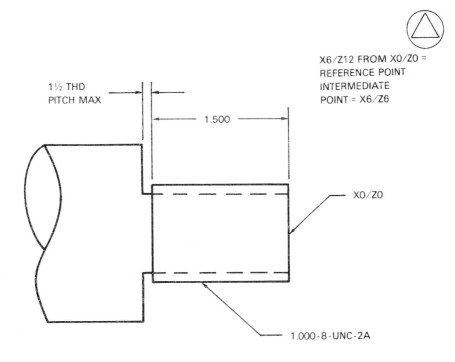

1½ THD
PITCH MAX

1.500

X6/Z12 FROM X0/Z0 =
REFERENCE POINT
INTERMEDIATE
POINT = X6/Z6

X0/Z0

1.000-8-UNC-2A

MATERIAL: 4130 STEEL

FIGURE 14–21
Part drawing for review question #11

CHAPTER 15

Use of Computers in Numerical Control Programming

OBJECTIVES Upon completion of this chapter, you will be able to:

- Describe the three basic ways computers are used in numerical control programming.
- Describe offline programming.
- Explain the advantages of computer-aided programming.
- Describe the three types of statements used in a computer-aided program.
- Understand basic part geometry and tool motion statements in APT and COMPACT II.
- Describe two types of computer graphics programming systems and how they differ.
- Explain how graphics programming simplifies writing NC programs.

Computers are becoming commonplace on shop floors. This is particularly true in numerical control programming. Only a few years ago, computers for NC programming were limited to large companies, but with the advent of inexpensive computer memory, even the smallest mold shops can now afford them. Computers can be used to help write NC programs in three basic ways: in offline programming, computer-aided programming, and computer graphics programming. This chapter will introduce all three of these uses. It is not the purpose of this text to teach computer-aided or computer graphics programming but, rather, to make the student aware of the far-reaching effects that computers are having on manufacturing.

OFFLINE PROGRAMMING TERMINALS

Technically speaking, all uses of computers to assist programming are offline programming methods. *Offline programming* is programming that is performed away from the machine, not at the CNC computer keyboard. An offline programming terminal usually refers to a computer that is used as a text editor for writing programs. This type of programming station does not "aid" the programmer except by allowing the program to be entered into the computer exactly as if it were being entered via the MDI console. The advantage is that one program may be written while another program is being run on the machine. The program being written is simply saved on a computer disk, magnetic tape, punched tape, or a combination of these three.

COMPUTER-AIDED PROGRAMMING

Computer-aided programming had its beginnings in the 1950s when point-to-point tape machinery made manual programming an enormously laborious task. Computer-aided programming languages act as translators. The programmer "talks" to the computer via the keyboard in a computer language designed specifically for numerical control programming. The main advantage of a computer-aided language is that a number of different numerical control machines can be programmed using the same language. The computer is told the machine and tools to be used, the part to be made, and the path the cutter will take. The computer takes all this information, and through another program, called a postprocessor, writes an NC program (usually in word address) to machine the part on the NC machine specified. This type of programming requires a large amount of computer memory; the computer commands vary, depending on the programming language used.

The language a particular company uses depends on the parts it produces, the machines it uses, the cost of the programming system, and the computers that are available. Following are some of the more common programming languages.

APT (Automatic Programmed Tools).   APT is the oldest and the largest of the computer-aided programming languages. It can be used on only large computers and will perform the mathematical calculations required for complex curved surfaces using four and five axes.

AD-APT (Adaption of APT).   This is a version of APT that can run on smaller computers. It uses about half the commands of APT and can be used for two-axis contouring with a third axis of linear motion.

AUTOMAP (Automatic Machining Program).   AUTOMAP is another adaption of APT. It will run on medium size computers, has a limited number of commands, and is used for programming straight lines and circles.

COMPACT II.   Compact II can accomplish the same tasks as APT but is limited to three-axis work. The main difference between them is that Compact II uses a somewhat more conversational set of commands, without some of the strict rules of syntax that must be observed in APT. The computer will also aid in debugging the finished program by interactive conversation with the programmer.

UNIAPT.   UNIAPT is very similar to APT but is designed to run on dedicated minicomputers — that is, small computers that are used for only one task. UNIAPT will handle four- and five-axis programming.

NUFORM.   NUFORM differs from most other computer-aided languages in that the programmer inserts codes or dimensions in their appropriate location within the NUFORM format. NUFORM uses numeric rather than alphabetic codes.

In this chapter, APT and COMPACT II will be used as examples. An APT or COMPACT II program has three parts: part geometry definitions, auxiliary function statements (tool changes, speeds, feedrates, etc.), and tool motion statements.

Part Geometry Definitions in APT

Parts are defined to the computer by describing features such as points, lines, planes, circles, and their relationships to each other. Following are some of the simpler APT geometry commands.

To Define a Point.   Points are the basis for defining other geometric features. Points may be defined by Cartesian coordinates or by relationship to other geometric features such as lines or circles. The following examples are all point definitions.

Point defined by Cartesian coordinates, Figure 15–1(a):
SYMBOL FOR POINT = POINT/X,Y,Z
Example: P1 = POINT/6,6,6
Meaning: P1 = a point X6.0000, Y6.0000, Z6.0000 from 0/0.

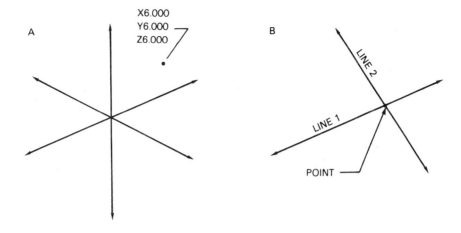

A

X6.000
Y6.000
Z6.000

B

LINE 2

LINE 1

POINT

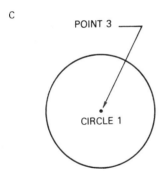

C

POINT 3

CIRCLE 1

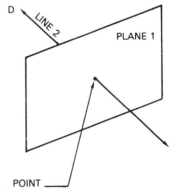

D

LINE 2

PLANE 1

POINT

FIGURE 15–1
Defining a point in APT

Point defined by the intersection of two lines, Figure 15−1(b):
SYMBOL FOR A POINT = POINT/INTOF, SYMBOL FOR A LINE, SYMBOL
   FOR A LINE
Example: P2 = POINT/INTOF,LN1,LN2
Meaning: P2 = a point located at the intersection of line LN1 and line LN2.

Point defined by the center of a circle, Figure 15−1(c):
SYMBOL FOR A POINT = POINT/CENTER, SYMBOL FOR A CIRCLE
Example: P3 = POINT/CENTER,CIR1
Meaning: P3 = a point located at the center of circle CIR1.

Point defined by the intersection of a line and a plane, Figure 15−1(d):
SYMBOL FOR A POINT = POINT/INTOF, SYMBOL OF A PLANE, SYMBOL
   OF A LINE
Example: P4 = POINT/INTOF,PN2,LN3
Meaning: P4 = a point located at the intersection of line LN3 with plane PN2.

To Define a Line.   Lines may be defined by using coordinates, other lines,
points, and circles. Lines extend to infinity and are calculated by the computer
mathematically from the information given it. It takes two points to define a line.

Lines defined by two points, Figure 15−2(a):
SYMBOL FOR A LINE = LINE/SYMBOL FOR A POINT, SYMBOL FOR A
   POINT
Example: LN1 = LINE/PT1,PT2
Meaning: LN1 = a line passing through point PT1 and PT2.

Lines defined by a point and parallel line, Figure 15−2(b):
SYMBOL FOR A LINE = LINE/SYMBOL FOR A POINT, PARLEL, SYMBOL
   FOR A LINE
Example: LN2 = LINE/PT2,PARLEL,LN1
Meaning: LN2 = a line passing through point PT2, parallel to line LN1.

Lines defined by a point and a perpendicular line Figure 15−2(c):
SYMBOL FOR A LINE = LINE/SYMBOL FOR A POINT, PERPTO, SYMBOL
   FOR A LINE
Example: LN3 = LINE/PT3,PERPTO,LN2
Meaning: LN3 = a line passing through point PT3, perpendicular to line LN2.

Lines defined by a point and tangency to a circle, Figure 15−2(d):
SYMBOL FOR A LINE = LINE/SYMBOL FOR A POINT, LEFT/RIGHT,
   TANTO, SYMBOL FOR A CIRCLE
Example: LN4 = LINE/PT2,LEFT,TANTO,CIR1
Meaning: LN4 is the line passing through point 2 and tangent to a circle with
   radius 1 on the left side.

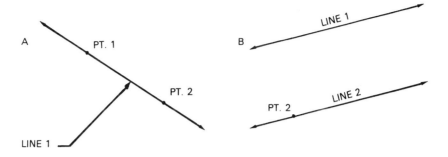

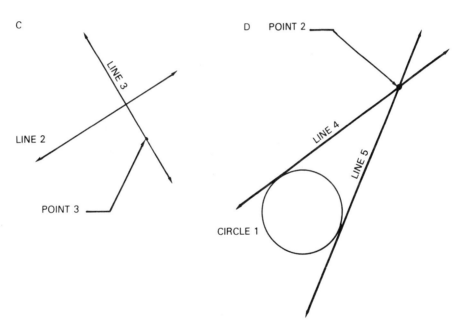

FIGURE 15–2
Defining a line in APT

Example: LN5 = LINE/PT3,RIGHT,TANTO,CIR1
Meaning: LN5 is the line passing through point 3 and tangent to a circle with
   radius 1 on the right side.

To Define a Circle.   A circle is defined by an arc section. The section of the
circle that is a part feature is programmed later as part of the tool motion statements.

Circle defined by the centerpoint and radius, Figure 15−3(a):
SYMBOL FOR A CIRCLE = CIRCLE/CENTER, SYMBOL FOR A POINT, RA-
   DIUS, RADIUS VALUE
Example: C1 = CIRCLE/CENTER,PT1,RADIUS,1.0
Meaning: C1 is a circle with center at point PT1 and a radius of 1.0 inch.

Circle defined by the centerpoint and a point on the circumference, Figure
   15−2(b):
SYMBOL FOR A CIRCLE = CIRCLE/CENTER, SYMBOL FOR THE POINT
   AT THE CENTER, SYMBOL FOR THE POINT ON THE CIRCUMFER-
   ENCE
Example: C2 = CIRCLE/PT2,PT4
Meaning: C2 is the circle with PT2 as its center and PT4 on its circumference.

Circle defined by the centerpoint and a tangent line, Figure 15−3(c):
SYMBOL FOR A CIRCLE = CIRCLE/SYMBOL FOR A POINT, TANTO, SYM-
   BOL FOR A LINE
Example: C3 = CIRCLE/CENTER,PT5,LN1
Meaning: C3 is the circle with point 5 as its centerpoint, tangent to line 1.

To Define a Plane.   Part surfaces are often defined by planes. Following are
some simple definitions of planes.

Defining a plane by points, Figure 15−4(a):
SYMBOL FOR A PLANE = PLANE/SYMBOL FOR A POINT, SYMBOL FOR
   A POINT, SYMBOL FOR A POINT
Example: TOP = PLANE/PT1,PT2,PT3
Meaning: TOP is the plane passing through points 1, 2, and 3.

Defining a plane by a point and a parallel plane, Figure 15−4(b):
SYMBOL FOR A PLANE = PLANE/SYMBOL FOR A POINT, PARLEL, SYM-
   BOL FOR A PLANE
Example: BOTTOM = PLANE/PT4,PARLEL,TOP
Meaning: BOTTOM is the plane passing through point 4 and parallel with plane
   TOP.

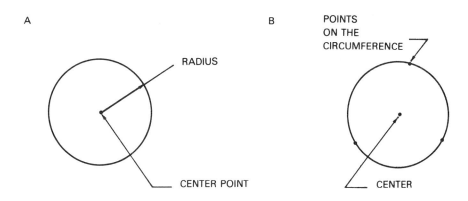

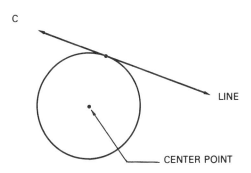

FIGURE 15-3
Defining a circle in APT

Defining a plane through two points, perpendicular to another plane, Figure 15-4(c):

SYMBOL FOR A PLANE = PLANE/PERPTO, SYMBOL FOR A PLANE, SYMBOL FOR A POINT, SYMBOL FOR A POINT

Example: PLN3 = PLANE/PERPTO,PLN2,PT3,PT5

Meaning: PLN3 is the plane perpendicular to plane PLN2, and passing through points PT3 and PT5.

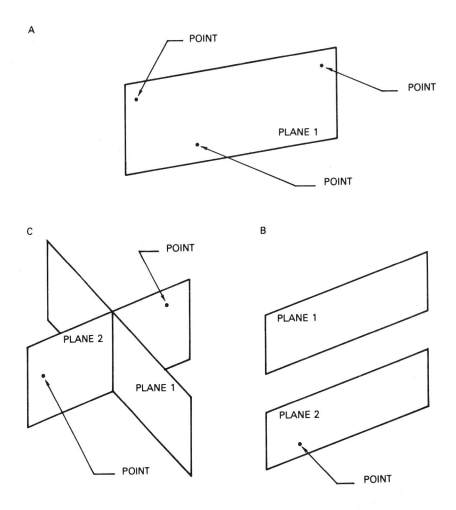

FIGURE 15–4
Defining a plane in APT

Geometric definitions can also contain modifiers, which help in clarifying a definition. These modifiers are:

XLARGE   XSMALL
YLARGE   YSMALL
ZLARGE   ZSMALL

LARGE means that the side with the greater value is specified. SMALL specifies the side of lesser value. The value is determined by the coordinate position of the feature in question. The statement LN2 = LINE/PARLEL,LN1,YLARGE,1.000 means that LN2 is line 2, parallel to line LN1 on the side where the Y coordinates are the largest and 1.000 inch away from LN1. YSMALL would mean that the line was 1.000 inch away on the side where the Y coordinates were smaller.

Other types of features can be used to define the geometric shape of a part, such as cylinders, vectors, surfaces, and angles. An APT dictionary, or a textbook on computer-aided programming, will contain a comprehensive list of APT geometric definitions.

Auxiliary Statements in APT

Before defining a cutter path using tool motion statements, the various tools to be used must be defined to the computer. The tool statements are auxiliary statements. The spindle speeds and feedrates used with the various tools are also auxiliary statements. Following are examples of tool, machine, spindle speed, and feedrate auxiliary statements.

Feedrates.

To define a feedrate in inches per minute:
FEDRAT/[feedrate value],IPM
Example: FEDRAT/20,IPM
Meaning: Feedrate is set to 20 inches per minute

To define a feedrate in inches per revolution:
FEDRAT/[feedrate value],IPR
Example: FEDRAT/.007,IPR
Meaning: Feedrate is set to .007 inch per revolution.

Tool Changes.

To change a tool when premeasured tools are not used:

LOADTL/[tool no.],ADJUST,[tool length offset register number]
Example: LOADTL/1,AJ,1
Meaning: Results in tape code to load tool number 1 into the spindle and call up tool length offset register number 1.

To change a tool when premeasured tools are used:
LOADTL/[tool no.],LENGTH,[tool length],ADJUST,[tool register]
Example: LOADTL/2,LENGTH,9,ADJUST,2

Meaning: Results in tape code to load tool number 2 into the spindle and call up tool length offset register number 2. All Z-axis coordinates will be modified by the 9-inch length of the tool.

Spindle Speeds.

To define a clockwise spindle speed:
SPINDL/[RPM],CLW
Example: SPINDL/2000,CLW
Meaning: Spindle speed is 2000 RPM in a clockwise direction.

To define a counterclockwise spindle speed:
SPINDL/[RPM],CCLW
Example: SPINDL/2300,CCLW
Meaning: Spindle speed is 2300 RPM in a counterclockwise direction.

Machine Statements.   Machine statements tell the computer where to find the instructions to write a program for a particular machine. The syntax is as follows:

MACHIN/[machine identifier]
Example: MACHIN/UNIV 1
Meaning: The machine to program is Universal machine #1.

Tool Statements.   Tool statements define to the computer the various tools that will be used during the program. There are two types of cutter definitions: simple and complex. Simple cutters have only a diameter and radius. Complex cutters have multiple features. Figure 15–5 shows both simple and complex cutters. The tool definition is done with a CUTTER statement.

Simple cutter:
CUTTER/d,r
Example: CUTTER/.75,.0625
Meaning: The cutter has a diameter of .750 inch and a corner radius of .0625 inch. The height of the cutter (h) is assumed to be 5.0 inches.

Complex cutter:
CUTTER/d,r,e,f,a,b,h
Example: CUTTER/.75,.0625,.3125,.0837,15,5,6
Meaning: The cutter has a diameter of .750 inch, a corner radius of .0625 inch, E distance is .3125 inch, F distance is .0837 inch, angle A is 15 degrees, angle B is 5 degrees, and the height of the cutter is 6.0 inches.

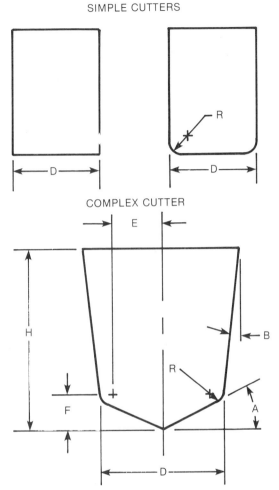

SIMPLE CUTTERS

COMPLEX CUTTER

FIGURE 15–5
Cutter dimensions

Tool Motion Statements in APT

Once the part geometry has been defined, the cutter path can be described to the computer using what are known as tool motion statements. Tool motion is controlled using the following commands: GOUP, GODOWN, GOFWD, GO-BACK, GORGT (right), and GOLFT (left). The relationship of these commands can be seen in Figure 15–6.

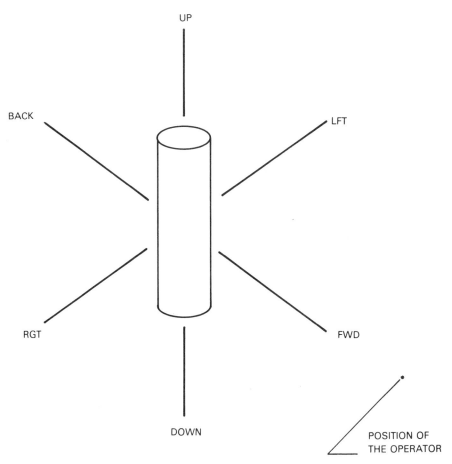

FIGURE 15–6
Tool motion commands in APT

Tool movement is also controlled by three surfaces known as the part surface, the drive surface, and the check surface (see Figure 15–7). The tool bottom is guided by the part surface, while the side of the tool is guided by the drive surface. Once initiated, tool motion continues until stopped by a check surface, which signals the end of the cut.

The six tool commands are used with one of four modifiers to define the check surface. These modifiers—TO, ON, PAST, and TANTO—are illustrated in Figure 15–8.

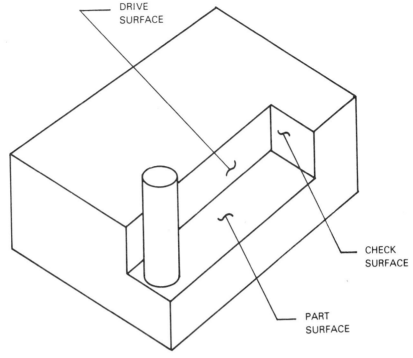

FIGURE 15–7
Controlling surfaces in APT

Figure 15–9 shows a rectangular piece defined by the APT language. The four sides have been labeled S1 through S4. Tool motion statements to move the tool from position A to B, and then mill the periphery of the part would be as follows:

GOTO/S1,PASTS2
GOFWD/S1,PASTS4
GORGT/S4,PASTS3
GORGT/S3,PASTS2
GORGT/S2,PASTS1

Figure 15–10 shows a part to be APT programmed. Figure 15–11 identifies the geometric figures used to define the part. Figure 15–12 is a simplified APT program written to mill the part periphery.

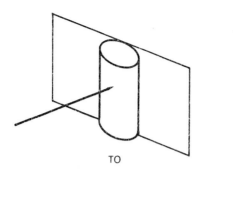

TO

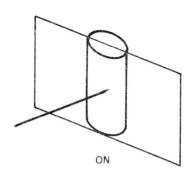

ON

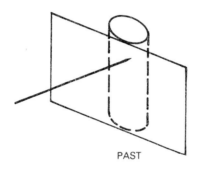

PAST

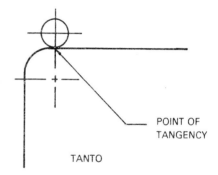

TANTO

POINT OF
TANGENCY

FIGURE 15-8
Modifiers for tool commands in APT

Part Geometry Definitions in COMPACT II

To Define a Point.   Points may be defined by Cartesian coordinates or by relationship to other geometric features such as lines or circles.

Point defined by Cartesian coordinates, Figure 15-1(a):
DPTn,nXB,nYB,nZB
Example: DPT1,6XB,6YB,6ZB
Meaning: Point 1 is the point X6.0000, Y6.0000, Z6.0000 from 0/0. The B specifies that a work coordinate system, called a BASE, is being referenced.

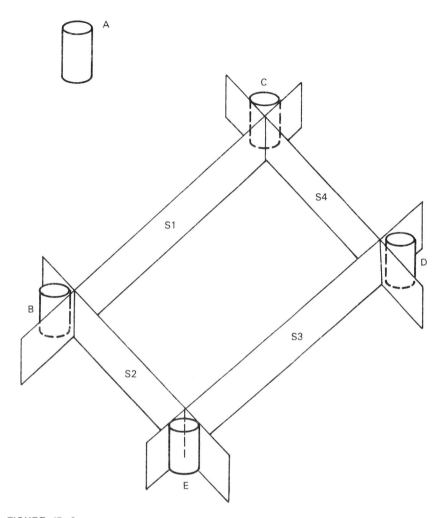

FIGURE 15-9

Point defined by the intersection of two lines, Figure 15-1(b):
DPTn,LNn,LNn
Example: DPT2,LN1,LN2
Meaning: Point 2 is the point formed by the intersection of line 1 and line 2.

Point defined by the center of a circle, Figure 15-1(c):
DPTn,CIRn,CNTR
Example: DPT3,CIR1,CNTR
Meaning: Point 3 is the point at the center of circle 1.

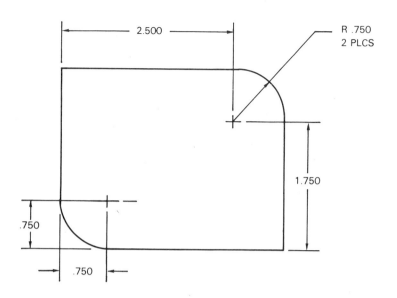

FIGURE 15–10
Part to be APT programmed

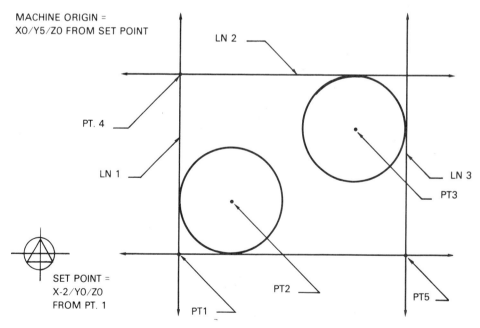

FIGURE 15–11
Geometry of part in Figure 15–10

To Define a Line.

Lines defined by two points, Figure 15−2(a):
DLNn,PTn,PTn
Example: DLN1,PT1,PT2
Meaning: Line 1 is the line passing through point 1 and point 2.

Lines defined by a point and parallel line, Figure 15−2(b):
DLNn,PTn,PARLNn
Example: DLN2,PT2,PARLN1
Meaning: Line 2 is the line passing through point 2 parallel to line 1.

Lines defined by a point and a perpendicular line, Figure 15−2(c):
DLNn,PTn,PERLNn
Example: DLN3,PT3,PERLN2
Meaning: Line 3 is the line passing through point 3 perpendicular to line 2.

Lines defined through a point and tangent to a circle, Figure 15−2(d):
DLNi,PTi,CIRi,MODIFIER
Example: DLN4,PT2,CIR1,YL
Meaning: Line 4 is the line passing through point 2, tangent to circle 1 on the Y
large (right) side.

To Define a Circle.

Circle defined by the centerpoint and radius, Figure 15−3(a):
DCIRn,PTn,R
Example: DCIR1,PT1,1.0R
Meaning: Circle 1 is the circle with center at point 1 and a radius of 1.0 inches.

Circle defined by three points on the circumference, Figure 15−3(d):
DCIRn,PTn,PTn,PTn
Example: DCIR2,PT2,PT3,PT4
Meaning: Circle 2 is the circle with points 2, 3, and 4 on its circumference.

Auxiliary Statements in COMPACT II

Five auxiliary statements are used to initiate and set up the COMPACT II program.

The machine statement MACHIN is used first. MACHIN,UNIVER1 would signal the computer that a machine called UNIVER (Universal) 1 was being used. This tells the computer which postprocessor to use.

The next statement is called the identification statement. This statement identifies the program so that it can be cataloged and retrieved at a later date.

```
MACHIN/UNIVERSAL-1
$$
$$GEOMETRY
$$
PL1       =PLANE/XYPLAN,-1

SETPT     =POINT/-2,0,0
PT1       =POINT/0,0,-1
PT2       =POINT/.75,.75,-1
PT3       =POINT/2.75,1.75,-1
PT4       =POINT/0,2.5,-1
PT5       =POINT/3.25,0,-1

CIR1      =CIRCLE/CENTER,PT2,RADIUS,.75
CIR2      =CIRCLE/CENTER,PT3,RADIUS,.75

LN1       =LINE/PT1,PT4
LN2       =LINE/PT2,LEFT,TANTO,CIR2
LN3       =LINE/PT5,PARLEL,LN1
LN4       =LINE/PT1,PT5
$$
$$MOTION
$$
FROM/SETPT
LOADTL/1,ADJUST,1
CUTTER/.75
SPINDL/2000,CLW
FEDRAT/12,IPM
RAPID,GO/LN1,PL1,PAST,LN4
TLLFT,GOLFT/LN1,PAST,LN2
GORGT/LN2,TANTO,CIR2
GOFWD/CIR2,TANTO,LN3
GOFWD/LN3,PAST,LN4
GORGT/LN4,PAST,LN1
RAPID,GOTO/SP
FINI
```

FIGURE 15-12
APT program to mill the part in Figure 15-10

The word IDENT begins the statement. IDENT,EXAMPLE,PARTNO.1 would identify the program as example 1, part number 1.

The next statement is called the initialization statement (INIT). This statement tells the computer what modes will be used for input and output. INIT,INCH/IN,INCH/OUT tells the computer that all input and output dimensions are in inches.

The fourth statement in the program is the setup statement (SETUP). This statement identifies the machine origin, tool change location, tool travel limits, positioning system to be used, and any other special requirements that may be necessary for the particular machine.

The fifth statement is the base statement. The base statement establishes a work coordinate system. BASE 6XA,6YA,6ZA establishes the work coordinate system 6.0 inches in X, 6.0 inches in Y, and 6.0 inches in Z from the machine origin. The A following the axes indicates machine absolute system (the machine origin). In geometry definition statements, the suffix B is used to indicate base coordinate system (the work coordinate system).

Other statements used in COMPACT II are: ATCHG to initiate an automatic tool change; CON to turn the coolant on; STK to identify the amount of stock to be left for finishing; FRM to assign a feedrate in inches per minute; IPR to assign a feedrate in inches per revolution.

Tool Motion Statements in COMPACT II

The tool motion statements used in COMPACT II are similar to those used in APT. The modifiers ON, TO, and PAST are used in identical fashion. Other statements are:

MOVE—To designate a move in rapid traverse.
CUT—To designate a move at feedrate.
ICON—To identify an inside contour.
OCON—To identify an outside contour.
S—To identify the starting location of a cut.
F—To identify the finish location of a cut.

In Figure 15–13 the geometry of a part is illustrated. The COMPACT II statements to move the tool from home to #1 and then around the part periphery are:

MOVE,TOLN1,PASTLN4
CUT,PARLN1,PASTLN2
CUT,PARLN2,TAN,CIR1
OCON,CIR1,CW,S(90),F(0)

S(90) means the starting point is at 90 degrees; F(0) means the finish point is at 0 degrees.

The Postprocessor

The postprocessor is a separate program that translates the part program into the codes necessary for the particular machine defined to the computer through the machine statement. *Postprocessor* is really a short term for post-processing program. Most machinery programmed using computer-aided programming languages uses RS-274 (word address) format. The postprocessor converts the APT or COMPACT II program (part geometry, auxiliary, and tool motion statements) to a word address format numerical control program. This program may then be punched into a tape using either RS-244 or RS-358 cod-

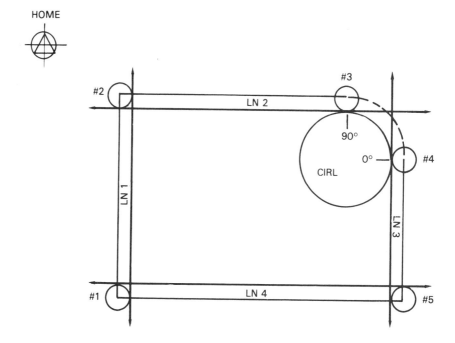

FIGURE 15-13

ing; or it may be transferred directly from the main computer to the machine's computer. The advantage of computer-aided programming languages over straight word address programming is the computer's ability to perform the calculations for very complex surfaces. Manual programming of parts requiring four and five axes is far more complicated than two- and three-axis programming. Figure 15-14 shows a part for which an APT program is written and postprocessed. Figure 15-15 shows the part geometry, and Figure 15-16 shows the APT program and the postprocessor output.

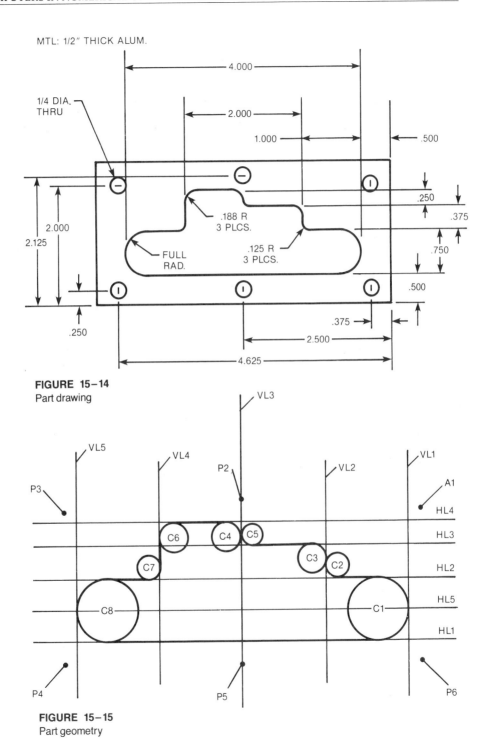

FIGURE 15–14
Part drawing

FIGURE 15–15
Part geometry

```
0001 PARTNO APT MILL EXAMPLE
0002 MACHIN/UNCM01,01
0003 MACHIN/PCPLOT,1
0004 $$
0005 $$MACRO TO PRINT A ROW OF ASTERIX ACROSS PAGE
0006 ASTRX  =MACRO/
0007 PPRINT ********************************************************
0008 TERMAC
0009 $$
0010 CALL/ASTRX
0011 PPRINT COORDINATE SYSTEM ORIGIN:
0012 PPRINT X0 = LOWER RIGHT CORNER OF PART
0013 PPRINT Y0 = LOWER RIGHT CORNER OF PART
0014 PPRINT Z0 = 2.000 INCHES ABOVE TOP OF PART
0015 CALL/ASTRX
0016 PPRINT TOOL LIST
0017 PPRINT TOOL 1:  3/8 DIA. 2 FLUTE CARBIDE END MILL
0018 PPRINT TOOL 2:  1/4 DIA. 2 FLUTE CARBIDE END MILL
0019 PPRINT TOOL 3:  NO. 4 X 90 DEG. C'DRILL
0020 PPRINT TOOL 4:  1/4 DRILL
0021 CALL/ASTRX
0022 PPRINT MACHINING SEQUENCE
0023 PPRINT TOOL 1:  SPEED 3800 RPM, FEEDRATE .003 IPR, CRO REGISTER D11
0024 PPRINT          ROUGH MILL INDSIDE CONTOUR OF PART.
0025 PPRINT
0026 PPRINT TOOL 2:  SPEED 4000 RPM, FEEDRATE .002 IPR, CRO REGISTER D21
0027 PPRINT          FINISH MILL INSIDE CONTOUR OF PART.
0028 PPRINT
0029 PPRINT TOOL 3:  SPEED 3500 RPM, FEEDRATE .004 IPR
0030 PPRINT          C'DRILL & C'SINL 1/4 HOLES TO .260 DIA. 6 PLCS.
0031 PPRINT
0032 PPRINT TOOL 4:  SPEED 3000 RPM, FEEDRATE .004 IPR
0033 PPRINT          DRILL 1/4 DIA. THRU 6 PLCS.
0034 CALL/ASTRX
0035 $$
0036 $$PART GEOMETRY
0037 $$
0038 PL1  =PLANE/XYPLAN,-2              $$TOP OF PART
0039 PL2  =PLANE/PARLEL,PL1,ZSMALL,.51  $$.01 BELOW BOTTOM OF PART
0040 CLR  =PLANE/PARLEL,PL1,ZLARGE,.05  $$.05 ABOVE TOP OF PART
0041 VL1  =LINE/YAXIS,-.5
0042 VL2  =LINE/PARLEL,VL1,XSMALL,1
0043 VL3  =LINE/YAXIS,-2.5
0044 VL4  =LINE/PARLEL,VL2,XSMALL,2
0045 VL5  =LINE/PARLEL,VL1,XSMALL,4

0046 HL1  =LINE/XAXIS,.5
0047 HL2  =LINE/PARLEL,HL1,YLARGE,.75
```

FIGURE 15–16 *(Continues to page 385)*

```
0048 HL3    =LINE/PARLEL,HL2,YLARGE,.375
0049 HL4    =LINE/PARLEL,HL3,YLARGE,.25
0050 HL5    =LINE/PARLEL,HL1,YLARGE,.375

0051 C1     =CIRCLE/XSMALL,VL1,YLARGE,HL1,YSMALL,HL2
0052 C2     =CIRCLE/XLARGE,VL2,YLARGE,HL2,RADIUS,.125
0053 C3     =CIRCLE/XSMALL,VL2,YSMALL,HL3,RADIUS,.188
0054 C4     =CIRCLE/XSMALL,VL3,YSMALL,HL4,RADIUS,.188
0055 C5     =CIRCLE/YLARGE,HL3,XLARGE,OUT,C4,RADIUS,.125
0056 C6     =CIRCLE/XLARGE,VL4,YSMALL,HL4,RADIUS,.188
0057 C7     =CIRCLE/XSMALL,VL4,YLARGE,HL2,RADIUS,.125
0058 C8     =CIRCLE/YLARGE,HL1,XLARGE,VL5,YSMALL,HL2

0059 ZSURF/PL1
0060     P1     =POINT/-.375,2
0061     P2     =POINT/-2.5,2.125
0062     P3     =POINT/-4.625,2
0063     P4     =POINT/-4.625,.25
0064     P5     =POINT/-2.5,.25
0065     P6     =POINT/-.375,.25
0066     PAT1 =PATERN/P1,P2,P3,P4,P5,P6
0067 $$
0068 $$MOTION STATEMENTS
0069 $$
0070 PPLOT/1
0071 FROM/0,4,0
0072 CALL/ASTRX
0073 PPRINT 3/8 END MILL - USES CRO REGISTER D11
0074 PPRINT ROUGH MILL INSIDE CONTOUR - LEAVE .01 STK. TO FINISH.
0075 LOADTL/1,ADJUST,1
0076 CUTTER/.375
0077 SPINDL/3800,CLW
0078 COOLNT/ON
0079 FEDRAT/.003,IPR

0080 PPRINT
0081 PPRINT POSITION TO START XY AND FEED TO DEPTH
0082 RAPID,GO/ON,VL2,CLR,ON,HL5
0083 THICK/0,.01,.01
0084 GO/ON,VL2,PL2
0085 CUTCOM/LEFT,11

0086 PPRINT
0087 PPRINT MILL .750 WIDE SLOT
0088 GO/HL2
0089     GOLFT/HL2,TANTO,C8
0090     GOFWD/C8,TANTO,HL1
0091     GOFWD/HL1,TANTO,C1
0092     GOFWD/C1,TANTO,HL2
0093     GOFWD/HL2,PAST,VL2

0094 PPRINT
0095 PPRINT MILL BALANCE OF SLOT
0096 GO/HL3
```

```
0097      GOLFT/HL3,PAST,VL3
0098      GORGT/VL3,HL4
0099      GOLFT/HL4,VL4
0100      GOLFT/VL4,ON,HL5

0101 PPRINT
0102 PPRINT RETRACT SPINDLE, CANCEL CUTTER COMP, CANCEL TOOL OFFSET
0103 RETRCT
0104 CUTCOM/OFF
0105 COOLNT/OFF
0106 RESET
0107 $$
0108 PPLOT/2
0109 CALL/ASTRX
0110 PPRINT 1/4 END MILL - USES CRO REGISTER D21
0111 PPRINT FINISH MILL SLOT
0112 LOADTL/2,ADJUST,2
0113 CUTTER/.250
0114 SPINDL/4000,CLW
0115 COOLNT/ON
0116 FEDRAT/.002,IPR

0117 PPRINT
0118 PPRINT POSTITON TO CENTER OF C1 AND FEED TO Z DRPTH
0119 THICK/0
0120 RAPID,GO/ON,(LINE/PARLEL,VL1,XSMALL,.375),CLR,ON,HL5
0121 NOPS,GO/PL2
0122 AUTOPS

0123 PPRINT
0124 PPRINT FINISH MILL CONTOUR OF I.D.
0125 CUTCOM/LEFT,21
0126 GO/HL2
0127      GOLFT/HL2,TANTO,C2
0128      GOFWD/C2,TANTO,VL2
0129      GOFWD/VL2,TANTO,C3
0130      GOFWD/C3,TANTO,HL3
0131      GOFWD/HL3,TANTO,C5
0132      GOFWD/C5,TANTO,C4
0133      GOFWD/C4,TANTO,HL4
0134      GOFWD/HL4,TANTO,C6
0135      GOFWD/C6,TANTO,VL4
0136      GOFWD/VL4,TANTO,C7
0137      GOFWD/C7,TANTO,HL2
0138      GOFWD/HL2,TANTO,C8
0139      GOFWD/C8,TANTO,HL1
0140      GOFWD/HL1,TANTO,C1
0141      GOFWD/C1,TANTO,HL2
0142      GOFWD/HL2,PAST,VL2

0143 PPRINT
0144 PPRINT RETRACT SPINDLE, CANCEL CUTTER COMP, CANCEL TOOL OFFSET
0145 RETRCT
0146 CUTCOM/OFF
```

```
0147 COOLNT/OFF
0148 RESET
0149 $$
0150 PPLOT/3
0151 CALL/ASTRX
0152 PPRINT NO. 4 C'DRILL - C'DRILL 6 PLCS. TO .260 DIA.
0153 LOADTL/3,ADJUST,3
0154 SPINDL/3500,CLW
0155 COOLNT/ON
0156 PREFUN/99,NEXT
0157 CYCLE/DRILL,.183,.004,IPR,.1
0158      GOTO/PAT1
0159 CYCLE/OFF
0160 COOLNT/OFF
0161 RESET
0162 $$
0163 DRAFT/OFF
0164 CALL/ASTRX
0165 PPRINT 1/4 DRILL - PECK DRILL .250 THRU 6 PLCS.
0166 LOADTL/4,ADJUST,4
0167 SPINDL/3000,CLW
0168 COOLNT/ON
0169 PREFUN/99,NEXT
0170 CYCLE/DEEP,.6,.003,IPR,.1,INCR,.25
0171      GOTO/PAT1
0172 CYCLE/OFF
0173 COOLNT/OFF
0174 RESET
0175 FINI


     NO DIAGNOSTICS ELICITED DURING TRANSLATION PHASE
     185 N/C SOURCE RECORDS (SYSIN)


          SECTION 1 ELAPSED CPU TIME IN MIN/SEC IS 0000/00.4433
          SECTION 2 ELAPSED CPU TIME IN MIN/SEC IS 0000/00.1866
```

```
UNCM01 03.000.I370 MACHIN/UNCM01, 1   DATE 88.319     PAGE   1
APT MILL EXAMPLE                                                (INCH)
INPUT   CLREC N3G2X34Y34R34Z34Q34I34J34K34A43P4F32S4T2D2H2M2
    7       9  APT MILL EXAMPLE
    7       9 $
    7       9  LEADER/   20.0
    7       9 $
    7       9  N001 G90 G00 G70 G80 G40$
    7       9  ***************************************************
   11      11  COORDINATE SYSTEM ORIGIN:
   12      13  X0 = LOWER RIGHT CORNER OF PART
   13      15  Y0 = LOWER RIGHT CORNER OF PART
   14      17  Z0 = 2.000 INCHES ABOVE TOP OF PART
    7      20  ***************************************************
   16      22  TOOL LIST
   17      24  TOOL 1:   3/8 DIA. 2 FLUTE CARBIDE END MILL
   18      26  TOOL 2:   1/4 DIA. 2 FLUTE CARBIDE END MILL
   19      28  TOOL 3:   NO. 4 X 90 DEG. C'DRILL
   20      30  TOOL 4:   1/4 DRILL
    7      33  ***************************************************
   22      35  MACHINING SEQUENCE
   23      37  TOOL 1:   SPEED 3800 RPM, FEEDRATE .003 IPR, CRO REGISTER D11
   24      39            ROUGH MILL INDSIDE CONTOUR OF PART.
   25      41
   26      43  TOOL 2:   SPEED 4000 RPM, FEEDRATE .002 IPR, CRO REGISTER D21
   27      45            FINISH MILL INSIDE CONTOUR OF PART.
   28      47
   29      49  TOOL 3:   SPEED 3500 RPM, FEEDRATE .004 IPR
   30      51            C'DRILL & C'SINL 1/4 HOLES TO .260 DIA. 6 PLCS.
   31      53
   32      55  TOOL 4:   SPEED 3000 RPM, FEEDRATE .004 IPR
   33      57            DRILL 1/4 DIA. THRU 6 PLCS.
    7      60  ***************************************************
   71      64  FROM/XYZ =      0.0000     4.0000     0.0000
    7      67  ***************************************************
   73      69  3/8 END MILL - USES CRO REGISTER D11
   74      71  ROUGH MILL INSIDE CONTOUR - LEAVE .01 STK. TO FINISH.
   75      73  TOOL TIME =    0.04
   75      73 $
   75      73  N002 G90 G00 X.0000 Y4.0000 T01$
   75      73  N003 S0500 M03$
   75      73  N004 G44 Z.0000 H01$
   77      77  N005 S3800$
   78      79  N006 M08$
   80      83
   81      85   POSITION TO START XY AND FEED TO DEPTH
   82      88  N007 G00 X-1.5000 Y.8750$
   82      88  N008 Z-1.9500$
                        TAPE    TIME    WARNING
                PAGE    6.38    0.15       0
                TOTAL   6.38    0.15       0
```

```
UNCM01 03.000.I370 MACHIN/UNCM01, 1   DATE 88.319    PAGE   2
APT MILL EXAMPLE                                              (INCH)
INPUT   CLREC N3G2X34Y34R34Z34Q34I34J34K34A43P4F32S4T2D2H2M2
    84       92 N009 G01 Z-2.5100 F11.40$
    85       94 N010 G17$
    86       96
    87       98   MILL .750 WIDE SLOT
    88      100 N011 G41 X-1.5000 Y.8760 J1.0000 D11$
    88      100 N012 G01 Y1.0525$
    89      102 N013 X-4.1250$
    91      107 N014 G03 X-4.1250 Y.6975 I-4.1250 J.8750$
    91      107 N015 G01 X-.8750$
    93      112 N016 G03 X-.8750 Y1.0525 I-.8750 J.8750$
    93      112 N017 G01 X-1.6975$
    94      114
    95      116   MILL BALANCE OF SLOT
    96      118 N018 Y1.4275$
    97      120 N019 X-2.6975$
    98      122 N020 Y1.6775$
    99      124 N021 X-3.3025$
   100      126 N022 Y.8750$
   101      128
   102      130   RETRACT SPINDLE, CANCEL CUTTER COMP, CANCEL TOOL OFFSET
   103      132 N023 G00 Z.0000$
   104      134 N024 G40$
   105      136 N025 M09$
   106      138 N026 G00 Z.0000$
   106      138 N027 G49$
   106      138 N028 M01$
     7      143 *************************************************
   110      145   1/4 END MILL - USES CRO REGISTER D21
   111      147   FINISH MILL SLOT
   112      149 TOOL TIME =     1.09
   112      149 $
   112      149 N029 G90 G00 X-3.3025 Y.8750 T02$
   112      149 N030 S0500 M03$
   112      149 N031 G44 Z.0000 H02$
   114      153 N032 S4000$
   115      155 N033 M08$
   117      159
   118      161   POSTITON TO CENTER OF C1 AND FEED TO Z DRPTH
   120      166 N034 G00 X-.8750$
   120      166 N035 Z-1.9500$
   121      168 N036 G01 Z-2.5100 F8.00$
   123      171
   124      173   FINISH MILL CONTOUR OF I.D.
   126      177 N037 G41 X-.8750 Y.8760 J1.0000 D21$
   126      177 N038 G01 Y1.1250$
                      TAPE    TIME    WARNING
              PAGE    3.70    1.21       0
              TOTAL  10.08    1.36       0
```

```
UNCM01 03.000.I370 MACHIN/UNCM01, 1    DATE 88.319     PAGE   3
APT MILL EXAMPLE                                                   (INCH)
INPUT   CLREC N3G2X34Y34R34Z34Q34I34J34K34A43P4F32S4T2D2H2M2
  127      179 N039 X-1.3750$
  129      184 N040 G02 X-1.6250 Y1.3750 I-1.3750 J1.3750$
  129      184 N041 G01 Y1.4370$
  131      189 N042 G03 X-1.6880 Y1.5000 I-1.6880 J1.4370$
  131      189 N043 G01 X-2.3814$
  133      194 N044 G02 X-2.6263 Y1.6997 I-2.3814 J1.7500$
  134      197 N045 G03 X-2.6880 Y1.7500 I-2.6880 J1.6870$
  134      197 N046 G01 X-3.3120$
  136      202 N047 G03 X-3.3750 Y1.6870 I-3.3120 J1.6870$
  136      202 N048 G01 Y1.3750$
  136      207 N049 G02 X-3.6250 Y1.1250 I-3.6250 J1.3750$
  138      207 N050 G01 X-4.1250$
  140      212 N051 G03 X-4.1250 Y.6250 I-4.1250 J.8750$
  140      212 N052 G01 X-.8750$
  142      217 N053 G03 X-.8750 Y1.1250 I-.8750 J.8750$
  142      217 N054 G01 X-1.6250$
  143      219
  144      221   RETRACT SPINDLE, CANCEL CUTTER COMP, CANCEL TOOL OFFSET
  145      223 N055 G00 Z.0000$
  146      225 N056 G40$
  147      227 N057 M09$
  148      229 N058 G00 Z.0000$
  148      229 N059 G49$
  148      229 N060 M01$
    7      234   *****************************************************
  152      236   NO. 4 C'DRILL - C'DRILL 6 PLCS. TO .260 DIA.
  153      238 TOOL TIME =      1.38
  153      238 $
  153      238 N061 G90 G00 X-1.6250 Y1.1250 T03$
  153      238 N062 S0500 M03$
  153      238 N063 G44 Z.0000 H03$
  154      240 N064 S3500$
  155      242 N065 M08$
  158      248 N066 G81 G99 X-.3750 Y2.0000 R-1.9000 Z-2.1830 F14.00$
  158      249 N067 X-2.5000 Y2.1250$
  158      250 N068 X-4.6250 Y2.0000$
  158      251 N069 Y.2500$
  158      252 N070 X-2.5000$
  158      253 N071 X-.3750$
  159      255 N072 G80$
  160      257 N073 M09$
  161      259 N074 G00 Z.0000$
  161      259 N075 G49$
  161      259 N076 M01$
    7      264   *****************************************************
                    TAPE     TIME    WARNING
             PAGE   5.63     1.37       0
             TOTAL 15.72     2.73       0
```

```
UNCM01 03.000.I370 MACHIN/UNCM01, 1    DATE 88.319      PAGE   4
APT MILL EXAMPLE                                                     (INCH)
INPUT   CLREC N3G2X34Y34R34Z34Q34I34J34K34A43P4F32S4T2D2H2M2
   165     266  1/4 DRILL - PECK DRILL .250 THRU 6 PLCS.
   166     268 TOOL TIME =      0.21
   166     268 $
   166     268 N077 G90 G00 X-.3750 Y.2500 T04$
   166     268 N078 S0500 M03$
   166     268 N079 G44 Z.0000 H04$
   167     270 N080 S3000$
   168     272 N081 M08$
   171     278 N082 G83 G99 X-.3750 Y2.0000 R-1.9000 Z-2.6000 Q.2500 F9.00$
   171     279 N083 X-2.5000 Y2.1250$
   171     280 N084 X-4.6250 Y2.0000$
   171     281 N085 Y.2500$
   171     282 N086 X-2.5000$
   171     283 N087 X-.3750$
   172     285 N088 G80$
   173     287 N089 M09$
   174     289 N090 G00 Z.0000$
   174     289 N091 G49$
   174     289 N092 M01$
   175     291 TOOL TIME =      0.56
   175     291 N093 G00 Y4.0000 T15$
   175     291 N094 M30$
   175     291 =$
                      TAPE     TIME    WARNING
              PAGE    2.13     0.56       0
              TOTAL  17.85     3.29       0


UNCM01 03.000.I370 MACHIN/UNCM01, 1    DATE 88.319      PAGE   5
APT MILL EXAMPLE                                                     (INCH)
INPUT   CLREC N3G2X34Y34R34Z34Q34I34J34K34A43P4F32S4T2D2H2M2
             TOOL SEQUENCE
INPUT   CLREC     TOOL  OFFSET  LENGTH Z
   75      73       1      1     0.0000
  112     149       2      2     0.0000
  153     238       3      3     0.0000
  166     268       4      4     0.0000
        TOTAL PART PROGRAM CPU TIME IN MIN/SEC IS 0000/02.3966
              **** END OF APT PROCESSING ****
```

FIGURE 15–16
APT program and postprocessor output for part shown in Figure 15–14

COMPUTER GRAPHICS PROGRAMMING

The newest form of NC programming is called CAM. CAM stands for computer aided manufacturing. When using a CAM system, the programmer either calls up an existing part drawing, or defines the part geometry to the computer. Next, the cutter path is drawn around the part and the necessary information on cut direction, tool, speeds, and feeds are input. This information is then converted into either an APT file, or a cutter centerline data file by a postprocessor. This data is then fed through a secondary postprocessor, which produces the necessary NC code for a given machine.

There are a number of different CAM systems on the market. The following is a brief explanation of several different types.

Digitizing Systems

Digitizing systems use an existing part drawing to obtain the geometry information. The drawing used must be drawn to a true scale of the finished part. The scale drawing is fed into a digitizer which is connected to the CAM system computer. A *digitizer* is a device consisting of a table with a probe or other sensor attached. The sensing device is passed over the drawing, converting drawing lines in the necessary mathematical information into electronic form which the computer needs to recreate the drawing. The cutter path is then defined by the programmer and the cutter, speed, and feed information input. The result is then postprocessed into the necessary tape code for the CNC machine.

Digitizing is one of the simplest CAM methods, but it is also the least accurate. The accuracy of the part geometry is dependent upon the accuracy of the scaled drawing which was digitized. Fortunately this shortcoming is minimized by the fact that drawings 30 times size or larger can be used.

Scanning Systems

Scanning systems use the part itself rather than a part drawing to obtain the geometric database to machine the part. Scanning is used when complex curves are to be machined which are difficult to draw and which do not fit a true mathematical model. Automobile bodies are an example of such curves. A *scanner* is a probing device connected to the CAM system computer which is passed over a model of the part. The probe feeds information concerning the part geometry into the computer which then calculates the points necessary to define the part shape. The cutter path, tool data, speeds, and feeds are then input. The information is then fed to the postprocessor which will convert the information into the necessary tape coding.

CAM Systems

Although the two methods mentioned above are referred to as CAM systems, CAM is also the term used to describe computer graphics programming. CAM systems can be run on a mainframe computer, or on a PC. Mainframe systems are found in larger plants primarily involved with four- or five-axis programming. Microcomputer-based CAM systems function well for three-axis. New systems can handle four-axis programming as well. Microcomputer systems are an economical choice for many small to midsized shops. Figure 15–17 illustrates a PC-based CAM system running on an APPLE Macintosh. Figure 15–18 depicts another microbased system.

In CAM (graphics) programming, the programmer defines the part geometry to the computer using one or more of several input devices. These devices may be the keyboard, a mouse, a digitizer, or a light pen. After the part geometry is defined, the cutter path is drawn around the part. Information on the cut direction, tool data, speeds, and feeds are then input. This information is then translated by a series of postprocessors into the necessary NC code. Figure 15–19 shows a cutter and tool path generated on a mainframe-based CAM system.

FIGURE 15–17
Mac EZ-CAM II *(Photo courtesy of Bridgeport Machine Inc.)*

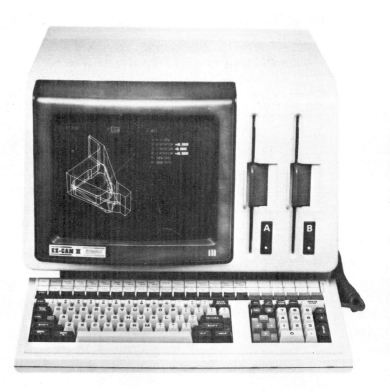

FIGURE 15–18
EZ-CAM II *(Photo courtesy of Bridgeport Machine Inc.)*

Once programmed and postprocessed, the cutter path can be plotted out on paper as a preliminary prove-out of the program. Figure 15–20 shows one type of plotter. The program plot can help the programmer spot errors that have been overlooked when the cutter path was on the computer monitor screen.

There are many types of CAM systems on the market. One need only glance at the pages of an industry trade magazine to appreciate the number that exist.

CAD/CAM Programming

CAD/CAM programming is one of the more sophisticated CAM programming systems. In a CAD/CAM system, the part geometry is obtained from the CAD (computer aided design) drawing itself. The cutter path, tool, speeds, and feeds are then defined and the program postprocessed into the NC tape code. There are two basic types of CAD/CAM systems: stand alone and modular.

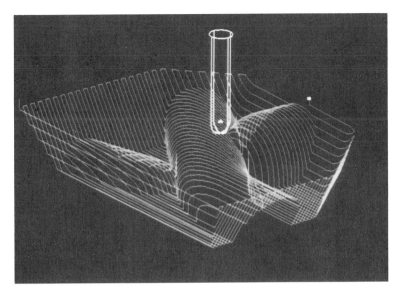

FIGURE 15–19
Cutter and tool path generated on a mainframe-based CAM system
(Photo courtesy of Battelle Inc.)

Stand alone systems contain both the CAD and the CAM modules in one system, furnished by one supplier. Generally, these systems are intended to function only with the supplied modules and will not link with other vendors products.

Modular systems consist of a CAD package made by one manufacturer, and a CAM package by another. The CAM packages are designed to link to the most popular CAD packages such as AutoCad, Micro CADAM, and GEOSPOT.

CIM

CAD/CAM programming coupled with DNC systems for program delivery to the individual machines form the fundamental building block for CIM (computer integrated manufacturing). CIM is not a true reality at the time of this publication, but is the wave of the future. In a CIM system, the entire manufacturing process is done with the aid of computers. Product design, manufacturing engineering, production control, quality assurance, procurement, accounting, and management functions would all be linked together in a shared database. The goal of all this is to eliminate paperwork, eliminate duplication of effort, and reduce overall costs while improving part quality and delivery schedules. Much is needed in the way of standardization to achieve a CIM environment, but much has already been accomplished.

FIGURE 15-20
Plotter *(Photo courtesy of Calcomp Inc.)*

For a company to be competitive in the coming years, a move toward computer integration will become a necessity. It is important therefore, for students desiring to make a career in the manufacturing fields to learn and improve upon their computer skills.

SUMMARY

The important concepts presented in this chapter are:

- Computers are used three basic ways in numerical control programming: in offline programming, computer-aided programming, and computer graphics programming.
- Offline programming terminals allow the NC program to be written away from the machine. They do not assist the programmer in any way other than eliminating the need to enter the program at the machine's computer console.
- Computer-aided programming permits the programming of parts for many machines using one programming language. The computer handles the necessary mathematical calculations for the cutter path. A postprocessor then translates this information into codes for a particular machine.
- Three types of statements are used in a computer-aided program: part geometry statements, auxiliary statements, and tool motion statements.
- There are a number of computer-aided programming languages. The one a company uses depends upon the parts it produces and the computers available.
- Graphics programming called CAM programming is being used to simplify the part programming process.
- There are four basic types of CAM systems: digitizing, scanning, CAM, and CAD/CAM.
- There are two basic types of CAD/CAM systems: stand alone and modular.

REVIEW QUESTIONS

1. What three ways are computers used in numerical control programming?
2. What is offline programming?
3. What does computer-aided programming allow the programmer to do?
4. What three types of statements are used in computer-aided programs?
5. What is CAM programming?
6. What are four types of CAM systems?
7. What is the difference between a scanning and a digitizing system?
8. What is the difference between a CAM and a CAD/CAM system?

CHAPTER 16

The Future of Numerical Control

OBJECTIVES Upon completion of this chapter, you will be able to:

- Explain why the use of CNC will increase in prototype and small lot job shops.
- Describe a flexible machining system.
- Describe a machining cell.
- Describe the responsibilities of the NC electronics technician, machine operator/ set-up operator, and part programmer.

Numerical control will play an increasingly important role in manufacturing in the coming years. CNC is already being applied to machine tools, punch presses, sheet metal brakes, electrical discharge machines, welding machinery, and inspection equipment. Smaller sized production shops, prototype operations, and large manufacturing concerns will all benefit from recent and continuing developments in numerical control, robotics, and computer technology.

NC IN PROTOTYPE AND JOB SHOPS

The ongoing development of less expensive numerical control systems will offer increasing options to companies that today cannot justify a numerical control system. The lower cost of acquiring machining and turning centers, coupled with the ease of programming and other features of the newest generation of CNC controllers, will result in the adoption of CNC machinery by more and more small job shops. Competition from foreign sources is forcing all companies to look for ways to improve quality while making the changes in design that market conditions so often require. CNC machinery can fulfill both requirements: (1) The repeatability of CNC can improve the overall quality of parts produced; and (2) since CNC uses software programs to produce part shapes, what would have been major retooling becomes the editing and revising of the part program.

Recently, a new type of DNC retrofit system was introduced specifically geared to prototype shop requirements. In this system a microcomputer such as the popular Apple II is used as the controller. The executive program for the system is software, not firmware, requiring little or no modification of the computer. This system is a surprisingly low cost way for a shop to acquire a DNC system. It is not designed for the demands of a manufacturing operation but will handle the one-of-a-kind parts made in prototype, die, or moldmaking shops. Designed for used on vertical mills, it is a three-axis contouring system capable of circular, helical, and linear interpolation.

A problem common to all companies is the shortage of skilled machinists. In smaller companies, the shortage of general machinists, tool and die makers, and mold makers is most acute. In coming years the shortage of skilled prototype machinists and instrument makers is likely to be felt by scientific and research organizations that have their own prototype shops. In addition, increasingly complex part geometries are being required for new technology applications. CNC offers solutions to all these problems.

CNC IN MANUFACTURING

The most exciting developments in NC applications are taking place in large scale manufacturing. In many industries, computer integration of the entire manufacturing process is believed to be possible in the coming decades. The computer capability for *computer integrated manufacturing (CIM)* already exists, but software bases and computer standards to allow networking of design, manufacturing, purchasing, inventory, and marketing functions must be developed and refined. In a CIM system, these various functions are interconnected, using the instant access to information that the computer allows, to eliminate duplication of effort, reduce inventories, reduce part handling, and provide a higher percentage of chip-making time. Although not yet a reality, one of the major building blocks in a CIM system is currently being produced and used by some industries—the flexible machining system (FMS).

Flexible Machining System

A *flexible machining system (FMS)* is a system of CNC machines, robots, and part transfer vehicles that can take a part from raw stock or casting and perform all necessary machining, part handling, and inspection operations to make a finished part or assembly. It is an entire unmanned, software-based, manufacturing/assembly line. An FMS consists of four major components: the CNC machines, coordinate measuring machines, part handling and assembly robots, and part/tool transfer vehicles. Figure 16–1 illustrates a small flexible

FIGURE 16–1 *(Photo courtesy of Cincinnati Milacron)*

machining system. This system employs a turning center, a horizontal machining center, and a vertical machining center. A single track-guided robot is used as both a load/unload robot and a transfer vehicle.

The main element in an FMS is the CNC machining or turning center. The automatic tool changing capability of these machines allows them to run untended, given the proper support system. Tool monitoring systems built into the CNC machine are used to detect and replace worn tools. The major obstacles in an FMS are not the machining centers but the support systems for the machines, such as part load/unload and part transfer.

Inspection in an FMS is accomplished through the use of coordinate measuring machines. These operate much like CNC machinery in that they are programmed to move to different positions on a workpiece. Instead of using a rotating spindle and a cutting tool, a coordinate measuring machine is equipped with electronic gaging probes which measure features on a workpiece. The results of the gaging are compared to acceptable limits programmed into the machine.

Robots are frequently used in an FMS to load and unload parts from the machines. Since robots are programmed pieces of equipment that lack the ability to make judgments, special workholding fixtures are employed on the transfer vehicles to orient the workpiece so that the robot can handle it correctly. Specially designed machine fixtures and clamping mechanisms are employed to ensure correct placement and clamping of the part on the machine. All part handling must be accomplished in a specific orderly fashion, with coordination of the part transfer vehicle, the robot, and the CNC machines. Future robots will probably employ some type of artificial intelligence which will enable them to make limited judgments as to workpiece orientation and take the necessary corrective actions.

The third critical component of an FMS is the tool and workpiece transfer vehicles. These vehicles shuttle workpieces from machine to machine. They also shuttle tool magazines to and from the machinery to maintain an adequate supply of sharp cutting tools at each CNC machine. Transfer vehicles employed in current flexible manufacturing systems are of four major types: automatic guided vehicles (AGV), wire guided vehicles, air cushion vehicles, and hardware guided vehicles.

Automatic guided vehicles rely on onboard sensors and/or a program to determine the path they take. There is no hardware connecting them to the system. An advantage of AGVs is that they can be reprogrammed to take different routes, eliminating the need to run tracks or wires for each route change. The corresponding disadvantage of AGVs is that they are the most difficult of the part delivery vehicles to make function, because of the lack of hardware connection.

A *wire guided vehicle* uses a wire buried in the floor to define its path. A sensor on the vehicle detects the location of the wire. A major advantage of wire guided vehicles is the ability to use the wire as opposed to an AGV without the need to have a hardware system such as an overhead wire or track on the floor. The disadvantage of wire guided vehicles is the necessity of installing new wire in the floor if a route change is required.

An *air cushion vehicle* is guided by some external hardware device, such as an overhead wire, but glides on a cushion of air rather than a track system. When using air cushioned vehicles, particular attention to chip removal and control must be built into the FMS. Chips in the path of an air cushion vehicle will stop its progress. These vehicles are generally used for straight paths.

Hardware guided vehicles are the most reliable but least flexible of the transfer vehicles. A track on the floor or an overhead guide rail controls the vehicle path. The advantages of these vehicles are their reliability and the ease of coordinating them with the rest of the system. The major disadvantage is, of course, the need to run new rail or track whenever a vehicle route change or new route is deemed necessary. A large FMS may employ several different types of vehicles, depending on the requirements of different parts of the manufacturing line.

FIGURE 16-2   *(Photo courtesy of Cincinnati Milacron)*

Machining Cells

Large flexible machining systems are often a collection of smaller coordinated units called machining cells. A *machining cell* is a system consisting of one or more CNC machines and a parts handling device, such as a robot. The cell performs a machining operation or a specific sequence of operations. Flexible machining systems are not in widespread use at present although their numbers are increasing. Stand-alone machining cells, however, are widely employed by manufacturers, frequently with a view to incorporating them into an FMS at a later date. Figure 16-2 shows a machining cell consisting of a turning center and a grinder. A robot on an overhead gantry services the two machines.

EMPLOYMENT OPPORTUNITIES IN NC

A number of skilled positions have been created by numerical control. The most common jobs are NC electronics technician, machine operator/setup operator, and part programmer.

Electronics Technician

Numerical control and computer numerical control equipment are electrical systems interfaced to a machine tool. The electronics necessary for a CNC machine to function are complex. The NC electronics technician is a skilled technician who specializes in the maintenance of numerical control equipment. The NC technician must be well trained in digital electronics and possess a knowledge of the cycles and functions of NC machinery. The technician must be able to troubleshoot and correct problems that occur in the electronic circuitry of various NC machines.

NC technicians generally acquire their skills through a two-year junior college program in digital electronics. Additional education in numerical control is often provided by the employer in the form of NC manufacturers' technical school classes and seminars.

Machine Operator/Setup Operator

The machine operator/setup operator is responsible for preparing an NC machine to run a program and for setting up the fixtures, tools, and workpieces. The operator must possess a knowledge of general machine shop practices and techniques, as well as the cycles and functions of an NC machine. The operator is responsible for overriding programmed speeds and feeds if required during machining. The operator also assigns the tool length offsets to the appropriate tool registers and may be called upon to single-step a program through its first cycle. The operator must also be trained in the use of precision measuring instruments as he or she is often responsible for measuring the parts as they are finished.

Machine operators/setup operators acquire their training either by years of running other types of manufacturing equipment and then transferring to an NC operator's position, or through a two-year junior college program. Factory seminars and other coursework may be provided by the employer as required.

Part Programmer

The part programmer is a highly skilled individual responsible for writing the programs that run on numerically controlled equipment. He or she must be trained in general machine shop practice, mathematics, and the use of computers. Based on the part drawing, the programmer selects a machine to machine the part and devises a machining strategy, listing the tools to be used and the coordinates necessary to accomplish the operations. This information is then assembled into a part program written for the particular machine selected.

An NC programmer may acquire training through a two-year junior college, a four-year engineering technology degree program, or by transferring from positions as journeyman machinists or tool and die makers. NC programmers take additional coursework and factory seminars as required by the employer. The educational requirements for a programmer vary with the employer.

SUMMARY

The important concepts presented in this chapter are:

- The use of CNC will increase in prototype and small job shops due to the arrival of lower cost controllers containing many advanced programming features.
- A flexible machining system is an unmanned manufacturing/assembly line that can take a part from raw stock and perform all the necessary operations to produce a finished part or assembly.
- A machining cell is a system of one or more CNC machines and part handling robots that performs a specific sequence of operations.
- An NC electronics technician is responsible for maintaining the electronics of an NC or CNC system.
- An NC operator/setup operator is responsible for preparing a machine prior to running a program and monitoring the machine during the program execution.
- An NC part programmer is responsible for creating the part program.

REVIEW QUESTIONS

1. For what reason will the use of CNC increase in one-of-a-kind and prototype shops?
2. What is a flexible machining system?
3. What are the four major components of an FMS?
4. What are the four types of part/tool transfer vehicles?
5. What is a machining cell?
6. What are the responsibilities of the CNC electronics technician?
7. What are the responsibilities of the CNC machine operator/setup operator?
8. What are the responsibilities of the CNC part programmer?

APPENDIX 1

EIA Codes

PREPARATORY FUNCTIONS

G00—Denotes rapid traverse for point-to-point positioning.

G01—Linear interpolation.

G02—Circular interpolation clockwise.

G03—Circular interpolation counterclockwise.

G04—Dwell.

G05–07—Unassigned.

G08—Acceleration at a smooth rate.

G09—Deceleration at a smooth rate.

G10–16—Unassigned.

G13–16—Axis selection codes.

G17—XY plane selection.

G18—ZX plane selection.

G19—YZ plane selection.

G20–32—Unassigned.

G33—Thread cutting, constant lead.

G34—Thread cutting, increasing lead.

G35—Thread cutting, decreasing lead.

G36–39—Unassigned.

G40—Cutter diameter compensation cancel.

G41—Cutter diameter compensation left.

G42—Cutter diameter compensation right.

G43—Cutter compensation inside corner (used to adjust for differences in programmed and actual cutter size).

G44—Cutter compensation outside corner (used to adjust for differences in programmed and actual cutter size).

G45–49—Unassigned.

G50–59—Used with adaptive controls.

G60–69—Unassigned.

G70—Inch programming.

G71—Metric programming.

G72—Three-dimensional circular interpolation clockwise.

G73—Three-dimensional circular interpolation counterclockwise.

G74—Multiquadrant circular interpolation cancel.

G75—Multiquadrant circular interpolation.
G76–79—Unassigned.
G80—Cycle cancel.
G81—Drill cycle.
G82—Drill cycle with dwell.
G83—Intermittent or deep hole drilling cycle.
G84—Tapping cycle.
G85–89—Boring cycles.
G90—Absolute positioning.
G91—Incremental positioning.
G92—Register preload code.
G93—Inverse time feedrate.
G94—Inches (millimeters) per minute feedrate.
G95—Inches (millimeters) per revolution feedrate.
G96—Unassigned.
G97—Revolutions per minute spindle speed.
G98–99—Unassigned.

MISCELLANEOUS FUNCTIONS

M00—Program stop.
M01—Optional (planned) stop.
M02—End of program.
M03—Spindle on clockwise.
M04—Spindle on counterclockwise.
M05—Spindle off.
M06—Tool change.
M07—Coolant on (flood).
M08—Coolant on (mist).
M09—Coolant off.
M10—Automatic clamp.
M11—Automatic unclamp.
M12—Synchronize multiple axes.
M13—Spindle clockwise and coolant on.
M14—Spindle counterclockwise and coolant on.
M15—Rapid motion positive direction.
M16—Rapid motion negative direction.
M17–18—Unassigned.
M19—Spindle orient and stop.
M20–29—Unassigned.
M30—End of tape, will rewind tape automatically.

M31—Interlock bypass.
M32–39—Unassigned.
M40–46—Gear changes if used, otherwise unassigned.
M47—Continues program execution from the start of program.
M48—Cancel M47.
M49—Deactivate manual speed or feed override.
M50–57—Unassigned.
M58—Cancel M59.
M59—RPM hold.
M60–99—Unassigned.

OTHER ADDRESSES

A—Rotary motion about the X axis.
B—Rotary motion about the Y axis.
C—Rotary motion about the Z axis.
D—Angular dimension around a special axis. Also used for a third feed function.
E—Angular dimension around a special axis, or special feed function.
H—Unassigned.
I—X axis arc centerpoint.
J—Y axis arc centerpoint.
K—Z axis arc centerpoint.
L—Unassigned.
O—Used on some controllers in place of N address for sequence numbers.
P—Special rapid traverse code, or a third axis parallel to the X axis.
Q—Special rapid traverse code, or a third axis parallel to the Y axis.
R—Special rapid traverse code, or a third axis parallel to the Z axis. Also used for radius designation.
U—Secondary axis parallel to X.
V—Secondary axis parallel to Y.
W—Secondary axis parallel to Z.

Machinist Shop Language Commands

COMMANDS

This is a list of Machinist Shop Language commands used with the CNC machine in this text.

A (absolute)—Specifies absolute positioning.

ARC—If used by itself, institutes the cutting of an arc. If used with CW or CCW, tells the computer that an arc is to be cut in a clockwise or counterclockwise direction. Following an ARC/direction command the computer will look for information describing the arc in the following two events.

AUX (auxiliary)—Allows changes to be made in normal control functions. The direction of the X, Y, and Z axes may be changed and mirror imaging may be instituted, for example. AUX codes act like the miscellaneous functions in word address format.

CALL—Executes a subroutine. CALL 1 for example, tells the machine to carry out the instructions in subroutine 1.

CCW—Specifies a counterclockwise arc rotation.

CW—Specifies a clockwise arc rotation.

DO—Do loop. Anything that is repeated over equal intervals of space (a row of holes, for example) may be placed in a do loop. Do 5 tells the machine to perform the operation that follows 5 times.

DWELL—Halts execution of the program until the start button is manually depressed. When start is pressed, the program continues, starting at the next event.

END—There are three uses for the END command. (1) In a do loop, END signals the end of the loop. (2) In a subroutine, END signals the end of the subroutine. (3) In a program, END signals the end of the program.

F (feed)—Tells the machine to make tool movements at the programmed feedrate.

FEED—Assigns a feedrate.

G (G code)—A preparatory function, G code calls up certain "canned" or standard cycles contained within the computer for such operations as drilling, boring, and reaming.

I (incremental)—Specifies incremental positioning.

R (rapid)—Tells the machine to make tool movements at rapid traverse.

SUBR (subroutine)—Like a miniprogram within a program, sections of a program that are to be repeated are often placed in a subroutine to eliminate having to program the same information twice. The subroutine is instituted by using the CALL command.

TOOL—Like a dwell, halts the program so that a tool can be inserted in the spindle. If the machine is equipped with three axes, TOOL also acts to assign certain tool length and/or cutter diameter values.

V (variable)—Assigns values to certain program variables such as canned cycle feedrates and feed engagement points.

PREPARATORY FUNCTIONS (G CODES)

Following is a list of preparatory functions used in conjunction with Machinist Shop Language.

G40—Cutter diameter compensation cancel.

G41—Cutter diameter compensation left.

G42—Cutter diameter compensation right.

G51—Institute polar rotation.

G52—Polar rotation cancel.

G53—Institute scaling.

G54—Scaling cancel.

G76—Hole milling.

G77—Circular pocket milling.

G78—Rectangular pocket milling.

G79—Bolt circle pattern.

G80—Canned cycle cancel.

G81—Basic drilling cycle.

G82—Counter-boring/spot-facing cycle (feed in, timed dwell, rapid out).

G83—Peck drilling cycle (feed in, rapid out, feed in, etc.).

G85—Boring cycle (feed in, feed out).

G86—Boring in one direction cycle (feed in, rapid out).

G87—Chip breaking cycle (feed in, retract .050, feed in, etc.).

G89—Flat bottom boring cycle (feed in, timed dwell, feed out).

VARIABLE (V) CODES

Following is a list of variable codes commonly available in Machinist Shop Language.

V11—X axis polar center (must be absolute dimension).

V12—Y axis polar center (must be absolute dimension).

V13—Polar rotation index angle (must be incremental). A negative number indicates clockwise rotation, a positive number counterclockwise rotation.

V14—Radius for polar moves (value must be positive).

V15—Angle for polar moves or angle of first hole in a bolt circle pattern.

V16—Angle of last hole in a bolt circle pattern or X axis scaling value.

V17—Number of holes to be machined in a bolt circle or Y axis scaling value.

V18—Diameter of bolt circle or Z axis scaling value.

V20—Feedrate for G80 series canned cycle.

V21—Buffer zone for G80 series canned cycles. Must be .100 for G83 or G87.

V22—Dwell time when using G82 or G89.

V23—Maximum peck when using G83 or G87.

V40—Z axis start height for pecked milling.

V41—Length of pocket on X axis (must be incremental).

V42—Width of pocket on Y axis (must be incremental) or number of rotations for helical interpolation.

V43—Depth of pocket on Z axis.

V44—Pocket corner radius or diameter of circle if circular pocket milling.

V45—Stepover value for pocket milling.

V46—Maximum depth of cut.

V47—Stock left for finish pass.

V48—Finish pass feedrate.

V49—Tool diameter for pocket milling (cutter comp cannot be active for pocket milling).

AUXILIARY (AUX) CODES

Following is a complete list of auxiliary codes commonly used in Machinist Shop Language.

AUX 100—Reverses sign of X axis.

AUX 200—Reverses sign of Y axis.

AUX 300—Reverses sign of X and Y axes.

AUX 400—Reverses sign of Z axis.

AUX 500—Reverses sign of X and Z axes.

AUX 600—Reverses sign of Y and Z axes.

AUX 700—Reverses sign of X, Y, and Z axes.

AUX 800—Turns off mirror image.

AUX 1000—Causes machine to continue to the next move before reaching its target (used only with contouring operations).

AUX 1101—Absolute zero shift.

AUX 1110—Turns off software limits.

AUX 1111—Turns on software limits.

AUX 1400—Feed percentage override for feedrate moves.

AUX 1401—Feed percentage override for feed and rapid moves.

AUX 1900—Single-step event mode.

AUX 1901—Single-step axis movement mode.

AUX 2000—Cancels AUX 1000.

AUX 2500—Sets control to use Z axis.

AUX 2600—Sets control to allow manual use of Z axis.

APPENDIX 3

Word Address Codes Used in Text Examples

PREPARATORY FUNCTIONS (G CODES) USED IN MILLING

Following is a list of preparatory functions used in CNC milling examples in this text. Other codes commonly used on General Numeric controllers are also listed.

G00—Rapid traverse positioning.
G01—Linear interpolation (feedrate movement).
G02—Circular interpolation clockwise.
G03—Circular interpolation counterclockwise.
G04—Dwell.
G10—Tool length offset value.
G17—Specifies X/Y plane.
G18—Specifies X/Z plane.
G19—Specifies Y/Z plane.
G20—Inch data input (on some systems).
G21—Metric data input (on some systems).
G22—Safety zone programming.
G23—Cross through safety zone.
G27—Reference point return check.
G28—Return to reference point.
G29—Return from reference point.
G30—Return to second reference point.
G40—Cutter diameter compensation cancel.
G41—Cutter diameter compensation left.
G42—Cutter diameter compensation right.
G43—Tool length compensation positive direction.
G44—Tool length compensation negative direction.
G45—Tool offset increase.
G46—Tool offset decrease.
G47—Tool offset double increase.
G48—Tool offset double decrease.

G49—Tool length compensation cancel.
G50—Scaling off.
G51—Scaling on.
G73—Peck drilling cycle.
G74—Counter tapping cycle.
G76—Fine boring cycle.
G80—Canned cycle cancel.
G81—Drilling cycle.
G82—Counter boring cycle.
G83—Peck drilling cycle.
G84—Tapping cycle.
G85—Boring cycle (feed return to reference level).
G86—Boring cycle (rapid return to reference level).
G87—Back boring cycle.
G88—Boring cycle (manual return).
G89—Boring cycle (dwell before feed return).
G90—Specifies absolute positioning.
G91—Specifies incremental positioning.
G92—Program absolute zero point.
G98—Return to initial level.
G99—Return to reference (R) level.

MISCELLANEOUS (M) FUNCTIONS USED IN MILLING AND TURNING

Following is a list of miscellaneous functions used in the milling and turning examples in this text. Other M functions common to General Numeric and Fanuc controllers are also listed.

M00—Program stop.
M01—Optional stop.
M02—End of program (rewind tape).
M03—Spindle start clockwise.
M04—Spindle start counterclockwise.
M05—Spindle stop.
M06—Tool change.
M08—Coolant on.
M09—Coolant off.
M13—Spindle on clockwise, coolant on (on some systems).
M14—Spindle on counterclockwise, coolant on.
M17—Spindle and coolant off (on some systems).

M19—Spindle orient and stop.
M21—Mirror image X axis.
M22—Mirror image Y axis.
M23—Mirror image off.
M30—End of program, memory reset.
M41—Low range.
M42—High range.
M48—Override cancel off.
M49—Override cancel on.
M98—Jump to subroutine.
M99—Return from subroutine.

PREPARATORY FUNCTIONS (G CODES) USED IN TURNING

Following is a list of preparatory functions used in CNC milling examples in this text. Other codes commonly used on Fanuc controllers are also listed.

G00—Rapid traverse positioning.
G01—Linear interpolation (feedrate movement).
G02—Circular interpolation clockwise.
G03—Circular interpolation counterclockwise.
G04—Dwell.
G10—Tool length offset value setting.
G17—Specifies X/Y plane.
G18—Specifies X/Z plane.
G19—Specifies Y/Z plane.
G20—Inch data input (on some systems).
G21—Metric data input (on some systems).
G22—Stored stroke limit on.
G23—Stored stroke limit off.
G27—Reference point return check.
G28—Return to reference point.
G29—Return from reference point.
G30—Return to second reference point.
G40—Tool nose radius compensation cancel.
G41—Tool nose radius compensation left.
G42—Tool nose radius compensation right.
G50—Programming of work coordinate system.
G68—Mirror image for double turrets on.
G69—Mirror image for double turrets off.

G70—Inch programming (some systems) or finish cycle.

G71—Metric programming (some systems) or stock removal in turning code.

G72—Stock removal in facing code.

G73—Pattern repeat.

G74—Z axis peck drilling.

G75—Groove cutting cycle, X axis.

G76—Multipass thread cutting.

G90—Absolute positioning.

G91—Incremental positioning.

G94—Per minute feed (some systems).

G95—Per revolution feed (some systems).

G98—Per minute feed (some systems).

G99—Per revolution feed (some systems).

APPENDIX 4

Codes in Common Use with Tape Machinery

PREPARATORY FUNCTIONS (g CODES)

(Note: On tape machinery, lowercase letters are generally used.)

g01—Linear interpolation.

g02—Circular interpolation clockwise.

g03—Circular interpolation counterclockwise.

g78—Mill cycle stop. A milling code used to position a spindle before lowering it. Upon receiving a g78 the spindle moves at rapid traverse to the programmed x/y coordinates, then rapids down to the feed engagement point, then feeds down to final depth at feedrate. The spindle is then clamped (either manually or automatically).

g79—Mill cycle. Usually (though not always) used following a g78. Upon receiving a g79 the spindle moves to the programmed x/y coordinates at feed rate, then rapids and subsequently feeds down to depth. If used following a g78, the spindle moves to the programmed coordinates at feedrate, since the spindle is already down.

g80—Cancel cycle.

g81—Drill cycle.

g84—Tapping cycle.

g85—Boring cycle.

MISCELLANEOUS (m) FUNCTIONS

m00—Program stop.

m02—End of program.

m03—Spindle on clockwise.

m04—Spindle on counterclockwise.

m05—Spindle off.

m06—Tool change.

m07—Flood coolant on.

m08—Mist coolant on.

m09—Coolant off.

m10—Clamp spindle.

m11—Spindle unclamp.

m13—Spindle on clockwise, flood coolant on.

m14—Spindle on counterclockwise, flood coolant on.

m17—Spindle on clockwise, mist coolant on.

m18—Spindle on counterclockwise, mist coolant on.

m26—Pseudo tool change. Used primarily for clamp changes. On some machines the spindle positions at rapid traverse to the tool change location but no tool change takes place. On other machines the spindle retracts to its "home" position.

m30—End of tape. Rewinds the tape to the start on machines where m02 does not do so.

m50−59—Z axis cam selection. Selects which of nine cams is to control z axis motion. m50 specifies no cam while m51−59 specify cams 1−9 respectively. Some machines use a 'w' function instead of an 'm' function to select cams.

APPENDIX 5

Safety Rules for Numerical Control

SAFETY RULES FOR OPERATING MACHINES

1. Use common sense in all situations.
2. Wear safety glasses at all times on the shop floor.
3. Wear safety shoes.
4. Keep long hair covered when operating or standing near a machine.
5. Do not wear jewelry (including rings), neckties, long sleeves, or loose clothing while operating machines.
6. Keep the floor free of obstructions.
7. Clean oil and grease spills immediately.
8. Do not play with compressed air or engage in general horseplay around machinery.
9. Do not use compressed air to clean machine slides. Chips blown under the machine ways will cause premature wear.
10. Do not perform grinding operations near NC machinery. Grinding grit will cause premature machine slide wear.
11. Platforms around machinery should be kept clean and have antislip surfaces.
12. Use caution when lifting heavy parts, tooling, or fixturing. Lift with the legs, not the back.
13. Keep tools and other parts off the machine.
14. Keep hands away from the spindle while it is revolving, and away from other moving parts of the machine.
15. Use a cloth or gloves when handling tools by their cutting edges.
16. Use caution when changing tools.
17. Use caution to avoid inadvertently bumping any NC controls.
18. Do not operate controls unless you have been instructed in their use.
19. Keep electrical panels in place. Electrical work should be performed only by qualified service personnel.
20. Make sure safety guards and devices are in place and working before operating the machine.

21. Do not remove chips from the machine or workpiece with hands or fingers. Use a brush. Do not remove chips with the spindle running.
22. Respect the programmer's knowledge of the machine.

SAFETY RULES FOR PROGRAMMERS

1. Never assume! When in doubt check the manual.
2. Do not attempt to program a machine without access to the programming manual for the machine tool and controller.
3. Cancel all modal commands in the first line of the program to ensure commands are not active when the program is cycled the first time.
4. Be sure all modal commands have been canceled at the end of the program so that no codes are active at the start of the next program.
5. Use a buffer zone between the part and the feed engagement point for all tool moves into the workpiece.
6. Respect the machine operator's knowledge of the machine.
7. When on the shop floor:
   a. Wear safety glasses at all times.
   b. Wear safety shoes.
   c. Remove neckties or tuck them inside your shirt.
   d. Keep hands away from moving machine parts.
   e. Keep long hair safely secured or covered.

APPENDIX 6

Useful Machining Formulas and Data

MACHINING FORMULAS

To determine spindle RPM:

$$RPM = \frac{(CS \times 4)}{D}$$

Where: CS is the material cutting speed in surface feet per minute, and D is the diameter of the part or cutter revolving in the spindle.

To determine feedrates:

    1. Milling feedrates:

$$FEED = RPM \times T \times N$$

Where: T is the chip load per tooth and N is the number of teeth on the cutter.

    2. Lathe feedrates (commonly .002 − .025 inch per revolution):

$$FEED\ (in./rev) = I/RPM$$

Where: I is the feedrate in inches per minute.

$$FEED\ (in./min) = RPM \times r$$

Where: r is the feedrate in inches per revolution.

To determine lead of a thread:

$$LEAD = P \times I$$

Where: P is the pitch of the thread and I is the number of leads on the thread.

To determine pitch of a thread:

$$PITCH = 1/N$$

Where: N is the number of threads per inch.

To determine tap drill diameter of a thread (Unified threads):

$$MD - \left[\frac{1.08254 \times \%}{N}\right]$$

Where: MD is the major diameter of the thread, % is the percentage of thread engagement desired, and N is the number of threads per inch.

To determine length of a drill point:

$$\text{DRILL POINT} = .3 \times \text{DRILL DIAMETER}$$

To determine depth of countersink to achieve a given diameter:

$$\text{DEPTH} = A - (B \times C)$$

Where: A is the diameter of countersink desired, B is the diameter of the hole, and C is a constant as follows:

.35 for a 110-degree countersink
.50 for a 90-degree countersink
.57 for a 82-degree countersink
.35 for a 60-degree countersink

To determine circumference of a circle:

$$\text{CIRCUMFERENCE} = \text{DIAMETER} \times \text{PI}$$

To determine diameter of a circle:

$$\text{DIAMETER} = \text{CIRCUMFERENCE} \times .31831$$

To determine area of a circle:

$$\text{AREA} = \text{PI} \times \text{RADIUS}^2$$

$$\text{AREA} = \frac{1}{2}C \times \frac{1}{2}D$$

Where: C is the circumference and D is the diameter.

To determine surface area of a sphere:

$$\text{SURFACE} = \text{DIAMETER}^2 \times \text{PI}$$

To determine volume of a sphere:

$$\text{VOLUME} = \text{DIAMETER}^3 \times .5236$$

CUTTING SPEED DATA

The following rates are averages for high-speed steel cutters. For carbide cutters, double the cutting speed value.

Cutting speeds for lathes:

MATERIAL	CUTTING SPEED
Tool steel	50
Cast iron	60
Mild steel	100
Brass, soft bronze	200
Aluminum, magnesium	300

Cutting speeds for drills:

MATERIAL	CUTTING SPEED
Tool steel	50
Cast iron	60
Mild steel	100
Brass, soft bronze	200
Aluminum, magnesium	300

Cutting speeds for milling:

MATERIAL	CUTTING SPEED
Tool steel	40
Cast iron	50
Mild steel	80
Brass, soft bronze	160
Aluminum, magnesium	200

FEEDRATE DATA

Feeds for drilling:

DRILL SIZE	FEEDRATE
$< \frac{1}{8}$	.001 – .002
$\frac{1}{8} - \frac{1}{4}$	.002 – .004
$\frac{1}{4} - \frac{1}{2}$	.004 – .007
$\frac{1}{2} - 1.000$	.007 – .015
> 1.000	.025

Feeds per tooth for milling:

MATERIAL	FACE MILLS	SIDE MILLS	END MILLS
Low carbon steel	.010	.005	.005
Medium carbon steel	.009	.005	.004
High carbon steel	.006	.003	.002
Stainless steel	.006	.004	.002
Cast iron	.012	.006	.006
Brass and bronze	.013	.008	.006
Aluminum	.020	.012	.010

CUTTER CENTERLINE FORMULAS

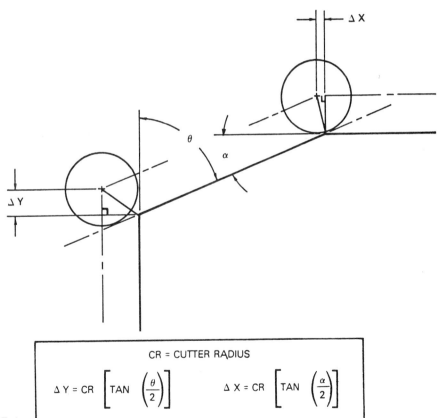

CR = CUTTER RADIUS

$$\Delta Y = CR \left[TAN \left(\frac{\theta}{2} \right) \right] \qquad \Delta X = CR \left[TAN \left(\frac{\alpha}{2} \right) \right]$$

FIGURE 1

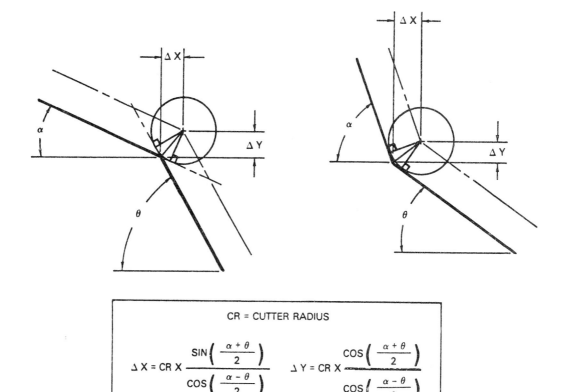

FIGURE 2
Two lines intersecting, not parallel to a machine axis

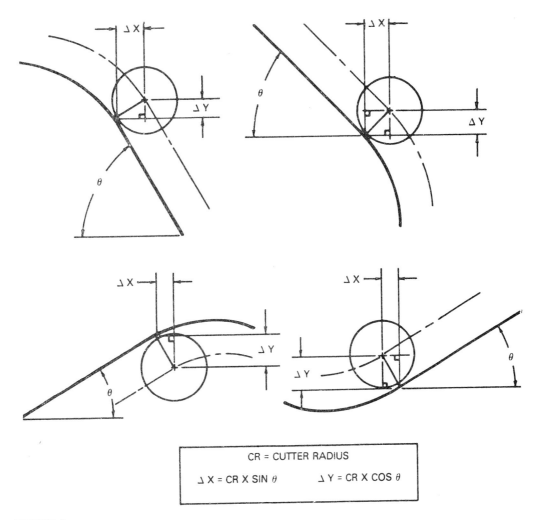

FIGURE 3
Line not parallel to a machine axis tangent to a circle

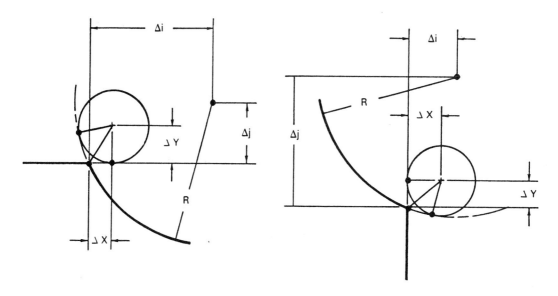

CR = CUTTER RADIUS

$$\bot X = \Delta i - \sqrt{(R - CR)^2 - (\Delta j - CR)^2}$$

$$\bot Y = CR$$

$$\bot X = CR$$

$$\bot Y = \Delta j - \sqrt{(R - CR)^2 - (\Delta i - CR)^2}$$

FIGURE 4
Intersection of a circle and a line parallel to a machine axis

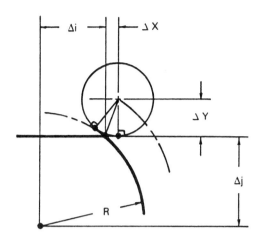

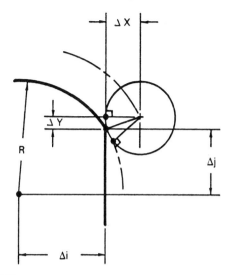

CR = CUTTER RADIUS

$$\lrcorner X = \sqrt{(R - CR)^2 - (\Delta j - CR)} - \Delta i$$
$$\lrcorner Y = CR$$

$$\lrcorner X = CR$$
$$\lrcorner Y = \sqrt{(R - CR)^2 - (\Delta i + CR)^2} - \Delta i$$

FIGURE 5
Intersection of a circle and a line parallel to a machine axis

APPENDIX 7

Lathe Canned Cycle Example

The program contained in this section is the courtesy of Hardinge Brothers Inc. Refer to Figures 1 and 2.

TOOLING USED

TURRET STATION	TOOL OFFSET	TOOLING USED
1	1	Number 4 centerdrill
3	3	55 deg. × .030r O.D. turning tool
4	4	Number 7 drill
5	5	¼—20 tap
6	6	35 degrees × .015r O.D. turning tool
8	8	O.D. threading tool

SEQUENCE OF OPERATIONS

1. Centerdrill—1500 RPM, feedrate .009 IPR, .250 deep.
2. Rough face and turn—55 degree tool, G71 cycle, 400 constant surface speed with a maximum RPM of 4200, depth of cut .125, stock for finish pass .006, feedrate .008 IPR.
3. Tap drill—1900 RPM, feedrate .007 IPR, .875 deep.
4. Tap—250 RPM, G32 cycle, feedrate .049 IPR.
5. Finish face and turn—35 degree tool, G70 cycle, 450 constant surface speed with a maximum RPM of 5000, feedrate .004 IPM.
6. O.D. thread—G76 cycle, 1000 RPM, 8 passes used, 59 degree infeed angle.

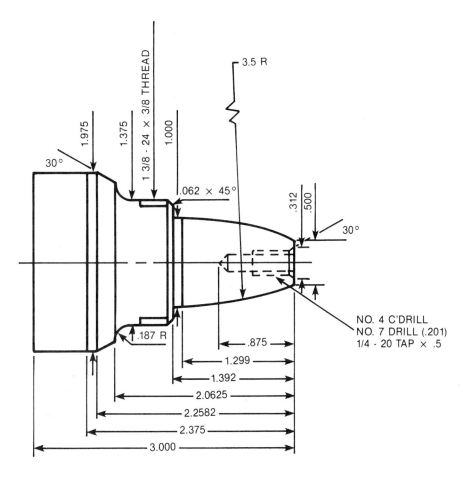

FIGURE 1
(Courtesy of Hardinge Brothers Inc.)

PROGRAM NOTES

The programmer of this part chose not to use sequence numbers on all lines. The uses of sequence numbers is up to the programmer, they are not required by the controller except where necessary for canned cycle or subroutine use.

```
%
O401 (SAMPLE PART PROGRAM)
      G20
      G65 P9150 D1.5
N1    (T0101 #4 CENTER DRILL)
      G97 S1000 M13
      M98P1
      G4 T0101
      X0 Z.100 S1500
      G99 Z-.25 F.009
      M98P2

N3    (T0303 ROUGH O.D. R.030 T3)
      G97  S1000 M13
      M98P1
      G4 T0303
      X2.16 Z.2
      G50 S4200
      G96 S400
      G42 X2.155 Z.1 F50.
      G99
      G71 P100 Q300 U.012 W.006 D12500 F.008
N100  G00 X.25
      G1 G99 Z0 F.004
      X.5
      G3 X1. Z-1.299 R3.5
      G1 Z-1.302
      X1.375 K-.062
      Z-2.0625 R.1875
      X1.75
      X1.975 Z-2.2582
      Z-2.375
N300  X2.155
      M98P2

N4    (T0404 #7 DRILL)
      G97 S1000 M13
      M98P1
      G4 T0404
      X0 Z.1 S1900
      G1 G99 Z-.875 F.007
      M98P2

N5    (T0505 1/4-20 TAP)
      G97 S250 M13
      M98P1
      G4 T0505
      X0 Z.25
      G32 Z-.406 F.049  (TAPPED .094 SHORT FOR DRIVER PULLOUT)
      G4 U.5
      G32 Z.25 F.05 M14
      M98P2
```

```
N6    (TO808 1 3/8-24 O.D. THREAD)
      G97 S1000 M13
      M98P1
      G4 TO808
      X1.4261 Z.1 S2880
      Z-1.4261 F50.
      G76 X1.3238 Z-1.767 K.0256 E.04166 D00904 A59
      M98P2
      TO100
      M30
      %
```

FIGURE 2
(Courtesy of Hardinge Brothers Inc.)

The uses of leading zeros is also optional at the programmer's discretion. In this example, G01 is written as G1, G02 as G2, etc.

0401

- The "O" number. Each program resident in the controller must have its own unique "O" number.

G65 P9150 D1.5

- Unique to Hardinge using Fanuc controls. This code sets the dwell time parameter.

M98P1

- Jump to subroutine call. This subprogram contains the startup code common to all tools. The subprogram is not part of the program printout.

M98P2

- Jump to subroutine call. This subprogram contains the tool cancel code common to all tools. The subprogram is not part of the program printout.

CANNED CYCLE NOTES

G71 P100 Q300 U.012 W.006 D12500 F.008

- G71 is the rough face and rough turn cycle.
- P100 is the sequence number (N100) that denotes the beginning of the finish pass lines.
- Q300 is the sequence number (N300) that denotes the end of the finish pass lines.

- U.012 tells the MCU to leave .012 stock on the X-axis for finishing. The lathe is diameter programmed, therefore, this will leave .006 stock per side to be finished.
- W.006 tells the MCU to leave .006 stock on the Z-axis for finishing.
- D12500 tells the MCU that the depth of each roughing pass is the .125 diameter stock removal.
- F.008 is the feedrate.

The G71 cycle will calculate backwards from the finished pass lines to generate the appropriate number of passes. The tool path shape is identical to the finish pass.

G32 Z − .406 F.049

- G32—This type of cycle was covered in Chapter 14. It is the single pass threading cycle. When used for tapping, no X infeed is necessary. It is the same as using a G01 except with G32 the speed and feed operator overrides will not function.
- F.049 is the feedrate. The tap is fed in at .001 IPR slower than the actual lead. This allows the tapping holder to pull out slightly. When the spindle is reversed at the bottom of the tapped hole, there is some travel in the holder to prevent tap breakage.

G70 P100 Q300

- G70 is the finish face and turn cycle.
- P100 is the sequence number (N100) of the finish pass lines.
- Q300 is the sequence number (N300) of the finish pass lines.

G70 will simply jump backwards in the program to the finish pass lines used with the G71 cycle and execute them. Note the G42 prior to the G70 line. Tool nose radius comp (TNR) is being used to allow for tool wear.

G76 X1.3238 Z − 1.767 K.0256 E.04166 D00904 A59

- This cycle was covered in Chapter 14. The A59 sets the infeed angle to 59 degrees. Other programmers prefer to use A60.

APPENDIX 8

Sample Programs

SAMPLE PART PROGRAM

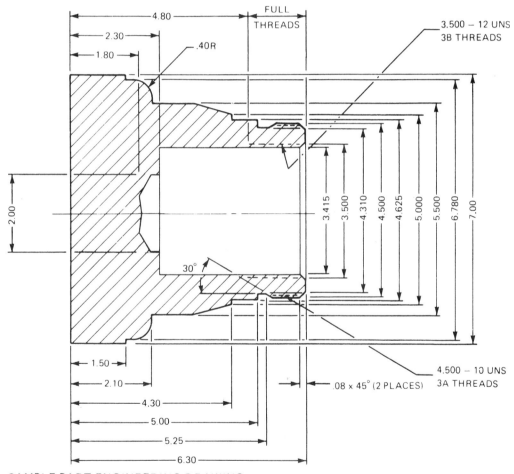

SAMPLE PART ENGINEERING DRAWING

Courtesy of Cincinnati Milacron

OD TOOLING

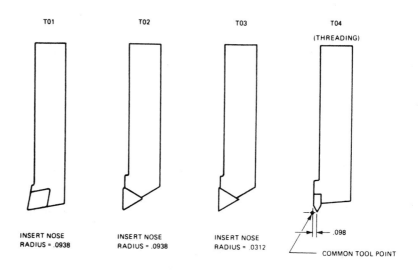

T01   T02   T03   T04
                                                                    (THREADING)

INSERT NOSE   INSERT NOSE   INSERT NOSE   .098
RADIUS = .0938   RADIUS = .0938   RADIUS = .0312

                                                                    COMMON TOOL POINT

ID TOOLING

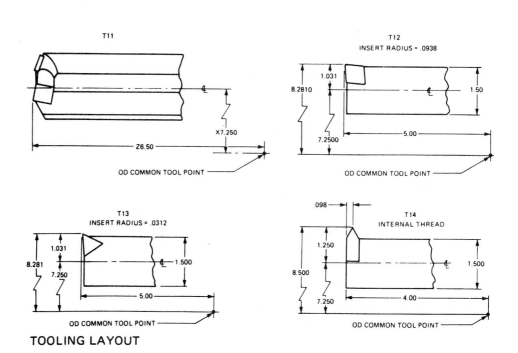

T11

X7.250

Z6.50

OD COMMON TOOL POINT

T12
INSERT RADIUS = .0938

1.031

8.2810   1.50

7.2500

5.00

OD COMMON TOOL POINT

T13
INSERT RADIUS = .0312

1.031

8.281   1.500

7.250

5.00

OD COMMON TOOL POINT

.098

T14
INTERNAL THREAD

1.250

8.500   1.500

7.250

4.00

OD COMMON TOOL POINT

TOOLING LAYOUT

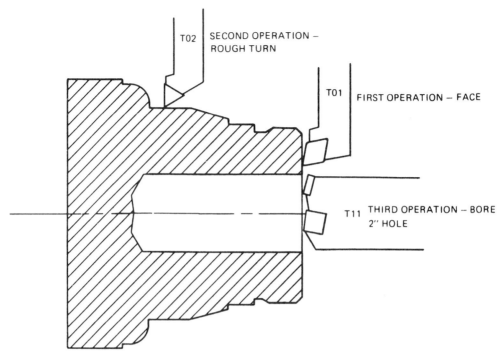

FIRST, SECOND AND THIRD OPERATIONS

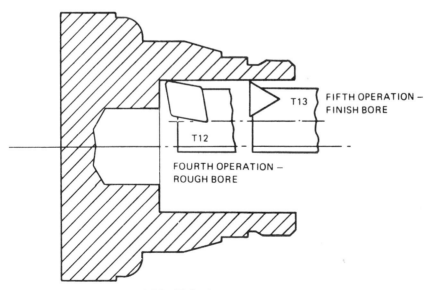

FOURTH AND FIFTH OPERATIONS

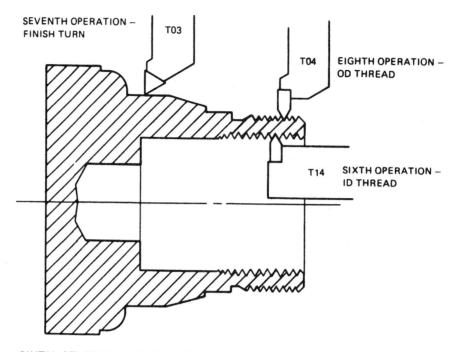

SIXTH, SEVENTH AND EIGHTH OPERATIONS

SAMPLE PART PROGRAM NO. 1

First Operation	O10	G90
	N20	G97 S100 M42
	N30	G70 M03
	N40	G00 X50000 Z85000 T0100 M06
	N50	G95
	N60	G92 S2500
	N70	G96 R50000 S600
	N80	G00 X37000 Z63500 M08
	N90	G01 X−940 F150
	N100	Z65500 F600
	N110	X37000
	N120	Z63000
	N130	X−940 F150
	N140	G00 Z65000
	N150	X45000
Second Operation	O160	G90
	N170	G97 S351 M41
	N180	G70 M13
	N190	G00 X45000 Z65000 T0200 M06
	N200	G95
	N210	G92 S2500
	N220	G96 R45000 S600
	N230	G00 X32600
	N240	G01 Z19690 F150
	N250	G03 X34100 Z16062 I28962 K16062
	N260	G01 X37000
	N270	G00 Z65000
	N280	X30100
	N290	G01 Z21200
	N300	X32100
	N310	G00 Z65000
	N320	X27600
	N330	G01 Z21100
	N340	X28962
	N350	G03 X34000 Z16062 I28962 K16062
	N360	G01 Z15100
	N370	X36000
	N380	G00 X29600 Z65000
	N390	X25100
	N400	G01 Z42199
	N410	X27600 Z32986
	N420	G00 Z65000
	N430	X22600
	N440	G01 Z55989
	N450	X21650 Z51659
	N460	Z50100
	N470	X23225
	N480	Z43100
	N490	X26000
	N500	G00 Z130000
Third Operation	O510	G90
	N520	G97 S600 M41
	N530	G70 M14
	N540	G00 X26000 Z130000 T1100 M06

```
                              N550  G95
                              N560  G00  X−72500
                              N570  G01  Z83000  F150
                              N580  G00  Z148000
Fourth Operation   O590  G90  M05
                              N600  G97  S1000  M42
                              N610  G70  M13
                              N620  G00  X−72500  Z148000  T1200  M06
                              N630  G95
                              N640  G92  S2500
                              N650  G96  R10310  S600
                              N660  G00  X−71560  Z115000
                              N670  G01  Z73100  F150
                              N680  X−73750
                              N690  G00  Z115000
                              N700  X−70310
                              N710  G01  Z73100
                              N720  X−72500
                              N730  G00  Z115000
                              N740  X−69060
                              N750  G01  Z73100
                              N760  X−71250
                              N770  G00  Z115000
                              N780  X−67810
                              N790  G01  Z73100
                              N800  X−70000
                              N810  G00  Z115000
                              N820  X−66560
                              N830  G01  Z73100
                              N840  X−68750
                              N850  G00  Z115000
                              N860  X−65835
                              N870  G01  Z73100
                              N880  X−68023
                              N890  G00  Z148000
Fifth Operation     O900  G90
                              N910  G97  S846  M42
                              N920  G70  M13
                              N930  G00  X−68023  Z148000  T1300  M06
                              N940  G95
                              N950  G92  S2500
                              N960  G96  R14787  S800
                              N970  G00  X−64752  Z115000
                              N980  G01  X−65735  Z112017  F100
                              N990  Z73000
                              N1000 X−73210
                              N1010 G00  Z148000
Sixth Operation    O1020 G90  M05
                              N1030 G97  S400  M41
                              N1040 G70  M14
                              N1050 G00  X−73210  Z148000  T1400  M06
                              N1060 X−67805  Z103000
                              N1070 G91
                              N1080 G33  X−1500  Z−18668  K8333
                              N1090 G00  Z18613
                              N1100 X1600
```

```
                          N1110 G33 X−1500 Z−18668 K8333
                          N1120 G00 Z18629
                          N1130 X1570
                          N1140 G33 X−1500 Z−18668 K8333
                          N1150 G00 Z18640
                          N1160 X1550
                          N1170 G33 X−1500 Z−18668 K8333
                          N1180 G00 Z18646
                          N1190 X1550
                          N1200 G33 X−1500 Z−18668 K8333
                          N1210 G00 Z18654
                          N1220 X1525
                          N1230 G33 X−1500 Z−18668 K8333
                          N1240 G00 Z18670
                          N1250 X1520
                          N1260 G33 X−1500 Z−18668 K8333
                          N1270 G00
                          N1280 G90
                          N1290 Z105000
                          N1300 X45000
      Seventh Operation   O1310 G90
                          N1320 G97 S780 M41
                          N1330 G70 M13
                          N1340 G00 X45000 Z105000 T0300 M06
                          N1350 G95
                          N1360 G92 S2500
                          N1370 G96 R45000 S800
                          N1380 G00 X19517 Z65000
                          N1390 G01 X22500 Z62017 F100
                          N1400 Z54373
                          N1410 X21550 Z52728
                          N1420 Z50000
                          N1430 X23125
                          N1440 Z43000
                          N1450 X24927
                          N1460 X27500 Z33399
                          N1470 Z21000
                          N1480 X29588
                          N1490 G03 X33900 Z16688 I29588 K16688
                          N1500 G01 Z15000
                          N1510 X37000
                          N1520 G00 X45000 Z65000
      Eighth Operation    O1530 G90
                          N1540 G97 S300 M41
                          N1550 G70 M14
                          N1560 G00 X45000 Z65000 T0400 M06
                          N1570 X22300 Z66029
                          N1580 G33 Z50300 K10000
                          N1590 G00 X24500
                          N1600 Z65946
                          N1610 X22150
                          N1620 G33 Z50300 K10000
                          N1630 G00 X24500
                          N1640 Z65879
                          N1650 X22030
                          N1660 G33 Z50300 K10000
```

```
N1670 G00  X24500
N1680 Z65835
N1690 X21950
N1700 G33  Z50300  K10000
N1710 G00  X24500
N1720 Z65807
N1730 X21900
N1740 G33  Z50300  K10000
N1750 G00  X24500
N1760 Z65791
N1770 X21870
N1780 G33  Z50300  K10000
N1790 G00  X24500
N1800 Z65780
N1810 X21850
N1820 G33  Z50300  K10000
N1830 G00  X24500
N1840 X50000  Z85000
N1850 M30
```

SAMPLE MILLING PROGRAM

MAT: 103/1020 STEEL
6"x4" x 1" OR 1/2" THICK

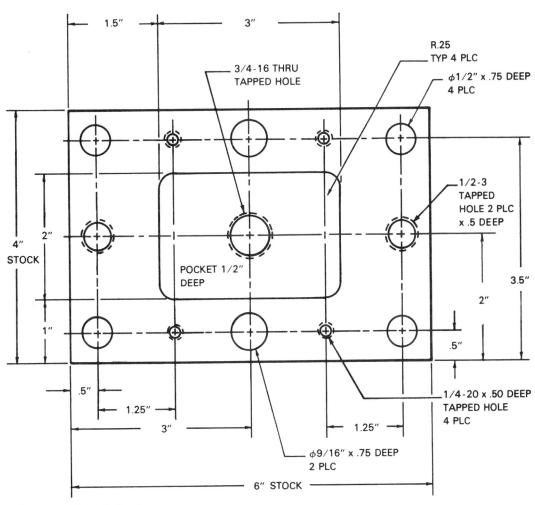

Courtesy of Bayer Industries

```
 1  O0001
 2  G20G40G49G80G90M03T01
 3  G00X-1.65Y2.75S0637
 4  G43Z0H01M08
 5  G01Z-.15F50.
 6  X6.25F22.93
 7  Y1.25
 8  X-1.65
 9  G00X0H00T02
10  X-1.65Y2.75S1490
11  G43Z0H02
12  G01Z-.16F50
13  X6.25F14.9
14  Y1.25
15  X-1.65
16  G00Z0H00T03
17  X3.Y2.S0407
18  G43Z0H03
19  G81G99Z-.725R0F6.11
20  G80Z0H00T04
21  X3.Y2.S0637
22  G43Z0H04
23  G01Z-.59F10.
24  G41X4.115F6.37D24
25  Y2.65
26  X2.385
27  Y1.35
28  X4.49
29  Y2.99
30  X1.51
31  Y1.01
32  X4.49
33  Z-.58F10.
34  G00G40X3.Y2.
35  Z0H00T05
36  X3.Y2.S1192
37  G43Z05H05
38  G01Z-.58F15.
39  Z-.6F11.92
40  G41X3.375D25
41  Y2.325
42  X2.625
43  Y1.675
44  X3.75
45  Y2.625
46  X2.25
47  Y1.375
48  X4.125
49  Y2.85
50  X1.875
51  Y1.15
52  X4.5
53  Y3
54  X1.5
```

```
55 Y1.
56 X4.5
57 Z-.59F15.
58 G00G40X3.Y2.
59 Z0H00T06
60 X3.Y2.S0444
61 G43Z0H06
62 G81G98Z-1.25R-.5F6.66
63 G80Z0H00T07
64 X3.Y2.S0162
65 G43Z0H07
66 G84G99Z-1.35R-.4F9.62
67 G80Z0H00T08
68 X.5Y.5S1222
69 G43Z0H08
70 G81G99Z-.25R0F6.11
71 X1.75
72 X3.
73 X4.25
74 X5.5
75 Y2.
76 Y3.5
77 X4.25
78 X3.
79 X1.75
80 X.5
81 Y2.
82 G80Z0H00T09
83 X.5Y.5S0611
84 G43Z0H009
85 G81G99Z-.75R0F7.33
86 Y3.5
87 X5.5
88 Y.5
89 G80Z0H00T10
90 X3.Y.5S0543
91 G43Z0H10
92 G81G99Z-.77R0F6.52
93 Y3.5
94 G80Z0H00T11
95 X1.75Y3.5S1520
96 G43Z0H11
97 G83G99Z-.66R0Q.25F9.12
98 X4.25
99 Y.5
100 X1.75
101 G80Z0H00T12
102 X1.75Y.5S0326
103 G43Z.2H12
104 G84G99Z-.4R.2F15.49
105 X4.25
106 Y3.5
107 X1.75
108 G80X0H00T13
```

```
109  X.5Y2.S0724
110  G43Z0H13
111  G83G99Z-.73R0Q.375F8.69
112  X5.5
113  G80Z0H00T14
114  X5.5Y2.S0255
115  G43Z.2H14
116  G84G99Z-.45R.2F18.63
117  X.5
118  G80Z0H00T15
119  X.5Y2.S0795
120  G43Z0H15
121  G81G99Z-.375R0F12.5
122  Y3.5
123  X1.75Z-.23
124  X3.Z-.4
125  X4.25Z-.23
126  X5.5Z-.375
127  Y2.
128  Y.5
129  X4.25Z-.23
130  X3.Z-.4
131  X1.75Z-.23
132  X.5Z-.375
133   G80Z0H00M09
134  X-4.Y6.M05
135  M30
```

SAMPLE LATHE PROGRAM

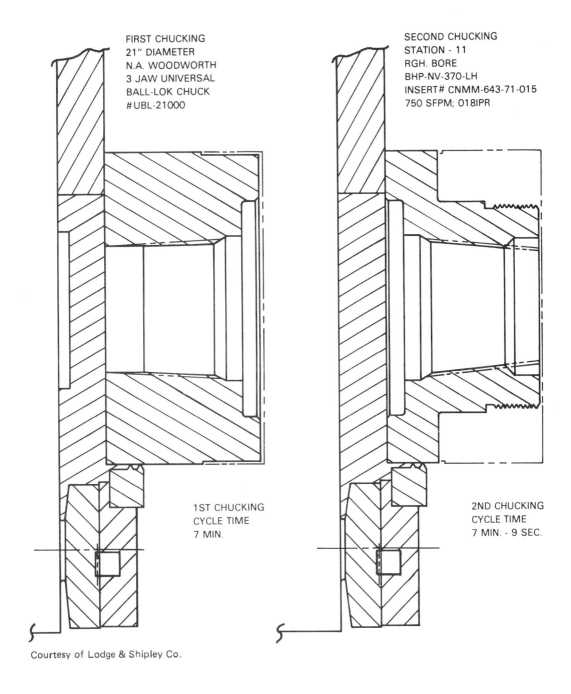

FIRST CHUCKING
21" DIAMETER
N.A. WOODWORTH
3 JAW UNIVERSAL
BALL-LOK CHUCK
#UBL-21000

SECOND CHUCKING
STATION - 11
RGH. BORE
BHP-NV-370-LH
INSERT# CNMM-643-71-015
750 SFPM; 018IPR

1ST CHUCKING
CYCLE TIME
7 MIN.

2ND CHUCKING
CYCLE TIME
7 MIN. - 9 SEC.

Courtesy of Lodge & Shipley Co.

```
N0010G70M12
N002G9T0202
N003G92X15.Z22.782
N004G97S065M03
N0050G95
N0060G21X0.Z5.F.8
N007G01Z-00.1F.007M08
N008G21Z07.782F.8M09
N009X11.Z07.782
N01G9T0101M12
N011G92X06.7969Z20.2039S0712
N012G97S021M04
N0130G95
N014G21X05.2Z05.182F.8
N0150G96R5.1S06
N0160G01X1.7F.014M08
N0170Z5.212
N018G21X04.95F.8
N0190G01Z5.182F.022
N02G03X05.057Z05.075K00.107
N0210G01Z3.307
N022X05.15
N023G21X06.7969Z20.2039F.8M09
N024G9T0303
N0250M13
N026G92X12.4531Z12.5789S15
N027G97S071M04
N0280G95
N0290G21X1.9485Z5.295F.8
N0300G96R2.S075
N0310G01Z-.1F.018M08
N03540X1.8485
N0350G21Z5.272F.8
N0360X2.2509
N0370G01Z4.047F.018
N0380X2.1058Z3.7366
N0382X2.0293Z1.2875
N0384Z-.1
N0386X1.9293
N0390G21Z5.272F.8
N0400X2.4485
N0410G01Z4.557F.018
N0420X2.1985
N0430G21Z5.272F.8
N0440X2.6985
N0450G01Z4.557F.018
N0460X2.4485
N0470G21Z5.272F.8
N0480X2.9485
N0490G01Z4.557F.018
N0500X2.4485
N0510G21Z5.272F.8
N0520X3.1985
N0530G01Z4.55F.018
```

```
N0540X2.9485
N0550G21Z5.272F.8
N0560X3.433
N0570G01Z4.635F.018
N575G03X03.355Z04.557I00.78
N0580G01X3.1985
N059G21Z07.272F.8
N0592T0606
N0593G04X02.
N0595G92X03.1985Z07.272
N06G21X03.5406Z05.272
N0610G01Z5.172F.010
N0620X3.453Z5.0844
N0630Z4.625
N064G03X03.375Z04.547I00.78
N0642G01X02.3209
N065X02.2609Z04.487
N0651Z04.047
N0652X02.0958Z03.7266
N0653X02.0458
N0660G97S0350M03
N069G21X01.4531Z07.272F.8M09
N069G9T101
N0693G92X.967Z10.1941
N0694G21X2.0764Z5.5F.8
N0695G21X2.237M08
N0696G33X2.1209Z1.75I0039K.125
N0697G33X1.9597Z1.5888I.125K.125
N0698G21Z5.5F.8
N0699X2.2476
N0700G33X2.1309Z1.75I.0039K.125
N0701G33X1.9597Z1.5888I.125K.125
N0702G21Z8.F.8
N0703X11.967Z15.501
N0705G97S054M04
N071G9T0909M82
N0720
N073G92X06.781Z20.188S15
N074G97S054M04
N0750G95
N0760G21X3.48Z5.256F.8
N0770G96R3.5S1
N0780G01Z5.156F.008M08
N079X04.93
N08G03X05.021Z05.065K00.091
N0810G01Z3.281
N0820X5.2
N0825G97S038
N083G21X10.781Z35.188F.8M09
N0840T0000
N0845
N0850M00
N0855M12
N086G9T1101
```

```
N087G92X10.7969Z35.2039S0712
N088G97S021M004
N09G21X05.2Z05.057F.8
G0910G96R5.1S06
N092G01X1.8F.014M08
N0930Z5.147
N0940G21X4.567F.8
N0945M11
N0960G01Z1.807F.022
N097X04.95
N098G03X05.057Z01.7K00.107
N099G01X5.147F.8
N1G21Z05.147F.8
N1005M12
N1010X4.067
N1020G01Z1.807F.022
N1030X4.6
N1040G21Z5.147F.8
N0150X3.567
N1060G01Z1.807F.022
N1070X4.1
N1080G21Z5.147F.8
N0190X3.1644
N1100G01Z5.057F.022
N1110X3.317Z4.9094
N1120Z3.307
N1130X3.442
N1140Z1.885
N115G02X03.52Z01.1807I00.078
N1160G01X5.
N1170G97S023
N118G21X06.7969Z20.2039F.8M09
N119G9T2111
N1205M13
N121G92X12.4531Z16.0789S15
N122G97S07M04
N1230G95
N1240G21X2.3241Z5.147F.8
N125G96R2.3S075
N126G01Z04.422F.018M08
N1270X1.9465Z4.0444F.01
N1280G21Z5.147F.8
N1330X2.433
N1340G01Z4.432F.018
N1350X2.3
N136G21Z07.147F.8
N1362T0808
N1363G04X02.
N1365G92X02.3Z07.077
N137G21X02.56Z05.147
N1380G01Z5.04F.01
N139G02X02.453Z04.94K00.107
N1400G01Z4.422
N141X02.2741
```

```
N1420G21Z5.24F.8
N1425G97S076
N143X12.4531Z16.0089M09
N144G9T1909
N145G92Z06.781Z20.188S15
N146G97S076M04
N1470G95
N1480G21X2.48Z5.131F.8
N1490G96R2.5S1
N1500G01Z5.031F.01M08
N1510X3.2029
N1520X3.281Z4.9529
N1530Z3.281
N1540X3.345
N155G03X03.406Z03.22K00.061
N1560G01Z1.875
N157G02X03.5Z01.781I00.94
N158G01X04.93
N159G03X05.021Z01.69K00.91
N1600G01Z1.59
N1610X5.031
N1615G97S029
N162G21X06.781Z20.188F.8M09
N1621G9T0707
N1622G92X06.75Z20.728
N1623G97S035M04
N1624G95
N1625G21X03.4Z03.25F.8
N1626G96R03.4S06M04
N1627G01X03.125F.006M08
N1628G21X03.35F.8M8
N1629Z03.43
N163G01X03.17Z03.25F.006
N1631G21X03.4F.8
N1632G21X06.75Z10.728M09
N1634G9T0404
N164G92X06.75Z20.157
N1645M12
N1650G97S0290M03
N1660G95
N1670G21X3.35Z5.5F.8
N1600M08
N1690G83X-.122Z-.007H01
N1700G33Z3.375K.0625
N1710G84X-.007Z-.004H01
N1720G84X-.0055Z-.0032H01
N1730G84X-.0046Z-.0027H01
N1740G84X-.0041Z-.0024H01
N1750G84X-.0038Z-.0022H01
N1760G84X-.0034Z-.002H01
N1770G84X-.0Z-.0H01
N178G21X10.75Z35.157F.8M09
N1790T0000
N1795
N1800M30
```

GLOSSARY

The majority of this glossary is from Luggen, *Fundamentals of Numerical Control*, copyright 1984 by Delmar Publishers Inc. Reprinted with permission.

TERM AND DEFINITION	EXAMPLE

A AXIS The axis of circular motion of a machine tool member or slide about the X axis. (Usually called alpha.)

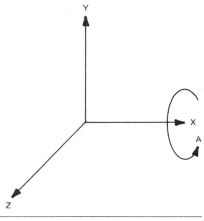

ABSOLUTE ACCURACY Accuracy as measured from a reference which must be specified.

ABSOLUTE READOUT A display of the true slide position as derived from the position commands within the control system.

ABSOLUTE SYSTEM A numerical control system in which all positional dimensions, both input and feedback, are given with respect to a common datum point. The alternative is the incremental system.

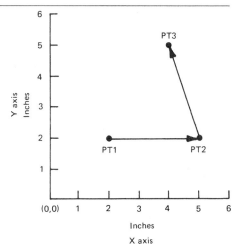

Coordinate Positions

Point	X value	Y value
PT1	2	2
PT2	5	2
PT3	4	5

In an absolute system, all points are relative to (0,0), and the absolute coordinates for each of the required points are programmed with respect to (0,0).

444

ACCURACY   1. Measured by the difference between the actual position of the machine slide and the position demanded. 2. Conformity of an indicated value to a true value, i.e., an actual or an accepted standard value. The accuracy of a control system is expressed as the deviation (the difference between the ultimately controlled variable and its ideal value), usually in the steady state or at sampled instants.

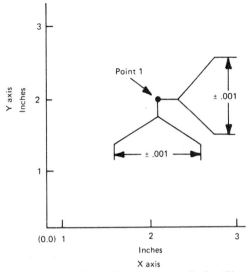

The position of point 1 in this example is X = 2 and Y = 2. If the machine accuracy is specified as ± .001, the X axis movement could be between X = 1.999 and X = 2.001. The Y axis movement could be between Y = 1.999 and Y = 2.001.

AD-APT   An Air Force adaptation of APT program language with limited vocabulary. It can be used on some small to medium sizes of U.S. computers for NC programming.

C1 = circle/center, PT1, radius, 2.5

Similar to the APT language except it does not possess the advanced contouring capabilities of APT.

ADAPTIVE CONTROL   A technique which automatically adjusts feeds and/or speeds to an optimum by sensing cutting conditions and acting upon them.

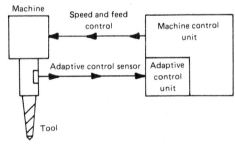

Sensors may measure variable factors, e.g. vibration, heat, torque, and deflection. Cutting speeds and feeds may be increased or decreased depending on conditions sensed.

ADDRESS   1. A symbol indicating the significance of the information immediately following. 2. A means of identifying information or a location in a control system. 3. A number which identifies one location in memory.

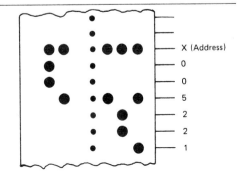

ALPHANUMERIC CODING A system in which the characters are letters A through Z and numerals 0 through 9.

APT and AD-APT statements use alphanumeric coding, e.g. GOFWD, CT12/PAST, 2, INTOF, L13

ANALOG 1. Applies to a system which uses electrical voltage magnitudes or ratios to represent physical axis positions. 2. Pertains to information which can have continuously variable values.

ANALYST A person skilled in the definition and development of techniques to solve problems.

APT (Automatic Programmed Tool) A universal computer-assisted program system for multiaxis contouring programming. APT III provides for five axes of machine tool motion.

Typical APT geometry definition statement:
C1 = CIRCLE/XLARGE, L12, XLARGE, L13, RADIUS, 3.5

Typical APT tool motion statement:
TLRGT, GORGT/AL3, PAST, AL12

ARC CLOCKWISE An arc generated by the coordinated motion of two axes, in which curvature of the tool path with respect to the workpiece is clockwise, when viewing the plane of motion from the positive direction of the perpendicular axis.

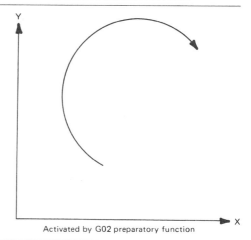

Activated by G02 preparatory function

ARC COUNTERCLOCKWISE An arc generated by the coordinated motion of two axes, in which curvature of the tool path with respect to the workpiece is counterclockwise, when viewing the plane of motion from the positive direction of the perpendicular axis.

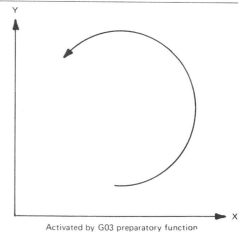

Activated by G03 preparatory function

ASCII (American Standard Code for Information Interchange) A data transmission code which has been established as an American standard by the American Standards Association. It is a code in which seven bits are used to represent each character. Formerly USASCII.

AUTO-MAP An abbreviation for AUTOmatic MAchining Programming. A computer-aided programming language which is a subset of APT. It is used for simple contouring and straight line programming.

AUTOMATION 1. The implementation of processes by automatic means. 2. The investigation, design, development, and application of methods to render processes automatic, self-moving, or self-controlling.

AUTOSPOT (Automatic System for Positioning of Tools) A computer-assigned program for NC positioning and straight-cut systems, developed in the U.S. by the IBM Space Guidance Center. It is maintained and taught by IBM.

AUX CODE Auxiliary function command in Machinist Shop Language. Used to control specific functions within a CNC program.

AUXILIARY FUNCTION A programmable function of a machine other than the control of the coordinate movements or cutter.

- Transferring a tool to the select tool position.
- Turning coolant ON or OFF.
- Starting or stopping the spindle.
- Initiating pallet shuttle or movement.

AXIS A principal direction along which the relative movements of the tool or workpiece occur. There are usually three linear axes, mutually at right angles, designated as X, Y, and Z.

AXIS INHIBIT A feature of an NC unit which enables the operator to withhold command information from a machine tool slide.

AXIS INTERCHANGE The capability of inputting the information concerning one axis into the storage of another axis.

AXIS INVERSION The reversal of plus and minus values along an axis. This allows the machining of a left-handed part from right-handed programming or vice versa.

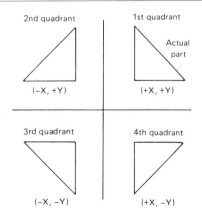

B (BETA) AXIS The axis of circular motion of a machine tool member or slide about the Y axis.

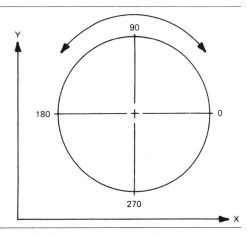

BACKLASH A relative movement between interacting mechanical parts as a result of looseness.

BCD (Binary-coded decimal) A system of number representation in which each decimal digit is represented by a group of binary digits forming a character.

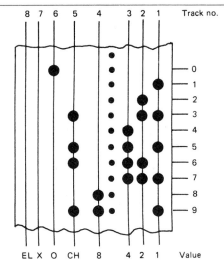

Numbers and letters are expressed by punched holes across the tape for the code or value desired.

BINARY CODE Based on binary numbers, which are expressed as either 1 or 0, true or false, on or off.

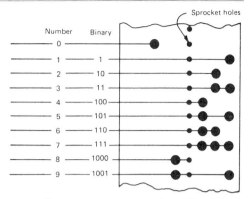

Most computers operate on some form of binary system where a number or letter can be expressed as ON (a hole) or OFF (no hole).

BIT (Binary digit) 1. Binary digit having only two possible states. 2. A single character of a language using exactly two distinct kinds of characters. 3. A magnetized spot on any storage device.

BLOCK A word, or group of words, considered as a unit. A block is separated from other units by an end of block character. On punched tape, a block of data provides sufficient information for an operation.

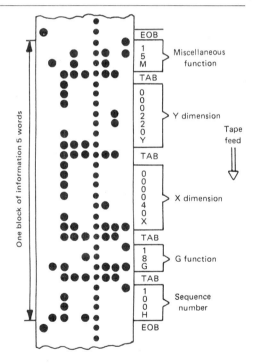

BLOCK DELETE   Permits selected blocks of tape to be ignored by the control system, at the operator's discretion with permission of the programmer.

This feature allows certain blocks of information to be skipped by programming a slash (/) code in front of the block to be skipped. One lot of parts with holes 1, 2, and 3 are required. On another lot, only holes 1 and 3 are required. The same tape could be used for both lots by activating the block delete switch on the second lot and eliminating hole 2. The (/) code would be in front of the block of information for hole 2.

BUFFER STORAGE   A place for storing information in a control system or computer for planned use. Information from the buffer storage section of a control system can be transferred almost instantly to active storage (that portion of the control system commanding the operation at the particular time). Buffer storage allows a control system to act immediately on stored information rather than wait for the information to be read into the machine from the tape reader.

BUG   1. A mistake or malfunction. 2. An integrated circuit (slang).

BYTE   A sequence of adjacent binary digits usually operated on as a unit and shorter than a computer word.

Eight bits equal one byte. A computer word usually consists of either sixteen or thirty-two bits (two or four bytes).

CAD   Computer-aided design

CAM   (Computer-aided manufacturing) The use of computers to assist in phases of manufacturing.

CAM-I   (Computer Aided Manufacturing International) The outgrowth and replacement organization of the APT Long Range Program.

CANCEL   A command which will discontinue any canned cycles or sequence commands.

CANNED CYCLE   A preset sequence of events initiated by a single command. For example, code G84 will perform tap cycle by NC.

CARTESIAN COORDINATES   A means whereby the position of a point can be defined with reference to a set of axes at right angles to each other.

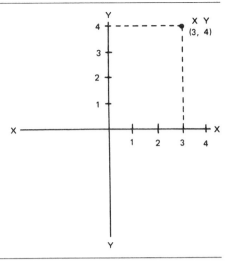

C AXIS   Normally the axis of circular motion of a machine tool member or slide about the Z axis.

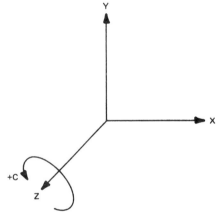

CHAD   Pieces of material removed in card or tape operations.

CHANNELS   Paths parallel to the edge of the tape along which information may be stored by the presence or absence of holes or magnetized areas. This term is also known as level or track. The EIA standard one-inch-wide tape has eight channels.

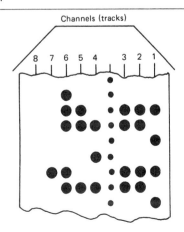

Channels (tracks)

CHARACTERS   A general term for all symbols, such as alphabetic letters, numerals, and punctuation marks. It is also the coded representation of such symbols.

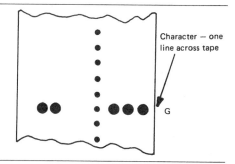

Character — one line across tape

G

CHIP   A single piece of silicon cut from a slice by scribing and breaking. It can contain one or more circuits but is packaged as a unit.

CIRCULAR INTERPOLATION   1. Capability of generating up to 360 degrees of arc using only one block of information as defined by EIA. 2. A mode of contouring control which uses the information contained in a single block to produce an arc of a circle.

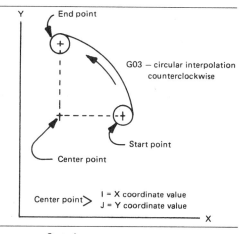

End point

G03 — circular interpolation counterclockwise

Start point

Center point

Center point〉 I = X coordinate value
J = Y coordinate value

CLOSED-LOOP SYSTEM   A system in which the output, or some result of the output, is measured and fed back for comparison with the input. In an NC system, the output is the position of the table or head; the input is the tape information which ordinarily differs from the output. This difference is measured and results in a machine movement to reduce and eliminate the variance.

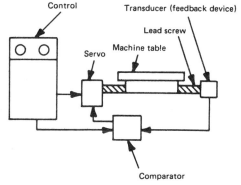

Control

Transducer (feedback device)

Lead screw

Machine table

Servo

Comparator

CNC   Computer numerical control

CODE   A system describing the formation of characters on a tape for representing information, in a language that can be understood and handled by the control system.

COMMAND A signal, or series of signals, initiating one step in the execution of a program.

COMMAND READOUT A display of the slide position as commanded from the control system.

COMPUTER NUMERICAL CONTROL A numerical control system utilizing an on-board computer as an MCU.

CONSTANT CUTTING SPEED The condition achieved by varying the speed of rotation of the workpiece relative to the tool, inversely proportional to the distance of the tool from the center of rotation.

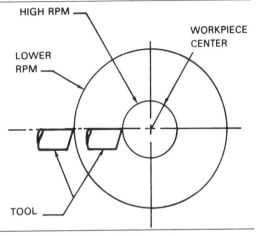

See *contouring control system.*

CONTINUOUS-PATH OPERATION An operation in which rate and direction of relative movement of machine members is under continuous numerical control. There is no pause for data reading.

CONTOURING CONTROL SYSTEM An NC system for controlling a machine (e.g., milling, drafting) in a path resulting from the coordinated, simultaneous motion of two or more axes.

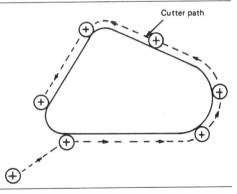

CPU Central processing unit of a computer. The memory or logic of a computer that includes overall circuits, processing, and execution of instructions.

CRT (Cathode Ray Tube) A device that represents data (alphanumeric or graphic) form by means of a controlled electron beam directed against a fluorescent coating in the tube.

CUTTER DIAMETER COMPENSATION   A system in which the programmed path may be altered to allow for the difference between actual and programmed cutter diameters.

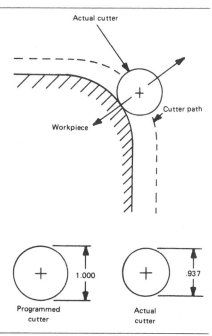

CUTTER OFFSET   The distance from the part surface to the axial center of a cutter.

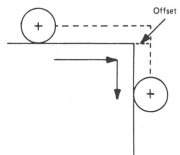

CUTTER PATH   The path defined by the center of a cutter.

CYCLE   1. A sequence of operations that is repeated regularly. 2. The time it takes for one such sequence to occur.

DATA   A representation of information in the form of words, symbols, numbers, letters, characters, digits, etc.

DATUM DIMENSIONING   A system of dimensioning based on a common starting point.

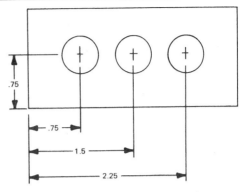

(Also known as absolute dimensioning)

DEBUG   1. To detect, locate, and remove mistakes from a program. 2. Troubleshoot.

DECIMAL CODE   A code in which each allowable position has one of ten possible states. (The conventional decimal number system is a decimal code.)

DELETE CHARACTER   A character used primarily to obliterate any erroneous or unwanted characters on punched tape. The delete character consists of perforations in all punching positions.

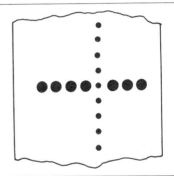

DELTA DIMENSIONING   A system used for defining part dimensions on a part drawing in which each dimension is referenced from the preceding one. Also known as incremental dimensioning in some shops.

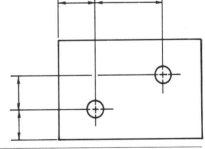

DIAGNOSTIC TEST   The running of a machine program or routine to discover a failure or potential failure of a machine element and to determine its location.

DIGIT   A character in any numbering system.

DIGITAL   1. Refers to discrete states of a signal (on or off). A combination of these makes up a specific value. 2. Relating to data in the form of digits.

DISPLAY   A visual representation of data.

DOCUMENTATION   Manuals and other printed materials (tables, magnetic tape, listing, diagrams) which provide information for use and maintenance of a manufactured product, both hardware and software.

DWELL   A timed or untimed delay in a program's execution. A timed dwell will resume the program after the programmed duration. An untimed dwell requires operator intervention to continue the program.

EDIT   To modify the form of data.

EIA STANDARD CODE   A standard code for positioning, straight-cut, and contouring control systems proposed by the U.S. EIA in their Standard RS-244. Eight-track paper (one-inch wide) has been accepted by the American Standards Association as an American standard for numerical control.

END OF BLOCK CHARACTER   1. A character indicating the end of a block of tape information. Used to stop the tape reader after a block has been read. 2. The typewriter function of the carriage return when preparing machine control tapes.

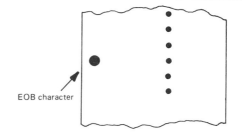

EOB character

END OF PROGRAM   A miscellaneous function (M02) indicating the completion of a workpiece. Stops spindle, coolant, and feed after completion of all commands in the block. Used to reset control and/or machine.

END OF TAPE   A miscellaneous function (M30) which stops spindle, coolant, and feed after completion of all commands in the block. Used to reset control and/or machine.

END POINT   The extremities of a span.

ERROR SIGNAL   Indication of a difference between the output and input signals in a servo system.

EXECUTIVE PROGRAM   A series of programming instructions enabling a dedicated minicomputer to produce a specific output control. For example, it is the executive program in a CNC unit that enables the control to think like a lathe or machining center.

FEED   The programmed or manually established rate of movement of the cutting tool into the workpiece for the required machining operation.

FEEDBACK   The transmission of a signal from a late to an earlier stage in a system. In a closed-loop NC system, a signal of the machine slide position is fed back and compared with the input signal, which specifies the demanded position. These two signals are compared and generate an error signal if a difference exists.

FEED FUNCTION   The relative motion between the tool or instrument and the work due to motion of the programmed axis.

FEEDRATE (CODE WORD)   A multiple-character code containing the letter F followed by digits. It determines the machine slide rate of feed.

FEEDRATE DIVIDER   A feature of some machine control units that gives the capability of dividing the programmed feedrate by a selected amount as provided for in the machine control unit.

FEEDRATE MULTIPLIER   A feature of some machine control units that gives the capability of multiplying the programmed feedrate by a selected amount as provided for in the machine control unit.

FEEDRATE OVERRIDE   A variable manual control function directing the control system to reduce the programmed feedrate.

Feedrate override is a percentage function to reduce the programmed feed rate. If the programmed feed rate was 30 inches per minute and the operator wanted 15 inches per minute, the feedrate override dial would be set at 50 percent.

FIXED BLOCK FORMAT   A format in which the number and sequence of words and characters appearing in successive blocks is constant.

FIXED CYCLE   See canned cycle.

FIXED SEQUENTIAL FORMAT   A means of identifying a word by its location in a block of information. Words must be presented in a specific order, and all possible words preceding the last desired word must be present in the block.

FLOATING ZERO   A characteristic of a machine control unit permitting the zero reference point on an axis to be established readily at any point in the travel.

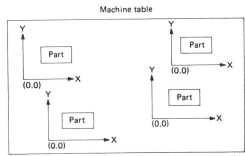

Machine table

The part or workpiece may be moved to *any* location on the machine table and zero may be established at that point.

FORMAT (TAPE)   The general order in which information appears on the input media, such as the location of holes on a punched tape or the magnetized areas on a magnetic tape.

FULL RANGE FLOATING ZERO   A characteristic of a numerical machine tool control permitting the zero point on an axis to be shifted readily over a specified range. The control retains information on the location of permanent zero.

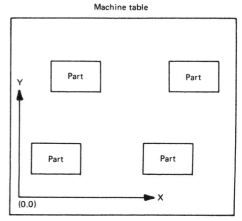

The part or workpiece may be shifted to any position on the machine table, but the actual position of permanent zero remains constant.

GAGE HEIGHT   A predetermined partial retraction point along the Z axis to which the cutter retreats from time to time to allow safe XY table travel. Also called the reference or rapid level.

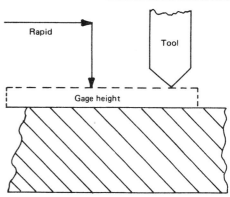

Gage height, usually .100 to .125, is a set distance established in the control or set by the operator. Gage height allows the tool, while advancing in rapid traverse, to stop at the established distance (gage height) and begin feed motion. Without gage height, the tool would rapid into the part causing tool damage or breakage and potential operator injury.

G CODE A word addressed by the letter G and followed by a numerical code defining preparatory functions or cycle types in a numerical control system.

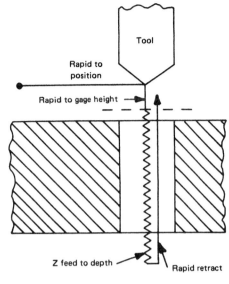

G81 — Drill Cycle

GENERAL PROCESSOR 1. A computer program for converting geometric input data into cutter path data required by an NC machine. 2. A fixed software program designed for a specific logical manipulation of data.

HARD COPY A readable form of data output on paper.

HARDWARE The component parts used to build a computer or control system, e.g. integrated circuits, diodes, transistors.

HARD-WIRED Having logic circuits interconnected on a backplane to give a fixed pattern of events.

HIGH-SPEED READER A reading device which can be connected to a computer or control so as to operate on line without seriously holding up the computer or control.

INCREMENTAL SYSTEM A control system in which each coordinate or positional dimension, both input and feedback, is taken from the past position rather than from a common datum point, as in the absolute system.

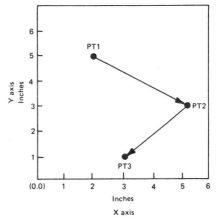

Coordinate positions

Point	X value	Y value
PT1	2	5
PT2	3	-2
PT3	-2	-2

In an incremental system, all points are expressed relative to the preceding point.

INDEX TABLE A multiple-character code containing the letter B followed by digits. This code determines the position of the rotary index table in degrees.

See B (Beta) axis.

INHIBIT To prevent an action or acceptance of data by applying an appropriate signal to the appropriate input.

INITIAL LEVEL The position of the spindle at the beginning of a canned cycle operation.

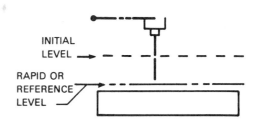

INPUT Transfer of external information into the control system.

INPUT MEDIA 1. The form of input such as punched cards and tape or magnetic tape. 2. The device used to input information.

INTERCHANGEABLE VARIABLE BLOCK FORMAT A programming arrangement consisting of a combination of the word address and tab sequential formats to provide greater compatibility in programming. Words are interchangeable within the block. Length of block varies since words may be omitted.

This is one of the most sophisticated tape formats in use today.

See *block.*

INTERCHANGE STATION The position where a tool of an automatic tool changing machine awaits automatic transfer to either the spindle or the appropriate coded drum station.

INTERMEDIATE TRANSFER ARM The mechanical device in automatic tool changing that grips and removes a programmed tool from the coded drum station and places it into the interchange station, where it awaits transfer to the machine spindle. This device then automatically grips and removes the used tool from the interchange station and returns it to the appropriate coded drum station.

INTERPOLATION 1. The insertion of intermediate information based on an assumed order or computation. 2. A function of a control whereby data points are generated between given coordinate positions.

INTERPOLATOR A device which is part of a numerical control system and performs interpolation.

ISO International Organization for Standardization.

JOG A control function which momentarily operates a drive to the machine.

LEADING ZEROES Redundant zeroes to the left of a number.

Leading zeroes

X + 0062500

LEADING ZERO SUPPRESSION See zero suppression.

LETTER ADDRESS The method by which information is directed to different parts of the system. All information must be preceded by its proper letter address, e.g., X, Y, Z, M.

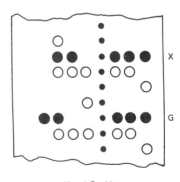

X and G address

An identifying letter inserted in front of each word.

LINEAR INTERPOLATION   A function of a control whereby data points are generated between given coordinate positions to allow simultaneous movement of two or more axes of motion in a linear (straight) path.

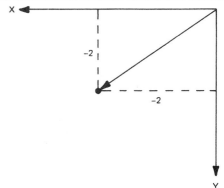

The control system moves X and Y axes proportionately to arrive at the destination point.

LOOP TAPE   A short piece of tape, with joined ends, which contains a complete program or operation.

MACHINING CENTER   Machine tools, usually numerically controlled, capable of automatically drilling, reaming, tapping, milling, and boring multiple faces of a part. Equipped with a system for automatically changing cutting tools.

MACRO   A group of instructions which can be stored and recalled as a group to solve a recurring problem.

An APT macro could be as follows:

DRILL1 = MACRO/X, Y, Z, Z1, FR, RR
    GOTO/POINT, X, Y, Z, RR
    GODLTA/–Z1, FR
    GODLTA/+Z1, RR
TERMAC

X, Y, Z, Z1, FR, and RR would be variables which would have values assigned when the macro is called into action. The variables would be as follows:
    X = X position
    Y = Y position
    Z = Z position (above work surface)
    Z1 = Z feed distance
    FR = feed rate
    RR = rapid rate

The call statement could be:
    CALL/DRILL1, X = 2, Y = 4, Z = .100, Z1 = 1.25, FR = 2, RR = 200

MAGIC-THREE CODING   A feedrate code that uses three digits of data in the F word. The first digit defines the power of ten multiplier. It determines the positioning of the floating decimal point. The last two digits are the most significant digits of the desired feedrate.

To program a feed rate of 12 inches per minute in magic-three coding:
1) count the number of decimal places to the left of the decimal. 12 = 2
2) Add magic "3" to the number of counted decimal places. (3 + 2 = 5)
3) write the F word address, the added digit, and the first two digits of the actual feed rate to be programmed. (F512)
4) F512 would be the magic "3" coded feed rate.

This method of feed rate coding is now almost obsolete.

MAGNETIC TAPE   A tape made of plastic and coated with magnetic material. It stores information by selective polarization of portions of the surface.

MANUAL DATA INPUT   A mode or control that enables an operator to insert data into the control system. The data are identical to information that could be inserted by tape.

MANUAL PART PROGRAMMING   The preparation of a manuscript in machine control language
and format to define a sequence of commands for
use on an NC machine.

Manual, or hand, programming is programming the actual codes, X and Y positions, functions, etc. as they are punched in the N/C tape.

H001 G81   X+37500 Y+52500   W01

MANUSCRIPT   A written or printed copy, in symbolic form, containing the same data as that punched on cards or tape or retained in a memory unit.

MEMORY   An organized collection of storage ele-
ments, e.g., disc, drum, ferrite cores, into which a
unit of information consisting of a binary digit can
be stored and from which it can later be retrieved.

A computer with a 64,000-word capacity is said to have a memory of 64 K.

MIRROR IMAGE   See axis inversion.

MODAL   Information that is retained by the system until new information is obtained and replaces it.

MODULE   An interchangeable plug-in item containing components.

NC   **(Numerical control)** The technique of controlling a machine or process by using command instructions in coded numerical form.

NULL   1. Pertaining to no deflection from a center or end position. 2. Pertaining to a balanced or zero output from a device.

NUMERICAL CONTROL SYSTEM   A system in which programmed numerical values are directly inserted, stored on some form of input medium, and automatically
read and decoded to cause a corresponding movement in a machine or process.

OFFLINE PROGRAMMING   The development of an NC part program away from the machine console to be transferred to the MCU at a later time.

OFFSET   A displacement in the axial direction of the tool which is the difference between the actual tool length and the programmed tool length.

OPEN-LOOP SYSTEM   A control system that has no means of comparing the output with the input for control purposes. No feedback.

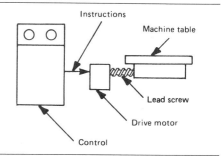

OPTIMIZE   To rearrange the instructions or data in storage so that a minimum number of transfers are required in the running of a program. To obtain maximum accuracy and minimum part production time by manipulation of the program.

OPTIONAL STOP   A miscellaneous function (M01) command similar to Program Stop except the control ignores the command unless the operator has previously pushed a button to validate the command.

OVERSHOOT   A term applied when the motion exceeds the target value. The amount of overshoot depends on the feedrate, the acceleration of the slide unit, or the angular change in direction.

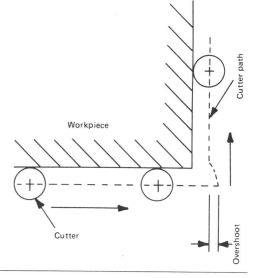

PARABOLA   A plane curve generated by a point moving so that its distance from a fixed second point is equal to its distance from a fixed line.

PARABOLIC INTERPOLATION   Control of cutter path by interpolation between three fixed points by assuming the intermediate points are on a parabola.

PARITY CHECK   1. A hole punched in one of the tape channels whenever the total number of holes is even, to obtain an odd number, or vice versa depending on whether the check is even or odd. 2. A check that tests whether the number of ones (or zeroes) in any array of binary digits is odd or even.

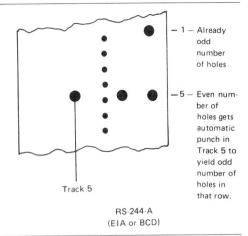

Track 5

1 — Already odd number of holes

5 — Even number of holes gets automatic punch in Track 5 to yield odd number of holes in that row.

RS-244-A
(EIA or BCD)

PART PROGRAM   A specific and complete set of data and instructions written in source languages for computer processing or in machine language for manual programming to manufacture a part on an NC machine.

PART PROGRAMMER   A person who prepares the planned sequence of events for the operation of a numerically controlled machine tool.

PERFORATED TAPE   A tape on which a pattern of holes or cuts is used to represent data.

PLOTTER   A device which will draw a plot or trace from coded NC data input.

POINT-TO-POINT CONTROL SYSTEM   A numerical control system in which controlled motion is required only to reach a given end point, with no path control during the transition from one end point to the next.

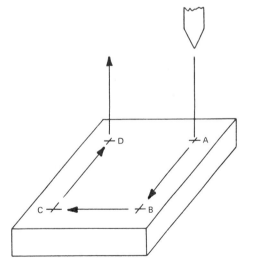

POSITIONING/CONTOURING   A type of numerical control system that has the capability of contouring, without buffer storage, in two axes and positioning in a third axis for such operations as drilling, tapping, and boring.

POSITIONING SYSTEM   See point-to-point control system.

POSITION READOUT   A display of absolute slide position as derived from a position feedback device (transducer) normally attached to the lead screw of the machine. See command readout.

POSTPROCESSOR   The part of the software which converts the cutter path coordinate data into a form which the machine control can interpret correctly. The cutter path coordinate data are obtained from the general processor and all other programming instructions and specifications for the particular machine and control.

PREPARATORY FUNCTION   An NC command on the input tape changing the mode of operation of the control. (Generally noted at the beginning of a block by the letter G plus two digits.)

Some preparatory functions are:

G84 — tap cycle
G01 — linear interpolation
G82 — dwell cycle
G02 — circular interpolation — clockwise
G03 — circular interpolation — counter clockwise

See *G code.*

PROGRAM   A sequence of steps to be executed by a control or a computer to perform a given function.

PROGRAMMED DWELL   The capability of commanding delays in program execution for a programmable length of time.

PROGRAMMER (PART PROGRAMMER)   A person who prepares the planned sequence of events for the operation of a numerically controlled machine tool. The programmer's principal tool is the manuscript on which the instructions are recorded.

Manual part programming instructions:

H001 G81	X+123750	Y+62500	W01
N002	X+105000		
N003		Y+51250	M06

Computer part programming instructions:

TLRGT, GORGT/HL3, TANTO, C1
        GOFWD/C1, TANTO, HL2
        GOFWD/HL2, PAST, VL2

PROGRAM STOP   A miscellaneous function (M00) command to stop the spindle, coolant, and feed after completion of the dimensional move commanded in the block. To continue with the remainder of the program, the operator must push a button.

QUADRANT   Any of the four parts into which a plane is divided by rectangular coordinate axes in that plane.

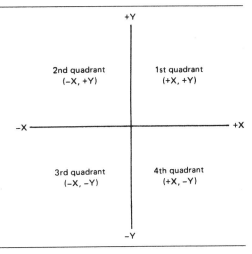

RANDOM   Not necessarily in a logical order of arrangement according to usage, but having the ability to select from any location and in any order from the storage system.

RAPID   Positioning the cutter and workpiece into close proximity with one another at a high rate of travel speed, usually 150 to 400 inches per minute (IPM) before the cut is started.

READER   A pneumatic, photoelectric, or mechanical device used to sense bits of information on punched cards, punched tape, or magnetic tape.

REGISTER   An internal array of hardware binary circuits for temporary storage of information.

REPEATABILITY   Closeness of, or agreement in, repeated measurements of the same characteristics by the same method, using the same conditions.

RESET   To return a register or storage location to zero or to a specified initial condition.

ROW (TAPE)   A path perpendicular to the edge of the tape along which information may be stored by the presence or absence of holes or magnetized areas. A character would be represented by a combination of holes.

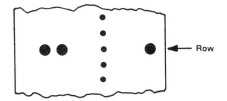

SEQUENCE NUMBER (CODE WORD) A series of numerals programmed on a tape or card and sometimes displayed as a readout; normally used as a data location reference or for card sequencing.

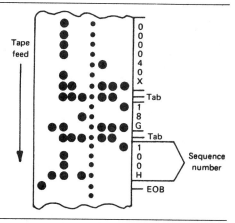

SEQUENCE READOUT A display of the number of the block of tape being read by the tape reader.

SEQUENTIAL Arranged in some predetermined logical order.

SIGNIFICANT DIGIT A digit that must be kept to preserve a specific accuracy or precision.

Significant digits

$$X + 00\underbrace{5250}0$$

Insignificant digits

SLOW-DOWN SPAN A span of information having the necessary length to allow the machine to decelerate from the initial feedrate to the maximum allowable cornering feedrate that maintains the specified tolerance.

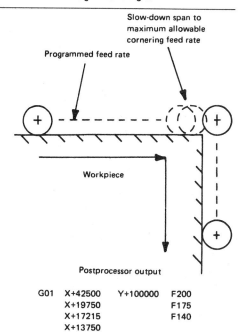

Postprocessor output

G01	X+42500	Y+100000	F200
	X+19750		F175
	X+17215		F140
	X+13750		
		Y+68750	F200

SOFTWARE   Instructional literature and computer programs used to aid in part programming, operating, and maintaining the machining center.

Examples of software programs are:

APT
FORTRAN
COBOL
RPG

SPAN   A certain distance or section of a program designated by two end points for linear interpolation; a beginning point, a center point, and an ending point for circular interpolation; and two end points and a diameter point for parabolic interpolation.

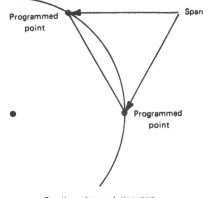

One linear interpolation span

SPINDLE SPEED (CODE WORD)   A multiple-character code containing the letter S followed by digits. This code determines the RPM of the cutting spindle of the machine.

STORAGE   A device into which information can be introduced, held, and then extracted at a later time.

STORAGE MEDIA   A device onto which information can be transferred and retained for later use. Storage media may also be used as input media, thereby serving a dual purpose.

TAB   A nonprinting spacing action on tape preparation equipment. A tab code is used to separate words or groups of characters in the tab sequential format. The spacing action sets typewritten information on a manuscript into tabular form.

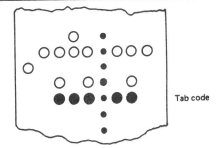

Tab code

TAB SEQUENTIAL FORMAT   Means of identifying a word by the number of tab characters preceding the word in a block. The first character of each word is a tab character. Words must be presented in a specific order, but all characters in a word, except the tab character, may be omitted when the command represented by that word is not desired.

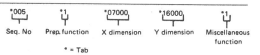

*005	*1	*07000	*16000	*1
Seq. No	Prep. function	X dimension	Y dimension	Miscellaneous function

* = Tab

The tab sequential format is, for the most part, obsolete.

TAPE   A magnetic or perforated paper medium for storing information.

TAPE LAGGER   The trailing end portion of a tape.
TAPE LEADER   The front or lead portion of a tape.

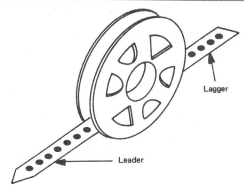

Reel tapes should have a leader and lagger of approximately three feet with just sprocket holes for tape loading and threading purposes.

TOOL FUNCTION   A tape command identifying a tool and calling for its selection. The address is normally a T word.

T06 would be a tape command calling for the tool assigned to spindle or pocket 6 to be put in the spindle.

TOOL LENGTH COMPENSATION   A manual input, by means of selector switches, to eliminate the need for preset tooling; allows the programmer to program all tools as if they are of equal length.

TOOL OFFSET   1. A correction for tool position parallel to a controlled axis. 2. The ability to reset tool position manually to compensate for tool wear, finish cuts, and tool exchange.

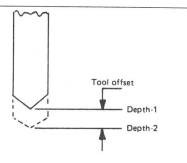

Tool offsets are used as final adjustments to increase or decrease depths due to cutting forces and tool deflection. In this case, a tool offset could be used to increase the drill depth from depth-1 to depth-2.

TRAILING ZERO SUPPRESSION   See zero suppression.

TURNKEY SYSTEM   A term applied to an agreement whereby a supplier will install an NC or computer system so that he has total responsibility for building, installing, and testing the system.

USASCII   United States of America Standard Code for Information Interchange. See ASCII.

VARIABLE BLOCK FORMAT (TAPE)   A format which allows the quantity of words in successive blocks to vary. Same as word address. Variable block means the length of the blocks can vary depending on what information needs to be conveyed in a given block. See *block*.

VECTOR   A quantity that has magnitude, direction, and sense; is represented by a directed line segment whose length represents the magnitude and whose orientation in space represents the direction.

VECTOR FEEDRATE   The feedrate at which a cutter or tool moves with respect to the work surface. The individual slides may move slower or faster than the programmed rate, but the resultant movement is equal to the programmed rate.

WORD   An ordered set of characters which is the normal unit in which information may be stored, transmitted, or operated upon.

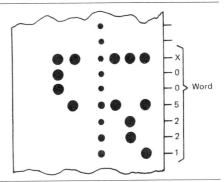

See *address* and *block*.

WORD ADDRESS FORMAT   The specific arrangement of addressing each word in a block of information by one or more alphabetical characters which identify the meaning of the word.

WORD LENGTH   The number of bits or characters in a word.

See *word*.

X AXIS   Axis of motion that is always horizontal and parallel to the workholding surface.

Y AXIS   Axis of motion that is perpendicular to both the X and Z axes.

Z AXIS   Axis of motion that is always parallel to the principal spindle of the machine.

ZERO OFFSET   A characteristic of a numerical machine tool control permitting the zero point on an axis to be shifted readily over a specified range. The control retains information on the location of the permanent zero.

See *full range floating zero* and *floating zero*.

ZERO SHIFT   A characteristic of a numerical machine tool control permitting the zero point on an axis to be shifted readily over a specified range. (The control does *not* retain information on the location of the permanent zero.)

See *floating zero*. Consult chapter 4 for additional details.

ZERO SUPPRESSION   Leading zero suppression: the elimination of insignificant leading zeroes to the left of significant digits usually before printing. Trailing zero suppression: the elimination of insignificant trailing zeroes to the right of significant digits usually before printing.

Leading zero suppression

X + 0043500

Insignificant digits

Could be written as:

X + 43500

Trailing zero suppression

X + 0043500

Insignificant digits

Could be written as:

X + 00435

RELATED SME TITLES

The Society of Manufacturing Engineers offers several related titles in CNC and Machine Trades.

Machinery's Handbook, E. Oberg, F. Jones,
  H. Horton
2500 pp, hardcover, 1988
Order Code: 1379–0802
$55.00 (SME Members: $49.00)

Speed and Feed Chart
Order Code: 825–0802
$15.00 (SME Members: $10.00)

Machine Controls (From TMEH Machining,
  Volume 1)
60 pp, softcover
Order Code: 1355–0802
$24.50 (SME Members: $21.50)

Fundamentals of Numerical Control, W. Luggen
287 pp, hardcover, 1988
Order Code: 1387–0802
$28.95 (SME Members: $26.95)

Troubleshooting Manufacturing Processes
545 pp, hardcover, 1988
Order Code: 1345–0802
$69.00 (SME Members: $64.00)

Cutting Fluids and Lubricants
350 pp, softcover, 1985
Order Code: 953–0802
$38.00 (SME Members: $33.00)

Nontraditional Machining Processes, 2nd Edition
255 pp, softcover, 1983
Order Code: 788–0802
$26.00 (SME Members: $23.00)

Machining Hard Materials
250 pp, hardcover, 1982
Order Code: 664–0802
$38.00 (SME Members: $32.00)

Machinists' Ready Reference Manual
466 pp, hardcover, 1988
Order Code: 1368-0802
$32.00 (SME Members: $27.00)

Modern Trends in Cutting Tools
300 pp, hardcover, 1982
Order Code: 686–0802
$38.00 (SME Members: $32.00)

*Improving Production with Coolants and
  Lubricants*
240 pp, hardcover, 1982
Order Code: 655–0802
$38.00 (SME Members: $32.00)

Machining Fundamentals
300 pp, softcover, 1980
Order Code: 514–0802
$29.50 (SME Members: $24.00)

Milling: Methods and Machines
250 pp, hardcover, 1980
Order Code: 669–0802
$38.00 (SME Members: $32.00)

For other titles in manufacturing, materials, design, quality control, automated systems, robots, and management, request a copy of the latest SME catalog from:

Society of Manufacturing Engineers
One SME Drive
P.O. Box 930
Dearborn, MI 48121

Call: 313/271–1500

INDEX